To

with my best wishes

Waste Heat Recovery in Process Industries

Waste Heat Recovery in Process Industries

Hussam Jouhara

WILEY-VCH

Author

Professor Hussam Jouhara
Chair of Thermal Engineering
Brunel University London
Kingston Lane
Uxbridge, UB8 3PH
United Kingdom

Cover Image: © ThomasVogel/Getty Images

Library of Congress Card No.: applied for

British Library Cataloguing-in-Publication Data
A catalogue record for this book is available from the British Library.

Bibliographic information published by the Deutsche Nationalbibliothek
The Deutsche Nationalbibliothek lists this publication in the Deutsche Nationalbibliografie; detailed bibliographic data are available on the Internet at <http://dnb.d-nb.de>.

Print ISBN: 978-3-527-34856-5
ePDF ISBN: 978-3-527-83002-2
ePub ISBN: 978-3-527-83001-5
oBook ISBN: 978-3-527-83000-8

Typesetting Straive, Chennai, India
Printing and Binding CPI Group (UK) Ltd, Croydon, CR0 4YY

Printed on acid-free paper

C110119_011221

To my family

Contents

Preface *xiii*

1 Thermodynamic Cycles *1*
1.1 Introduction to Thermodynamic Cycles *1*
1.2 Rankine Cycle *1*
1.2.1 Introduction *1*
1.2.2 Thermodynamic Diagrams *2*
1.2.3 The Carnot Cycle *10*
1.2.4 Ideal and Actual Rankine Cycles *12*
1.2.4.1 Ideal Cycle *13*
1.2.4.2 Superheated Rankine Cycle *15*
1.2.4.3 Actual Rankine Cycle *17*
1.2.4.4 Improvements to the Rankine Cycle *19*
1.2.4.5 Regenerative Rankine Cycles *22*
1.2.4.6 Cogeneration *26*
1.2.5 Other Configurations of the Rankine Cycle *29*
1.2.5.1 Supercritical Rankine Cycles *29*
1.2.5.2 Reverse Rankine Cycles *30*
1.2.6 Rankine Cycles in Power Plants *31*
1.2.6.1 Fossil Fuel Power Plants *31*
1.2.6.2 Nuclear Power Plants *32*
1.2.6.3 Overall Efficiency of a Power Plant *32*
1.2.6.4 Case Studies *33*
1.3 Organic Rankine Cycle *34*
1.3.1 Configurations of ORC *35*
1.3.1.1 Basic ORC Configuration *35*
1.3.1.2 ORC with Preheating *36*
1.3.1.3 Recuperative ORC *38*
1.3.1.4 Recuperative ORC with Preheating *39*
1.3.2 Organic Working Fluids *40*
1.3.3 Organic Working Fluid Selection *42*
1.3.4 Applications of the ORC *45*
1.3.4.1 Waste Heat Recovery *45*

1.4 Kalina Cycle 46
1.4.1 Cycle Fundamentals 46
1.4.1.1 Why Use Ammonia–Water Solution in Kalina Cycle? 48
1.4.2 Advantages and Drawbacks 49
1.4.2.1 Advantages 49
1.4.2.2 Drawbacks 50
1.4.3 Applications of the Kalina Cycle 50
1.4.3.1 The Different Configurations of the Cycle 51
1.4.4 Case Studies 53
1.5 Brayton Cycle 53
1.5.1 Regenerative Brayton Cycle (Regenerator) 57
1.5.1.1 Compressor Analysis 58
1.5.1.2 Turbine Analysis 58
1.5.1.3 Heat Supplied to the Cycle 59
1.5.2 Regenerative Brayton Cycle (Reheater and Intercooler) 59
1.5.2.1 Intercooling 60
1.5.2.2 Reheating 60
1.6 Chapter Summary 61
References 62

2 Waste Heat Recovery 67
2.1 Burner and Air Preheaters 67
2.1.1 Recuperators 67
2.1.1.1 Recuperative Burners 68
2.1.1.2 Classifying Recuperative Burners 71
2.1.1.3 Efficiency Improvement and Fuel Savings 72
2.1.2 Regenerators 74
2.1.2.1 Rotary Regenerators 74
2.1.2.2 Static Regenerators 75
2.1.2.3 Regenerative Burners 75
2.1.3 Burner Technology Comparison 76
2.1.4 NO_x Formation 77
2.1.5 Run-Around Coil 78
2.2 Heat Exchangers 79
2.2.1 Shell and Tube HEXs 79
2.2.1.1 Construction 80
2.2.1.2 Applications and Limitations 82
2.2.2 Plate Heat Exchanger 82
2.2.2.1 Spiral Plate Heat Exchanger 83
2.2.3 Heat Pipe Heat Exchanger 83
2.2.4 Compact HEX 85
2.3 Waste Heat Boilers 86
2.3.1 Different WHB Designs 87
2.3.2 WHB Methodologies 88
2.3.2.1 Feed Water Preheating Effect 88

2.3.2.2 Optimising Thermodynamic Cycles *89*
2.3.2.3 Heat Recovery Boiler with Water Spray Systems *91*
2.3.3 Failure Modes *92*
2.3.3.1 Failure Modes Analysis *92*
2.4 Heat Recovery Steam Generators *93*
2.4.1 Construction of Waste HRSG *94*
2.4.1.1 HRSG Design and Construction *95*
2.4.1.2 Evaporator *95*
2.4.1.3 Superheater *96*
2.4.1.4 Economiser *96*
2.4.1.5 Steam Drum *96*
2.4.1.6 Evaporator Types *96*
2.4.1.7 Horizontal Tube HEXs *98*
2.4.1.8 Natural Circulation HRSGs *98*
2.4.1.9 Assisted (or Forced) Circulation HRSGs *99*
2.4.1.10 Tube Materials *99*
2.4.1.11 The 'Pinch Point' and Other Effects *100*
2.5 Heat Pumps *100*
2.5.1 Fundamental Principles of Heat Pumps *100*
2.5.1.1 Cooling Mode *101*
2.5.1.2 Heating Mode *101*
2.5.2 Variation of Heat Pump System *102*
2.5.2.1 Air Source Heat Pump System *103*
2.5.2.2 Ground Source Heat Pump System *103*
2.5.2.3 Water Source Heat Pump System *105*
2.5.2.4 Water Loop Heat Pump System *105*
2.5.2.5 Exhaust Air System *106*
2.5.2.6 Hybrid Heat Pump *106*
2.5.2.7 Solar-Assisted Heat Pumps *106*
2.6 Direct Electrical Conversion Device *107*
2.6.1 TEG – Working Principle *108*
2.6.2 The Seebeck Effect *109*
2.6.3 The Peltier Effect *109*
2.6.3.1 Applications of the Peltier Effect *110*
2.6.4 Thomson Effect *110*
2.6.5 Joule Heating *111*
2.6.6 Theoretical Principle *112*
2.6.7 Figure of Merit *112*
2.6.8 Fermi Level *113*
2.6.9 Nano-Sizing *114*
2.6.10 Efficiency of TEG *115*
2.7 Thermal Storage *116*
2.7.1 Sensible Heat Storage *117*
2.7.2 Latent Heat Storage *120*
2.7.3 Thermochemical Storage *123*

2.7.4 Phase Change Materials 123
2.7.5 Organic Material 125
2.7.6 Inorganic PCMs 128
2.7.7 Eutectic PCMs 128
2.7.8 PCM Methodologies 129
2.7.8.1 Encapsulation of PCMs 129
2.7.8.2 Microencapsulated PCMs 129
2.7.8.3 Macroencapsulation of the PCMs 132
2.7.8.4 Nanomaterial PCMs 132
2.7.8.5 Shape Stabilisation 135
2.8 Design Development Methods 135
2.8.1 Introduction 135
2.8.2 Heat Exchangers 140
2.8.2.1 Local Heat Transfer 140
2.8.2.2 LMTD Method 147
2.8.2.3 Effectiveness-Number of Transfer Units (ε-NTU) Method 151
2.8.3 Regenerative and Recuperative Burners 152
2.8.3.1 Regenerative Burners 154
2.8.3.2 Recuperative Burners 156
2.8.4 Waste Heat Boilers 157
2.8.5 Air Preheaters 160
2.8.6 Heat Recovery Steam Generator 166
2.8.7 Heat Pumps 170
2.8.8 Direct Electrical Conversion Device 173
2.8.9 Thermal Storage 176
References 178

3 Low-Temperature Applications *191*
3.1 Refrigeration 191
3.2 Cryogenics 198
3.2.1 Loop Heat Pipe 199
3.3 HVAC 204
References 209

4 Medium-Temperature Applications *213*
4.1 Food Industry 213
4.1.1 Energy Use in the Industry 213
4.1.2 Case Study 1: Heat Recovery Potential of the Crisps Manufacturing Process 214
4.1.3 Case Study 2: Temperature and Energy Performance of Open Refrigerated Display Cabinets Using Heat Pipe Shelves 215
4.2 Ventilation 221
4.2.1 Applications 221
4.3 Solar Energy 223
4.4 Geothermal Energy 230

4.5 Automotive Industry *233*
4.5.1 Industrial Processes *235*
4.6 Aviation *237*
References *239*

5 High-Temperature Applications *245*
5.1 Steel Industry *245*
5.1.1 TEG Modules *246*
5.1.2 Heat Exchangers *246*
5.1.2.1 Application 1: Slag Particles Blast Furnace Retrofit *246*
5.1.2.2 Application 2: Flat Heat Pipe Heat Exchanger *247*
5.1.3 Recuperators *249*
5.1.3.1 Application 1: Heat Recuperator for Steel Slag *249*
5.2 Ceramic Industry *251*
5.2.1 Introduction *251*
5.2.2 Heat Exchangers *251*
5.2.2.1 Application 1: Radiative Heat Pipe *251*
5.2.2.2 Application 2: Multi-Pass Heat Pipe *252*
5.2.2.3 Application 3: Forced Convection Heat Pipe *253*
5.3 Cement Industry *254*
5.3.1 Gas Suspension Preheaters *255*
5.3.1.1 Application 1 *255*
5.3.1.2 Application 2 *256*
5.3.2 Heat Pipe Thermoelectric Generator *256*
5.4 Aluminium Industry *258*
5.4.1 Rotary Regenerator *258*
5.4.2 Heat Exchangers *258*
5.4.3 Heat Pumps *258*
5.4.4 Recuperators *260*
5.4.4.1 Radiative Recuperator *260*
5.4.4.2 Convective Recuperator *261*
5.4.4.3 Hybrid Recuperator *262*
5.4.5 Thermoelectric Device *262*
5.4.6 Regenerative Burner *262*
5.4.7 Preheating Scrap *264*
5.4.8 De-coating *265*
References *265*

Index *269*

Preface

In the twenty-first century, the topic of waste heat recovery from industrial processes is taking on an ever-increasing significance. There are essentially two reasons for this. The first is that there is an obligation to minimise greenhouse gas emissions and to aim for carbon-neutral operation. The second reason is economic. Waste heat can represent a huge financial penalty to an industrial process, but implementing waste heat recovery systems to re-use the energy can substantially reduce thermal inputs, and hence reduce costs.

This book begins with a study of the classical thermodynamic cycles: Rankine, organic Rankine, Kalina, and Brayton, on which modern systems for power generation are based. Accordingly, power plants developed from the Rankine and the Brayton cycles are used extensively in electricity generation today. However, these cycles may not be appropriate for future power generation because the moderate temperatures often associated with waste heat recovery make the organic Rankine cycle and Kalina cycle unsuitable for consideration in these circumstances.

The second chapter comprises an extensive examination of the many techniques available for the capture and re-use of waste heat rather than its employment in power cycles for energy conversion. The topics covered include air preheaters, regenerators, waste heat boilers, various heat exchanger designs, heat pumps, direct electrical conversion, and thermal energy storage. The chapter ends with methods of design development for the aforementioned topics.

The remaining chapters of the book examine the wide-ranging applications for waste energy utilisation at low temperature, medium temperature, and high temperature across specific industrial sectors. Some of these applications employ heat pipes, devices which are highly suitable in many cases for reliable and economic heat recovery due to their passive nature of operation. The final chapter closes with high-temperature applications in automotive engineering and aviation.

1

Thermodynamic Cycles

1.1 Introduction to Thermodynamic Cycles

The concept of a thermodynamic cycle is based on depicting thermodynamic processes which involve the transfer heat and work. This is achieved by altering temperature, pressure, as well as other state variables; the cycle ultimately returning to its initial state. The fundamental basis of these cycles is the first law of thermodynamics which states that 'energy cannot be created nor destroyed but only converted from one form to another'.

As shown in Figure 1.1, thermodynamic cycles are split into two primary classes – refrigeration cycles (also known as heat pump cycles) and power cycles such as the combustion engine cycle.

Cycles which transfer heat from low temperature to high temperature are classified as heat pump cycles, whereas cycles that convert heat input into mechanical work are designated as power cycles. And so, not inappropriately, thermodynamic power cycles provide the foundation for the operation of a heat engine. Power cycles are then further divided into groups depending on the type of heat engine. For cycles modelling internal combustion engines the groups are the Otto, Diesel, and Brayton cycles and for external combustion engines, they are the Rankine, Organic Rankine, and Kalina cycles.

1.2 Rankine Cycle

1.2.1 Introduction

The Rankine cycle is a closed thermodynamic cycle that converts thermal energy into mechanical energy and is based on the fundamentals of the Carnot cycle proposed by the French physicist Nicolas Léonard Sadi Carnot in 1824. The Rankine cycle was proposed by the Scottish physicist William John Macquorn Rankine in 1865, and contrary to Carnot cycle – which is only a theoretical cycle – the Rankine cycle has many applications, several of which will be discussed further within this section.

The main application of the Rankine cycle is to convert thermal energy into electrical energy and this thermodynamic cycle is based on the rotation of a turbine

Waste Heat Recovery in Process Industries, First Edition. Hussam Jouhara.

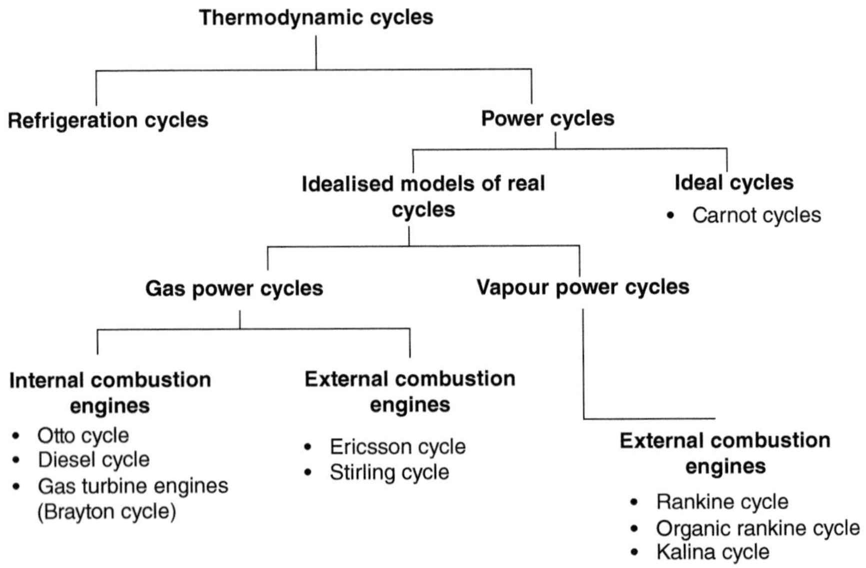

Figure 1.1 Map of all thermodynamic cycles.

fixed to a shaft and on to which an electrical generator is also located. In this way the mechanical energy of the turbine is converted into electrical energy, a more versatile and widely used form of energy. Rankine cycles and their different configurations are extensively used in thermal power plants such as nuclear, gas, oil, coal, geothermal, biomass, and concentrated solar power plants [1].

This section aims to present and explore in depth the Rankine thermodynamic cycle and following a general description of the cycle and a reminder of the Carnot cycle, different configurations will be presented and illustrated by means of diagrams and schematics. The driver for these various enhanced configurations always being to improve either the energetic or the exergetic efficiency of the cycle (or both). In this respect energy is a quantitative parameter and exergy a qualitative parameter, and we need to measure both of these in order to determine the total useful work we can extract from a thermodynamic system. The equations used to model such systems will be highlighted and explained as these formulas help us to understand the different parameters that influence the overall performance of a system. Real-life examples will also be used to clarify these explanations, and relevantly, a section of this document also explores the link between Rankine cycles and thermal power plants.

1.2.2 Thermodynamic Diagrams

In order to clearly understand the following sections of this document, it is important to be familiar with the basics of thermodynamics – the first and second laws as well as thermodynamic diagrams. Indeed, these diagrams are key aids in helping to develop a deeper understanding about how thermodynamic cycles function. The three main physical parameters used to describe a thermodynamic system are the temperature (°C or K), the pressure (bar or Pa), and volume (m^3) and these give rise to three significant diagrams which can then be created to visualise the state of a fluid under specific conditions: the *P–V* diagram, the *T–V* diagram, and the *P–T* diagram. *P–V* diagrams plot pressure with respect to specific volume, *T–V* diagrams plot temperature with respect to specific volume, and *P–T* diagrams plot pressure with respect to temperature. Now as most thermodynamic cycles operate by making use of a phase change of the working fluid, such diagrams are also useful in that they graphically display values (of pressure, volume, and temperature) at which these phase changes occur, i.e. for a given material, the points when changes of state from say a solid to a liquid (the point of melting), or from a liquid to a gas (the point of vapourisation) occur – or their reverse processes – freezing and condensing. In fact, we can display the three parameters of pressure, volume, and temperature as a 'surface', Figure 1.2.

Firstly, you will note that we say 'surface' not 'solid', and that's because at any particular temperature and pressure the condition of a material would only lie on the surface and not above nor below it. For example, take a point for say water in Figure 1.2, diagram (i), on the vapour section of the diagram. We might try to imagine a point above this – i.e. one with an increased pressure – but water vapour cannot exist at this higher pressure at that temperature and volume. Nor will it exist at a lower pressure. It can only exist at the values of pressure, temperature, and volume that are shown on the given surface [2].

So, what are the three single-phase regions we can see on the diagrams? They are the gas/vapour region, the liquid region, and the solid region, and between these we have the transition or two-state phase change regions where the solids are melting or freezing, and the liquids are evaporating or condensing.

Now when considering a vapour and a gas, because the two terms are often confused – and to some extent are synonymous in that they both refer to a physical state that is not solid or liquid – we perhaps explain the need to differentiate them. Technically, below the critical temperature, a vapour and its liquid are in

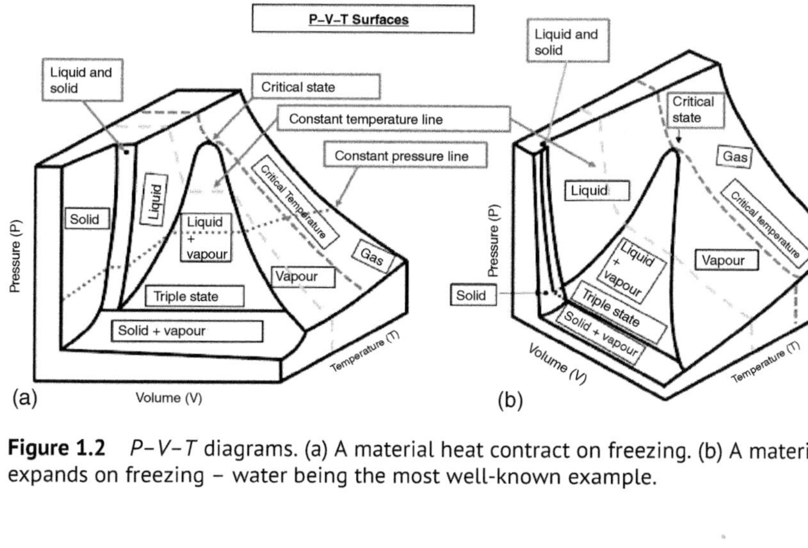

Figure 1.2 *P–V–T* diagrams. (a) A material heat contract on freezing. (b) A material that expands on freezing – water being the most well-known example.

'equilibrium' and a vapour can condense back to a liquid with a small change of pressure or temperature. Whereas a gas cannot condense to a liquid without changes in both pressure and temperature, and above the critical temperature line the liquid state does not exist at all! But what do we mean by being 'in equilibrium'?

Well, evaporation from the surface of a liquid or fluid takes place when the liquid molecules have enough momentum to overcome the cohesive forces within the fluid and escape into the space above it. To all intents and purposes, it becomes a gas, but because both the liquid and gas are present together it is more accurately known (in this state) as a vapour. If we now add heat to the liquid the momentum of the molecules increases, and more liquid evaporates. Similarly, a reduction in pressure above the liquid also reduces the momentum necessary for molecules to escape and so again evaporation is increased. But if the container which holds the liquid is sealed so that the evaporated vapour cannot escape, then the amount of liquid that can evaporate is finite. That is at any fixed value of temperature and pressure the amount of vapour will rise to a steady fixed amount and then no more can be added and the vapour is said to be 'fully saturated'. If any more molecules reach sufficient momentum to escape the surface then their number is exactly matched by those condensing back into the liquid, so an equilibrium or balance occurs between the two processes to limit the amount of vapour – or more correctly – the vapour pressure. Raise the temperature and the amount of vapour that can be maintained increases, and the 'fully saturated' line – the dome or bell-shaped line (in the earlier diagrams) illustrates the limit of this. So, to the right-hand side of the curve is where all the liquid will have evaporated and the liquid state cannot exist, whilst the left-hand side shows where all the solid has melted to liquid and the solid state can now longer exist. But under the curve is the region where both liquid and vapour exist together. Until that is, we reach the peak or 'critical point' or critical state where the distinction between the liquid and the gas no longer exists. Here the molecules are evaporating so rapidly that the density of the liquid and the vapour are equal and produce a 'supercritical fluid'; the point being that condensation of the gas will never occur above the critical point no matter how much pressure we apply.

But what else does the *P–V–T* diagram show us? For example, if we look at the liquid and solid regions, we will notice just how steep these surfaces are, and that's because, compared to gases, liquids and certainly solids are much harder to compress. So, at a particular temperature, if we increase the pressure it does not really change the volume or, conversely, the pressure will rise dramatically for a small change in volume. We can also note the transition phase between a solid and a vapour where 'sublimation' can occur. Dry ice is an example of such a material and is used where it is necessary to keep items cold but where any packaging will not get wet with liquid as the 'dry' ice warms up. We can also notice the triple state line where all three phases can coexist.

However, as useful as the *P–V–T* surface plot is to indicate the various states and transition phases of a substance they are a bit difficult to read-off from on to value scales. So, it is more usual to present thermodynamic information by means of two-axis diagrams that we briefly noted earlier. For example, the *P–V* diagram is the

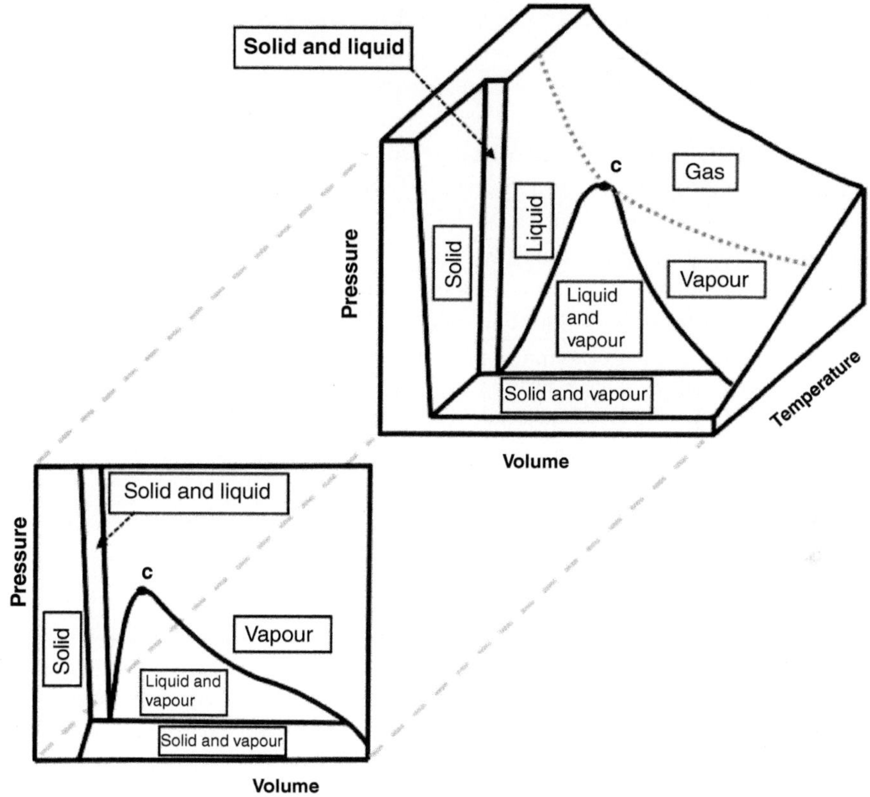

Figure 1.3 *P–V* perspective of a *P–V–T* diagram.

view we would obtain by looking at the surface from a pressure–volume perspective to see how a thermodynamic process changes with pressure and volume, Figure 1.3.

Usually we are only interested in the saturated liquid and saturated vapour areas, so our *P–V* diagrams are often 'truncated'. But in addition, our volume axis is more usually scaled in terms of *specific volume*, i.e. the cubic volume per kilogram of mass rather than just volume, because then, as an intensive property, the value does not depend on the size of the thermodynamic system. Hence, as we can see in Figure 1.4, the *P–V* diagram (also called a Clapeyron diagram) presents pressure as a function of the specific volume (m^3/kg), and on this diagram the left-hand side of the vapour dome shows the limit between the liquid region whilst the right-hand side shows the limit between the liquid/vapour region and the superheated vapour region. The diagram also denotes the plot of two isothermal lines or lines of constant temperature demonstrating that, inside the dome, the phase transition between liquid and vapour is a constant-pressure, constant-temperature process.

The other two diagrams that we can better visualise as two-dimensional graphs are the temperature–volume and the pressure–temperature diagrams as shown in Figure 1.5.

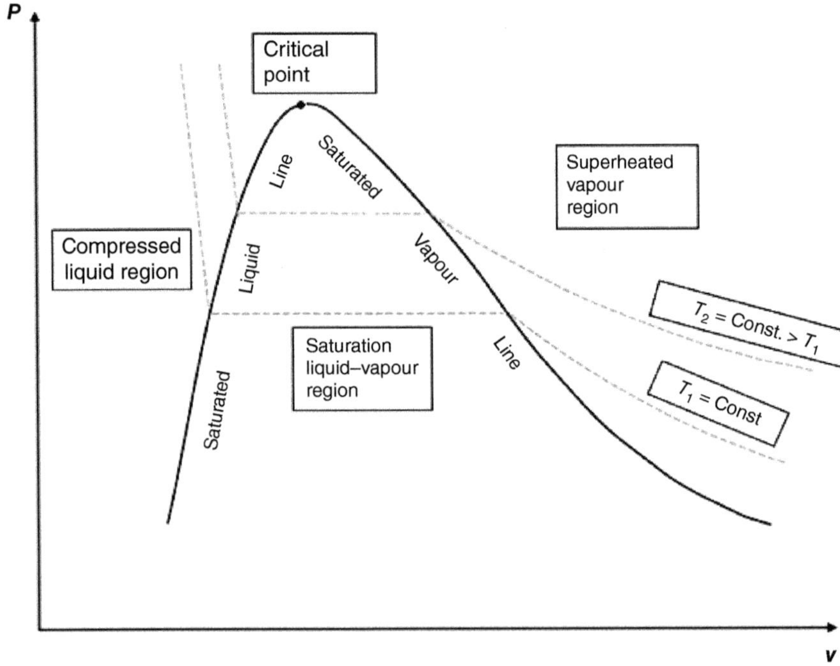

Figure 1.4 *P–V* diagram of a pure substance.

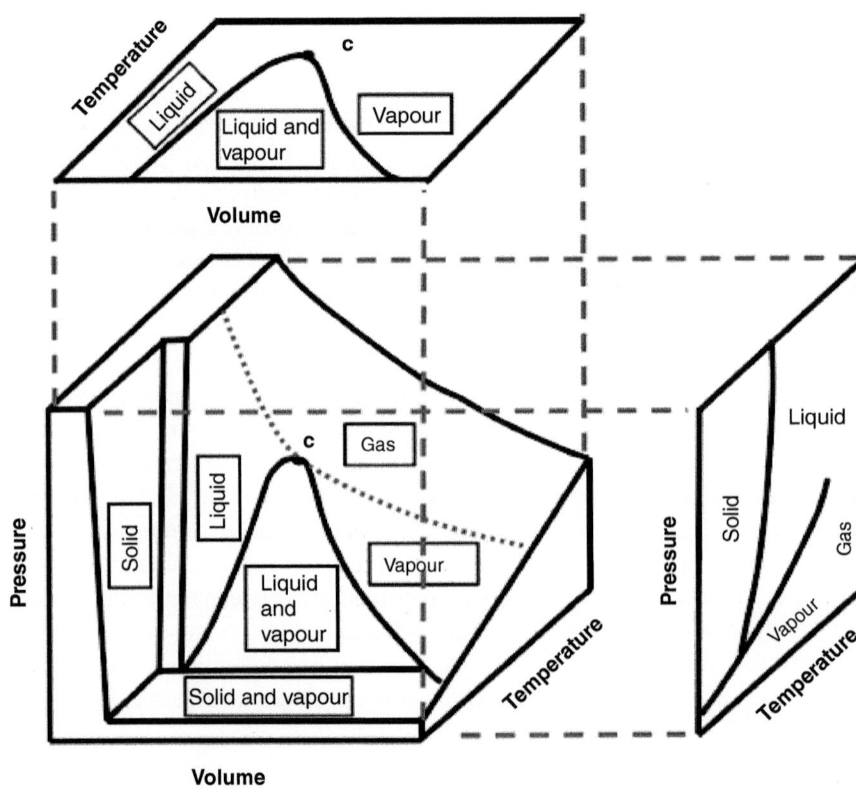

Figure 1.5 *T–V* and *P–T* perspectives of a *P–V–T* diagram.

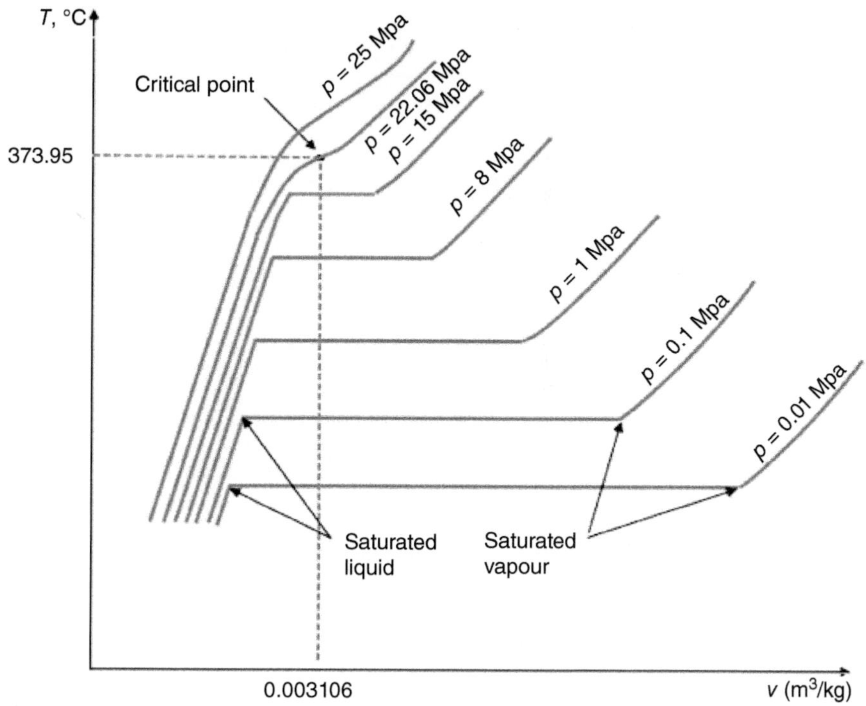

Figure 1.6 *T–V* diagram of a pure substance.

The vapour dome can also be seen on the *T–V* diagram in Figure 1.6, however, it is not directly visible on the *P–T* diagram in Figure 1.7 – which is also known as a phase diagram – in that for a given pressure and temperature the state of the substance is fully defined. As phase transitions occur under constant pressure and temperature for pure substances, alteration of the pressure or temperature will result in a phase change for the substance, the outcome being either a pure liquid or a pure vapour. For that reason, the liquid–vapour region is just a line on the *P–T* diagram and the vapour dome is not visible because, as we have already seen, a third dimension (the *P–V–T* diagram) is required to fully visualise the state of the fluid.

Two other thermodynamic variables are also very useful in describing a thermodynamic system from energetic and exergetic viewpoints, and these are the specific enthalpy h (kJ/kg) and the specific entropy s (kJ/kg K). Enthalpy (h) measures the total heat content of a system and is equal to the internal energy (i.e. the energy related to the molecular structure or state of the material and the degree of molecular activity) which is then added to the product of pressure and volume. In the analysis of systems that involve fluid flow, we often meet the combination of these two properties, defined from $h = u + PV$, where u represents the macroscopic energy of the non-flowing fluid and PV represents the energy needed to push and maintain the fluid flow around the system. But the second term, entropy, is defined as a measure of randomness of molecules, so what, we may ask, is its relevance to thermodynamics?

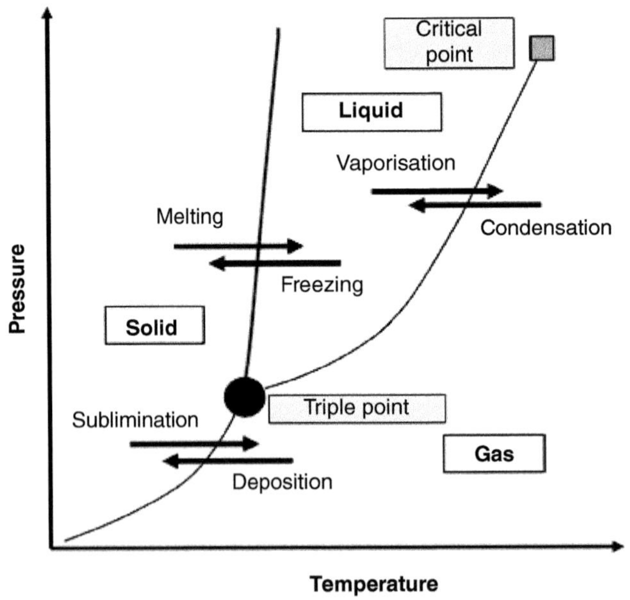

Figure 1.7 *P–T* diagram of a pure substance.

In fact, entropy is very relevant to thermodynamics and is the function of the quantity of heat that can be converted into work – and yes, it still does indicate the randomness of molecules. For example, suppose we put 100 J of energy into a system trying to convert this all into work can we then get 100 J back out? The answer of course is No! Some of that energy will be used in making more molecules move at random or increasing the momentum of those already in motion – that's why the entropy of a solid is less than that of a gas because gases have more molecules moving at random. Temperature increases the entropy of a material because as the temperature rises, the molecules spread out and take up more room – that is, they become more disordered. Pressure, on the other hand, has the opposite effect, it forces molecules closer together, so they become more ordered and entropy decreases. However, we should note that solids and liquids are nearly incompressible, so any entropy decrease is minimal.

So, in addition to those already mentioned, many other two- and three-dimensional diagrams can be plotted. For example, the *T–s* diagram (temperature as a function of specific entropy) and the *P–h* diagram (pressure as a function of specific enthalpy) are both very useful diagrams. An example of a temperature-specific entropy (*T–s*) diagram can be seen in Figure 1.8, and is often used to describe thermodynamic power cycles since the area drawn by the cycle on this form of diagram is equal to the work provided by it. Alternatively, a pressure–enthalpy (*P–h*) diagram, also called a Mollier diagram, might be used, as shown in Figure 1.9, to help describe receptor cycles such as refrigeration cycles.

These diagrams are also essential tools to aid understanding and for describing thermodynamic cycles, but in most cases when working with Rankine cycles the

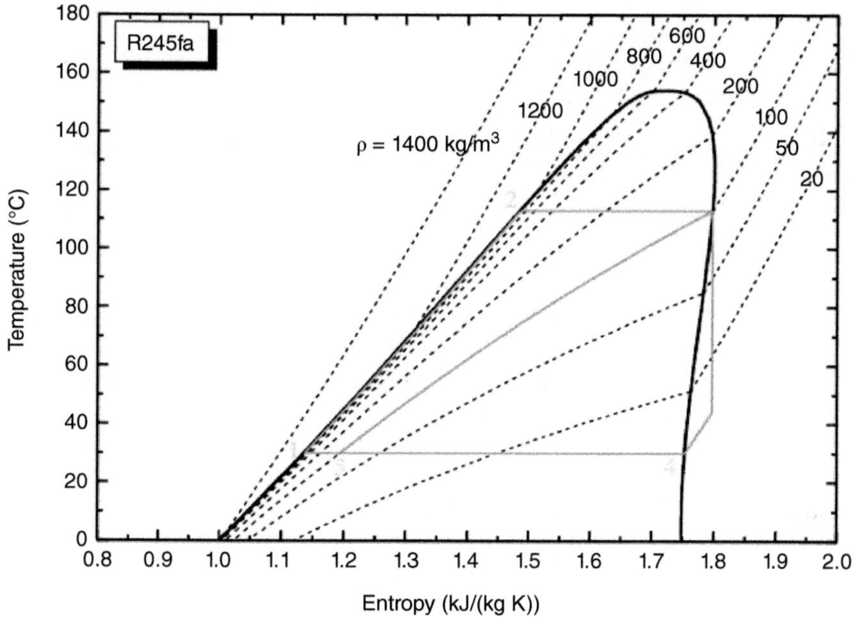

Figure 1.8 Example *T–s* diagram. Source: Based on Janie et al. [3].

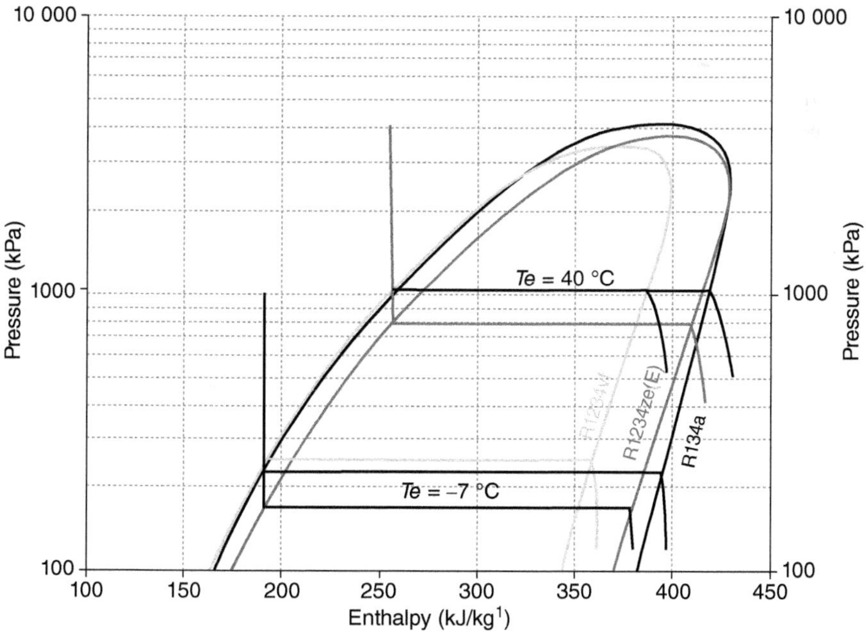

Figure 1.9 *P–h* diagram of refrigerants. Source: Sethi et al. [4].

T–s diagram is the most suitable. It will therefore be used extensively in the sections that follow.

1.2.3 The Carnot Cycle

Before we discuss the Rankine cycle and its application, we first need to consider its inspiration – the Carnot cycle [5].

Nicholas Léonard Sadi Carnot, the French mechanical engineer and physicist, is often described as the 'father of thermodynamics'. He published only one book, but in it he expressed the first successful theory of maximum heat engine efficiency and so laid the foundations, for what was then, an entirely new discipline. The Carnot cycle is therefore used as the common reference for every thermodynamic power cycle because even though the Carnot cycle is only a theoretical model, it does provide the standard as the most efficient engine possible. In other words, because the thermal efficiency of the Carnot cycle only depends on the absolute temperatures of the two heat reservoirs that drive the process, it gives us the upper limit of both energetic and exergetic thermodynamic cycle efficiency [6].

Most commonly, the working fluid used to explain the Carnot cycle is water and the cycle is divided into four phases as shown in Figure 1.10:

1. Isothermal expansion, (AB)
2. Isentropic (constant entropy) expansion, (BC)
3. Isothermal (constant temperature) compression, (CD)
4. Isentropic (constant entropy) compression, (DA)

Theoretically, these four processes are reversible, but in reality, reversible transformations do not exist because of, e.g. heat losses due to friction. The figure indicates a clockwise cycle and so is known as a Carnot 'power' cycle (power cycles are those which convert heat into mechanical work output whilst a 'heat pump' cycle transfers heat from lower to higher temperatures using mechanical work input).

Thermodynamic power cycles are the operating basis of heat engines, which currently supply much of the world's electric power and run almost all motor vehicles.

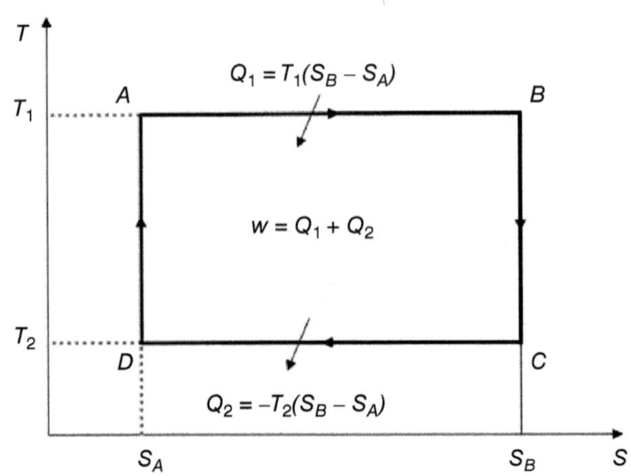

Figure 1.10 T–s diagram of the Carnot cycle.

Power cycles are classified according to the type of heat engine they model, e.g. we have the Otto cycle and the Diesel cycle for modelling internal combustion engines, the Brayton cycle for modelling gas turbines, and the Rankine cycle for modelling steam turbines.

Consider for a moment, the above *T–s* diagram as representing the Carnot cycle for an ideal or perfect gas turbine, i.e. one with no friction and the necessary heat transfers occurring without loss. What do the four stages of our cycle represent?

Essentially, we start with two 'heat reservoirs' with temperatures T_1 and T_2, respectively, and these have such a large thermal capacity that their temperatures are stable and unaffected by the cycle, which takes place as follows:

- Stage 1 (AB): Heat is transferred from the high-temperature reservoir (T_1) at constant temperature. During this phase, the entropy of the gas is increased by allowing it to expand and reduce its pressure and in doing so producing work on the surroundings – e.g. moving a piston or rotating a shaft within a turbine.
- Stage 2 (BC): During this stage of the cycle, it is assumed that the gas is thermally insulated from both the hot and cold reservoirs, so that they neither gain nor lose heat from the gas – i.e. the conditions are adiabatic (Greek ≡ without loss). Instead, the gas continues to expand (and do more work) further reducing the pressure and losing internal energy in an amount equal to the work done. Since this gas expansion occurs without any heat input, it cools to the 'cold' temperature T_2 but with the entropy unchanged.
- Stage 3 (CD): At this point the gas in now in thermal contact with the cold reservoir at temperature T_2, and work is then done on the gas to compress it – e.g. the piston is forced down or the gas passes into a compressor. In doing so, heat energy is transferred to the cold reservoir and the entropy of the system decreases by the same amount that it gained at Stage 1.
- Stage 4 (DA): It is assumed that the gas is thermally isolated from the hot and cold reservoirs (and any mechanical processes are frictionless) and further compression takes place, increasing the internal energy and so causing the temperature to rise back to T_1. This assumes that this temperature rise is solely due to the work added to the system and so the entropy remains unchanged. The gas is now back at the same state as the start of Stage 1.

As such, the work (kJ) provided by the thermodynamic cycle is given by:

$$W = \oint PdV = \oint TdS \tag{1.1}$$

In other words, the total work is given by integrating and thereby accumulating all the infinitesimally small changes in volume and pressure or entropy and temperature the gas makes on its journey around the cycle. Of course, the path taken by a real process differs somewhat from the ideal processes shown, but the idea here is to produce a perfect standard against which all other heat engines can be judged.

In a *T–s* diagram, the area under the upper portion of the figure represents the thermal energy absorbed during the cycle, whilst the area under the lower portion represents the thermal energy removed during the cycle – hence the area *inside* the cycle will represent the difference between these two. However, since the internal energy of the system must return to its initial value, this difference must also represent the total amount of work performed by the system over one cycle. So, in definitive terms on the diagram, the absolute value of *all* this work ($|W|$) corresponds to the area of the rectangle. Or:

$$|W| = (T_1 - T_2) \cdot (S_B - S_A) \tag{1.2}$$

If Q_H represents the total amount of thermal energy transferred from the hot reservoir to the system (kJ), and Q_C represents the total amount of thermal energy transferred from the system to the cold reservoir (kJ), both as given by Eqs. (1.3)–(1.5):

$$Q_H = T_1 \cdot (S_B - S_A) \quad \text{and} \quad Q_C = T_2 \cdot (S_B - S_A) \tag{1.3}$$

Then the efficiency of the Carnot cycle can be defined as:

$$\eta = \frac{|W|}{Q_H} = \frac{Q_H - Q_C}{Q_H} = \frac{(T_1 - T_2) \cdot (S_B - S_A)}{T_1 \cdot (S_B - S_A)} \tag{1.4}$$

where, T_1 is the hot source temperature and T_2 is the cold source temperature (both are expressed in [K] Kelvin), and S_B and S_A correspond, respectively, to the entropy (kJ/K) at points B and A. From which, the overall efficiency will be given by the following equation:

$$\eta = \frac{T_1 - T_2}{T_1} = 1 - \frac{T_2}{T_1} \tag{1.5}$$

1.2.4 Ideal and Actual Rankine Cycles

The Carnot cycle is just a theoretical cycle. It is not possible to build a real engine working under such conditions because of real component *irreversibility*. For example, the transformations result in heat dissipation through friction. So, these transformations are defined as irreversible because once they have occurred, it is impossible to return them to their initial state. A reversible transformation is a succession of infinitesimal changes, each occurring *without* any energy loss, and in practice, this kind of transformation does not exist in the real world even though each component is designed in order to approach an irreversible behaviour.

The Rankine cycle is very similar to the Carnot cycle, the difference being that the reversible isothermal transformations are now replaced by isobaric (constant pressure) ones. In most cases, the working fluid is water too.

The ideal Rankine cycle is composed of two isobaric transformations and two adiabatic and reversible isentropic (constant entropy) transformations as shown in Figure 1.11. The four main components required for the Rankine cycle are a pump, a boiler, a turbine, and a condenser as shown in Figure 1.12.

The four transformations involved in the cycle are:

- 1–2: Isentropic compression (pump). The fluid (in liquid state) is compressed while the specific entropy remains constant.

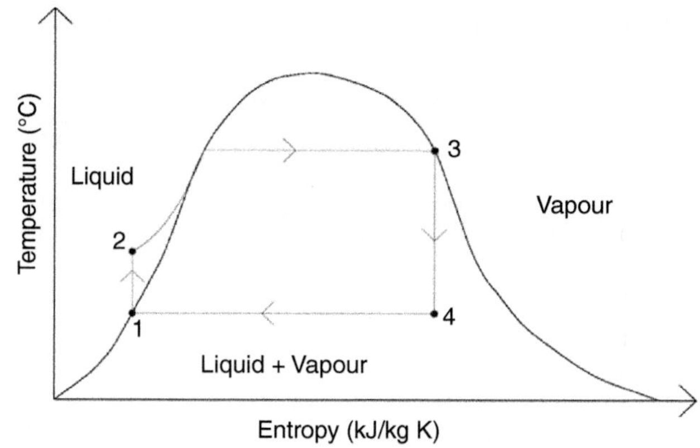

Figure 1.11 *T–s* diagram of the Rankine cycle.

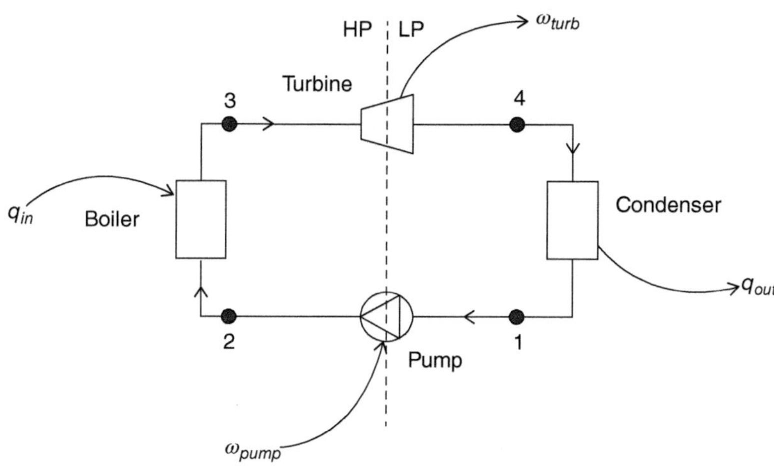

Figure 1.12 Schematic of the Rankine cycle.

- 2–3: Isobaric heat addition inducing vapourisation of the fluid (boiler), i.e. the fluid is heated under constant pressure and this results in a phase transition – the fluid changing from liquid to a mixture of liquid and vapour.
- 3–4: Isentropic expansion (turbine). The fluid is expanded through a turbine while the specific entropy again remains constant.
- 4–1: Isobaric heat rejection inducing liquefaction of the fluid (condenser), i.e. the fluid is cooled down under constant pressure and the vapour condenses.

1.2.4.1 Ideal Cycle

To describe the Rankine thermodynamic cycle, five physical variables are required:

1. The pressure P (Pa)
2. The temperature T (K)

3. The steam quality x, which represents the proportion of saturated vapour in a liquid/vapour mixture.

$$x = \frac{m_{vapour}}{m_{vapour} + m_{liquid}} \tag{1.6}$$

   where, m_{vapour} and m_{liquid} correspond, respectively, to the masses (kg) of the vapour and liquid contained in the mixture. (For example, at point number 1 on the *T*-*s* diagram, $x(1) = 0$, i.e. the fluid is a fully saturated liquid.)
4. The specific enthalpy of the fluid h (kJ/kg)
5. The specific entropy of the fluid s (kJ/kg K)

Knowing the values of each of these variables at every point of the diagram is very useful, and helpfully the five variables are linked to each other through physical relationships.

Looking at the diagram it may not be obvious, but the cycle is divided in two pressure areas: high pressure (HP) and low pressure (LP) – these pressures being expressed in Pascal (Pa). Points 2 and 3 are in the HP section and points 4 and 1 are in the LP section, and as the transformation between 2 and 3 is isobaric: $P_2 = P_3$. (And similarly, for the same reason, $P_1 = P_4$.) Then the work provided to the system corresponds to the area of the quadrilateral shape drawn by the four points, but to calculate this work, the specific enthalpy of the fluid must be used. From which we can then determine the *specific work* (ω) – which corresponds to the amount of work delivered per unit of mass (kJ/kg).

The thermodynamic system is exchanging energy (work or heat) with its environment, and we should note here the convention that: heat transferred *to* a fluid is a positive quantity; heat transferred *from* a fluid is a negative quantity. It is necessary to note this because a Rankine cycle uses part of its work to compress the water, so will need to deduct the power put into the pump from the power output of the turbine in order to obtain the net work. Accordingly, the required work and heat parameters are determined from:

ω_{turb}, the work given by the fluid to the turbine (kJ/kg). It is negative as the fluid is giving work:

$$\omega_{turb} = h_4 - h_3 \tag{1.7}$$

ω_{pump}, the work given by the pump to the fluid (kJ/kg). It is positive as the pump gives work to the fluid:

$$\omega_{pump} = h_2 - h_1 \tag{1.8}$$

q_{in}, the heat given by the boiler to the fluid (kJ/kg). It is positive as the boiler gives heat to the fluid:

$$q_{in} = h_3 - h_2 \tag{1.9}$$

q_{out}, the heat given by the fluid to the condenser (kJ/kg). It is negative as the fluid rejects heat:

$$q_{out} = h_1 - h_4 \tag{1.10}$$

So, as the net work (ω_{net}) (kJ/kg) provided by the cycle is governed by:

$$\omega_{net} = \omega_{turb} + \omega_{pump} \tag{1.11}$$

Then the thermal efficiency (η_{th}) of the ideal Rankine cycle can be defined by:

$$\eta_{th} = \frac{|\omega_{turb}|}{q_{in}} \tag{1.12}$$

And hence the overall efficiency (η) of the ideal Rankine cycle can be defined by:

$$\eta = \frac{|\omega_{net}|}{q_{in}} = \frac{|\omega_{turb}| - \omega_{pump}}{q_{in}} \tag{1.13}$$

Substituting Eqs. (1.4)–(1.6) then leads to:

$$\eta = \frac{|h_4 - h_3| - (h_2 - h_1)}{h_3 - h_2} = \frac{h_3 - h_4 - (h_2 - h_1)}{h_3 - h_2} \tag{1.14}$$

In practice however, compared to the output of the turbine, the work input to the pump is very small. In which case, if ω_{pump} can be neglected compared to the other coefficients, we arrive at the useful approximation of:

$$\eta \approx \frac{h_3 - h_4}{h_3 - h_2} \tag{1.15}$$

So to estimate the efficiency of the system, we can also use the first law of thermodynamics which states that the net energy of a system is equal to 0, i.e.

$$\omega_{turb} + \omega_{pump} + q_{in} + q_{out} = 0 \tag{1.16}$$

And further assuming ω_{pump} is neglectable compared to the other coefficients gives:

$$|\omega_{turb}| \approx q_{in} - |q_{out}| \tag{1.17}$$

From which the efficiency of the ideal Rankine cycle can now be estimated from the following formula:

$$\eta \approx \frac{|\omega_{turb}|}{q_{in}} \approx \frac{q_{in} - |q_{out}|}{q_{in}} = 1 - \frac{|q_{out}|}{q_{in}} \tag{1.18}$$

Practically the efficiency of the ideal Rankine cycle is necessarily lower than the efficiency of the theoretical Carnot cycle even using the same hot and cold sources. In addition, the cycle presented in this section is still an ideal cycle, and the actual cycle used, for instance, for power plant analysis is slightly different.

1.2.4.2 Superheated Rankine Cycle

In the conventional Rankine cycle, the isentropic expansion occurs in the wet steam domain (transformation 3–4 in Figure 1.11). To explain, when a boiler heats water, bubbles breaking through the water surface pull tiny droplets of water into the steam which then becomes partially wet ('wet steam') from the added liquid. As water is the working fluid, it can cause damage to the turbine due to corrosion. Therefore, to mitigate this in power plants, the vapour is superheated to remove any moisture that it contains. Steam dryness is important because it has a direct effect on the total

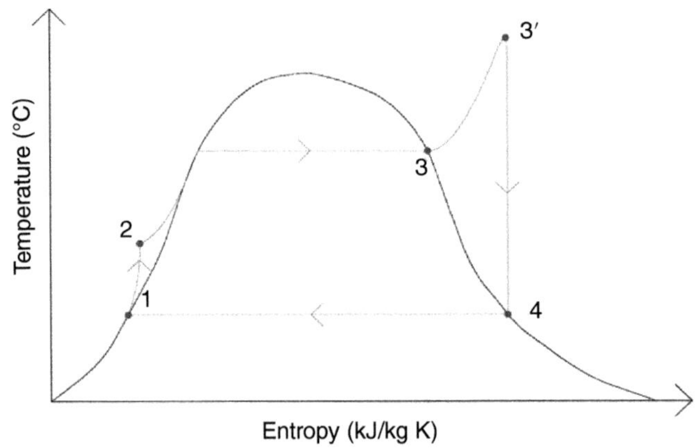

Figure 1.13 *T*–*s* diagram of the superheated Rankine cycle.

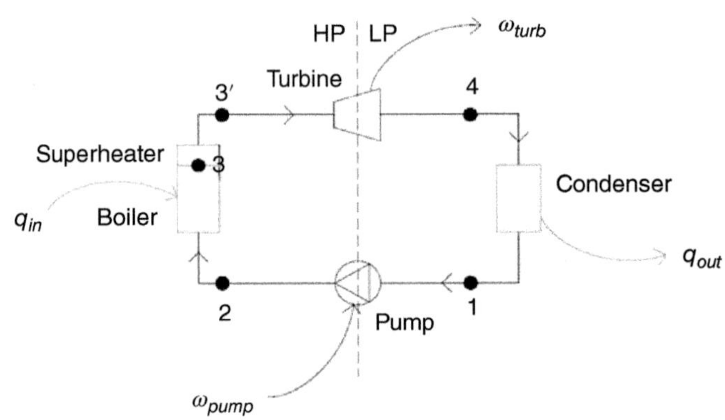

Figure 1.14 Schematic of the superheated Rankine cycle.

amount of transferable energy contained within the steam. For example, saturated steam (100% dry) contains 100% of the latent heat available at that pressure.

The result of this can be seen in the shape variation for the *T*–*s* diagram of the cycle as shown in Figure 1.13. In this case, the cycle is still a Rankine cycle and is sometimes called Hirn (or Rankine with superheating) cycle, so one extra component is required: the superheater as shown in Figure 1.14. However, in most cases, the boiler and the superheater are the same component.

Overheating is also interesting because it increases the amount of work given to the turbine. As we can see on the *T*–*s* diagram, the area of the superheated cycle is bigger than the area of the classical Rankine cycle. And in this way, for a same power output, the water flow rate can be lower for the Rankine–Hirn cycle than for a classical Rankine cycle, so improving the energetic efficiency of the system. Unfortunately, superheating also increases the irreversibility of the cycle and therefore, correspondingly, the exergetic efficiency of the system is regrettably reduced.

The main difference between the two cycles is that the heat given to the fluid by the boiler/superheater enhances the superheated cycle. On the *T*–*s* diagram, this extra heat addition corresponds to the transformation 3–3′. This way, the expansion of the fluid then occurs within a superheated steam domain and corresponds to the transformation 3′–4 on the diagram. Furthermore, this transformation is assumed to display a greater degree of isentropic change and so is shown as a vertical line on the diagram, contrary to the transformation we now see between 1 and 2.

The efficiency of the superheated cycle is given by the same equation as the classical Rankine cycle (Eqs. (1.13) and (1.14)) – the difference being in the position of point 3′ (instead of 3), i.e.

$$\eta = \frac{h_3' - h_4 - (h_2 - h_1)}{h_3' - h_2} \tag{1.19}$$

1.2.4.3 Actual Rankine Cycle

The *actual* Rankine cycle is the best cycle with which to model the transformations occurring within a power plant as it is the one which takes both system irreversibility and system imperfections into consideration.

In this cycle, friction caused by the moving fluid induces pressure losses or heat loss, and as a result the isobaric transformations described in the ideal Rankine cycle are not exactly isobaric in the actual cycle. Pressure drops in the condenser and in the turbine reduce the area of the cycle, as shown on the T–s diagram in Figure 1.15. Hence, the specific work of the cycle is lower for the actual cycle than for the ideal cycle. Furthermore, the isentropic transformations described in the ideal cycle are not reversible, as they are in the actual cycle: irreversibility deteriorates the performances of the pump and of the turbine. Hence, the exergetic efficiency is reduced.

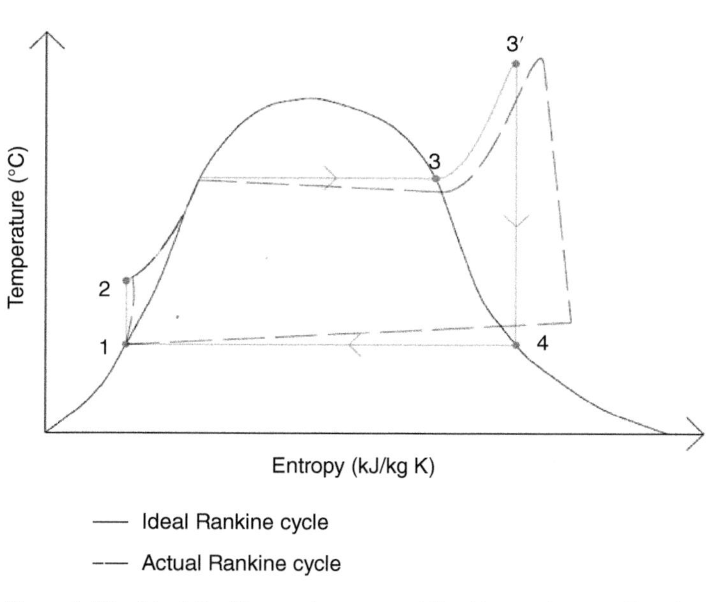

Figure 1.15 Ideal Rankine cycle vs. actual Rankine cycle on a T–s diagram.

In thermodynamics, *reversibility* defines the best theoretical work process, i.e. as the name suggests, one which can proceed in both the forward and the reverse directions – but with the key point that the system and its parameters are returned to their initial states at the end of the reverse process. *Irreversibility* is a concept that recognises that processes are not perfect, losses occur, and in thermodynamics these are quantified by:

- The difference between the work output (or input) of a reversible process and the actual work process
- The entropy a system generates during a process.

Fortunately, the efficiency of the cycle can be given by the same expression as the ideal Rankine cycle (Eq. (1.14)). However, the values of the specific enthalpy are not the same as those in the ideal Rankine cycle.

To make comparisons between the compression and expansion transformations of an actual cycle to the isentropic transformations of the ideal cycle, an isentropic efficiency needs to be introduced and this comes within the term the *adiabatic efficiency*, and a series exists to define isentropic efficiencies for turbines, compressors, and nozzles. For instance, the isentropic efficiency of the expansion transformation in the turbine is defined as follows:

$$\eta_{i,turb} = \frac{\text{Energy given to the turbine with the actual transformation}}{\text{Energy given to the turbine with the isentropic transformation}} \tag{1.20}$$

As an example, Figure 1.16 compares the actual and the isentropic expansion of the vapour in a turbine on a *h*–*s* diagram, and to accommodate the difference, an adaption is made to Eq. (1.20) in order to express this adiabatic efficiency using specific enthalpy as follows: If we start by assuming that vapour is an ideal gas, then it is possible to calculate the pressure, temperature, enthalpy, and entropy of the fluid using the Laplace's equation:

$$P^{1-\gamma} \cdot T^{\gamma} = \text{Constant} \tag{1.21}$$

In this equation, γ is the heat capacity ratio, a dimensionless coefficient.

Knowing the temperature and pressure values at the inlet and the pressure at the outlet of the turbine, it is then possible to calculate the temperature value that would be reached if the transformation was truly isentropic. Then once T_{2s} is calculated, and h_1, h_{2a}, and h_{2s} deduced from thermodynamic tables, the adiabatic efficiency of the turbine can be calculated. The isentropic efficiency of the pump is often also used to take the irreversibilities of the pump into consideration.

It is given by the following equation:

$$\eta_{i,pump} = \frac{\text{Isentropic pump work}}{\text{Actual pump work}} \tag{1.22}$$

And the process starts by assuming the water is incompressible, which then leads to:

$$\text{Isentropic pump work} = \Delta h = V \cdot \Delta P \tag{1.23}$$

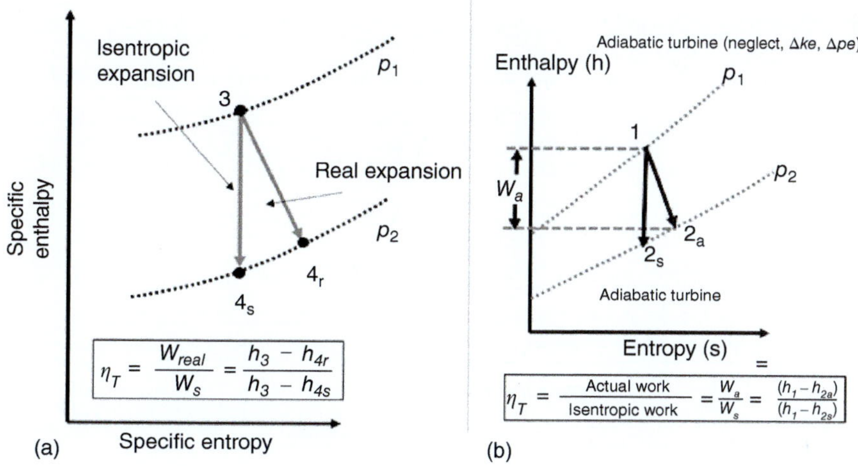

Figure 1.16 Adiabatic efficiency of a turbine.

In this equation, V is the specific volume of water (m^3/kg). It is constant along the transformation if the fluid is supposed incompressible. The pressure difference,ΔP, is expressed in Pascal (Pa). Again, knowing the actual inlet and outlet temperatures/pressures of the pump, it is possible to calculate the actual pump work, and this corresponds to the variation of enthalpy. All from which the isentropic efficiency of the pump can be deduced.

So, when using equations for the ideal cycle, it is important to be aware of the real efficiencies for all the components. For example, the turbine and the pump both also have a mechanical efficiency. All these imperfections will be described in depth in the power plant section so that we can better express the real link between the heat input and the electricity output produced by a plant.

1.2.4.4 Improvements to the Rankine Cycle

The Hirn cycle has a greater efficiency than the Rankine cycle and this is expected since the expression for Carnot's efficiency (Eq. (1.5)) illustrates that the higher the hot source temperature, the greater the efficiency. It is also true for the Rankine cycle (Eq. (1.18)).

Therefore, increasing the temperature of the hot source is a way of improving the efficiency of the Rankine cycle, but it also results in an increase in the boiler pressure. In the past, the boiler pressures were limited because of the materials used [7]. Over time however, the upper limit on boiler pressure rose and currently, with advances in metallurgy, it is possible to go over the critical pressure of water (220.6 bar) and some coal-fired power plants today are operating with boiler pressures around 300 bar.

Moreover, other improvements can also be made to improve the efficiency of the cycle, and these main improvements are presented within this section.

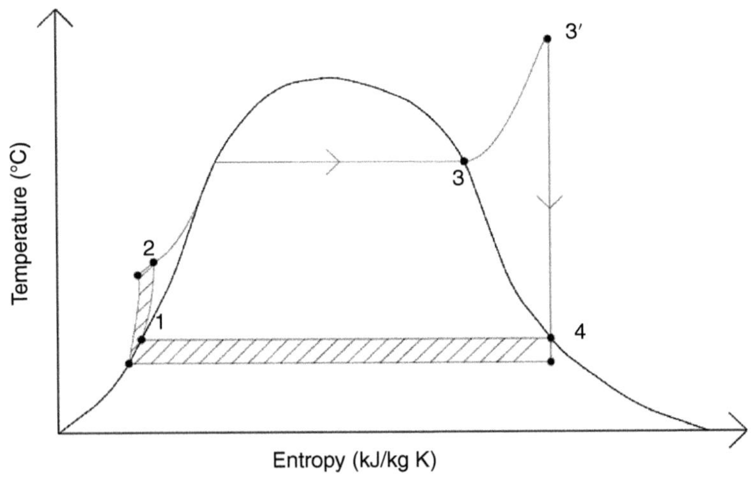

Figure 1.17 Effects of decreasing the LP on the Rankine cycle.

Decreasing the LP As we have noted, on our diagrams the lines drawn between the different points of the cycle represent thermodynamic processes and furthermore, the area enclosed by these lines corresponds to work provided by the system. This means that trying to increase the amount of work given by the system is the graphical equivalent to expanding this area, and one way to achieve this is to decrease the pressure of the system. As we can see, the lower the pressure, the lower the bottom line (condensation phase) of the cycle, i.e. the area of the cycle is increased (Figure 1.17).

In addition, if the pressure decreases, it leads to an increase in net work compared to the increase in heat input ($\Delta\omega_{net} > \Delta q_{in}$), then the efficiency of the system will also increase. However, as we see in Figure 1.17, decreasing the pressure of the system can result in introducing lower quality steam into the turbine, i.e. we run the risk or moving back into the wet steam domain – the further the point is from the saturated vapour curve, the lower the steam quality. Low steam quality can result in turbine deterioration due to corrosion and a subsequent decrease in mechanical efficiency. Hence, the pressure of the Rankine cycle must be set as low as possible in order to improve the energetic efficiency of the system – but not too low to still ensure a high steam quality.

Increasing the HP We might therefore conclude that increasing the pressure will also have a beneficial effect, but increasing the pressure actually keeps the work of the cycle roughly steady. Figure 1.18 shows the effect of this increase on a *T*–*s* diagram. Helpfully, the area of the cycle increases on the top part of the cycle (area shaded on the diagram) but unfortunately, it then also decreases on the right part of it (area shaded on the diagram) so the net gain is very small. Nonetheless, as the average temperature of the hot source increases, so increasing the pressure of the cycle has a positive effect on the cycle's efficiency. However, this increase then leads to a lower steam quality at the end of the expansion phase (point 4) which, as we have seen, can affect turbine operation.

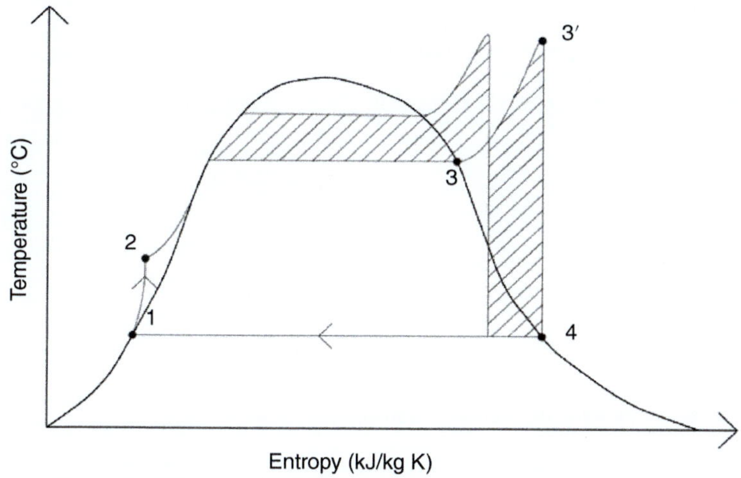

Figure 1.18 Effects of increasing the HP on the Rankine cycle.

To sum up, both the HP and the LP values must be carefully chosen so as to ensure that we achieve the highest possible energetic efficiency whilst still ensuring that we maintain high steam quality at the end of the expansion phase. Of course, it does not help that with modern plants, the LP and HP can vary as functions of time. But by using controllers which adjust these values (LP; HP) by taking into account external condition – such as outside temperature and pressure – we can ensure optimal performance.

Rankine Cycles with Reheat As we have seen, steam quality is a very important factor in the Rankine cycle. So, in order to avoid low steam quality at the end of the expansion phase, reheat cycles were introduced in the first part of the twentieth century. This category of cycle also involves having two turbines positioned on the same shaft as shown in Figure 1.19. The first turbine allows the expansion of the steam coming from the superheater from HP to a medium pressure (MP) (Figure 1.20), and this steam is then reheated and expanded in the second turbine from MP to LP. The main advantage of the reheat is to increase the steam quality to ensure a steadier thermodynamic efficiency.

Rankine cycles with reheat also ensure a higher overall mechanical efficiency by using two turbines for the cycle compared to only using one. In this configuration, the overall efficiency of the cycle is given by:

$$\eta = \frac{|\omega_{turb1}| + |\omega_{turb2}| - \omega_{pump}}{q_{inB} + q_{inR}} = \frac{|h_4 - h_3'| + |h_6 - h_5| - (h_2 - h_1)}{h_3' - h_2 + h_5 - h_4}$$
$$= \frac{h_3' - h_4 + h_5 - h_6 - (h_2 - h_1)}{h_3' - h_2 + h_5 - h_4} \quad (1.24)$$

where, q_{inB} corresponds to the heat (kJ/kg) provided by the boiler to the fluid and q_{inR} corresponds to the heat (kJ/kg) provided by the reheater to the fluid (In real-life power plant, the boiler and the superheater use the same heat source.)

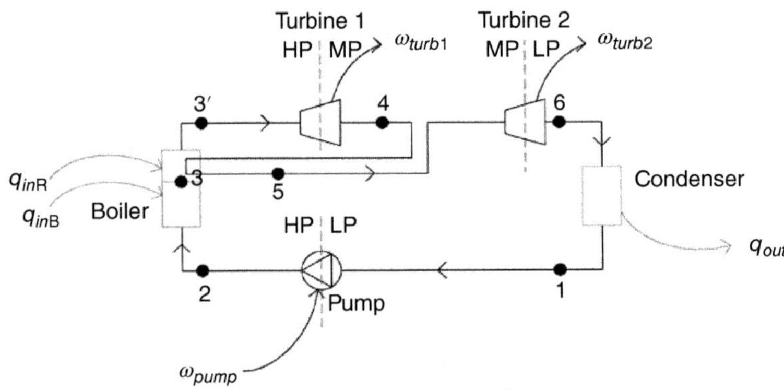

Figure 1.19 Schematic of the Rankine cycle with reheat.

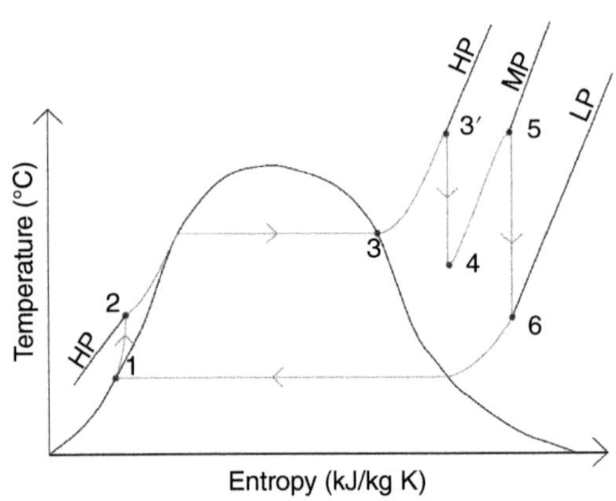

Figure 1.20 *T–s* diagram of a Rankine cycle with reheat.

1.2.4.5 Regenerative Rankine Cycles

To improve energetic and exergetic efficiency of the thermodynamic cycle, the regenerative Rankine cycles extracts a portion of the steam from the turbine to heat the fluid before it is sent to the boiler. This results in a small reduction in the work provided to the turbine but a substantial reduction in the heat input, because the heating phase creates entropy and thus irreversibility if this is all provided by the boiler. The portion of steam used therefore helps the boiler heat the fluid, reducing the fuel required by the boiler and hence both energetic and exergetic efficiencies improve. Furthermore, the average temperature of the heat input by the boiler is increased and this also results in a higher energetic efficiency [8].

Regenerative cycles require a new component. A feed water heater (FWH) also called a *regenerator* is required, its role being to heat the feed water before it enters the boiler. There are two different kinds of regenerative cycle: the open FWH cycle and the closed FWH cycle.

Open FWH cycles use an Open FWH. As seen in Figure 1.21, this component mixes the liquid delivered from pump 1 with the extracted steam arriving from turbine. It corresponds to point 6 on the *T–s* diagram shown in Figure 1.22.

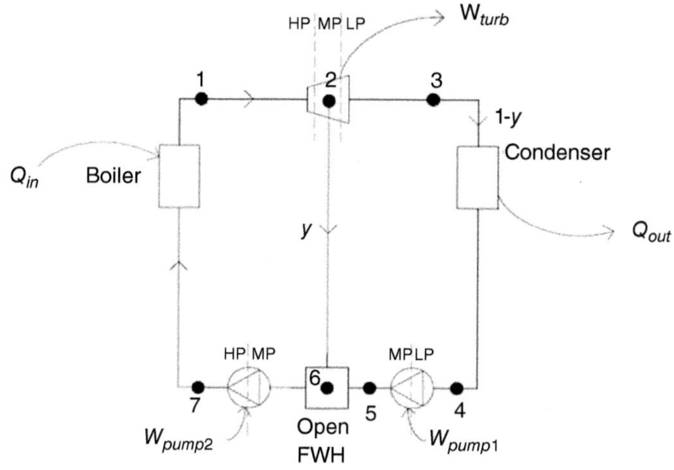

Figure 1.21 Schematic of a Rankine cycle with reheat and one open FWH.

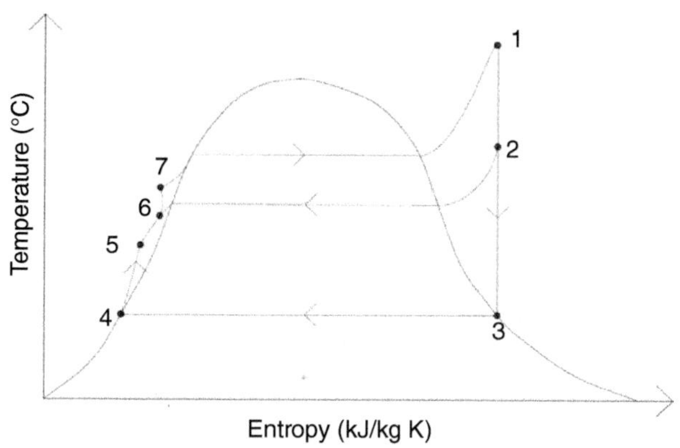

Figure 1.22 *T–s* diagram of a Rankine cycle with reheat and one open FWH.

As the mixing takes place at MP, pump 1 is only required to compress the fluid from LP to MP. Hence, after this mixing phase, a second pump is then required to increase the pressure to HP. This second compression phase corresponds to the transformation 6–7 on the *T–s* diagram (Figure 1.21).

Now if y represents the mass proportion of steam extracted from the compressor, and:

$$y = \frac{\dot{m}(2 \to 6)}{\dot{m}(1 \to 2)} \tag{1.25}$$

where $\dot{m}$ $(2 \to 6)$ corresponds to the mass flow rate (kg/s) between points 2 and 6, and $\dot{m}(1 \to 2)$ corresponds to the mass flow rate between points 1 and 2.

Then the overall efficiency of the open FWH cycle is given by:

$$\eta = \frac{|\omega_{turb}| - \omega_{pumps}}{q_{inB}} = \frac{(h_1 - h_2) + (1-y)(h_2 - h_3) - (1-y)(h_5 - h_4) - (h_7 - h_6)}{h_1 - h_7} \tag{1.26}$$

This efficiency is always greater than the efficiency of the ideal Rankine cycle given in Eq. (1.14). Indeed, the greater the number of regenerators, the higher the cycle efficiency.

So, writing the energy balance of the open FWH leads to:

$$y \cdot h_2 + (1 - y) \cdot h_5 = h_6 \tag{1.27}$$

Now the open FWH requires two pumps: one at its inlet and one at its outlet, but to avoid this multiplication of pumps in a power plant, it is possible to use a closed FWH. Moreover, closed FWHs are more reversible than open ones.

A closed FWH has the same role as the open, but there is no mixing of the extracted steam with the feed water. This way the extracted steam and the feed water can have different pressures, the FWH then operating as a shell and tube heat exchanger. It is more expensive and complex than an open FWH, but requires less area and maintenance is reduced because of the lower number of pumps.

In the closed FWH system the extracted steam is cooled and turned into condensate and two options are available to reinject this condensate in to the cycle as shown Figures 1.23 and 1.24.

The 'drain cascaded backward' technology is the cheapest way to reinject the condensate, and with this configuration a steam trap is first used to reduce the pressure from MP to LP, which then reintroduces the condensate to the condenser as shown in Figures 1.25 and 1.26. However, the overall adiabatic efficiency of a steam trap is lower than the average efficiency of a pump, so the *drain pump forward* technology offers better performance. In fact, in the steam trap, because the pressure drops suddenly, exergy is destroyed and so is not reversible. On the other hand, in the pump (of the pump forward technology), the pressure of the fluid is increased more slowly resulting in a process more closely resembling a reversible process and resulting in less exergy reduction.

Reiterating, the drain pumped forward technology is more expensive and requires more maintenance. In the configuration, a pump is used to increase the condensate

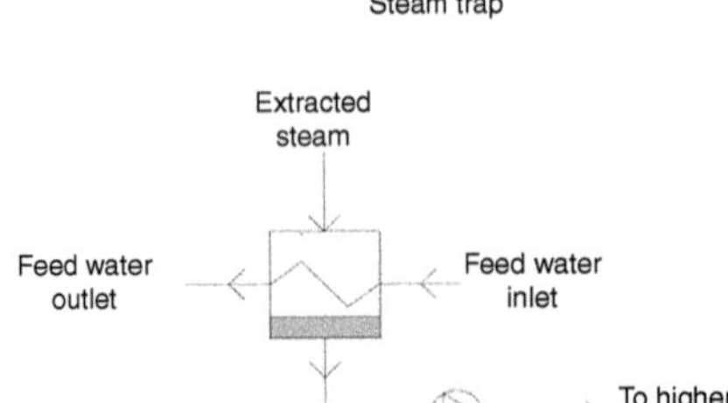

Figure 1.23 Schematic of the drain cascaded backward technology.

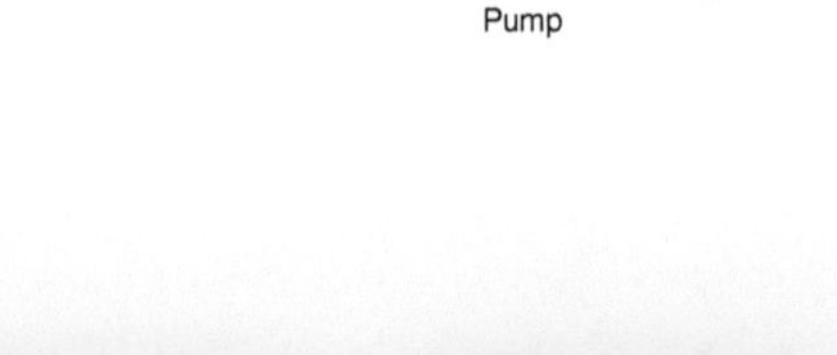

Figure 1.24 Schematic of the drain pump forward technology.

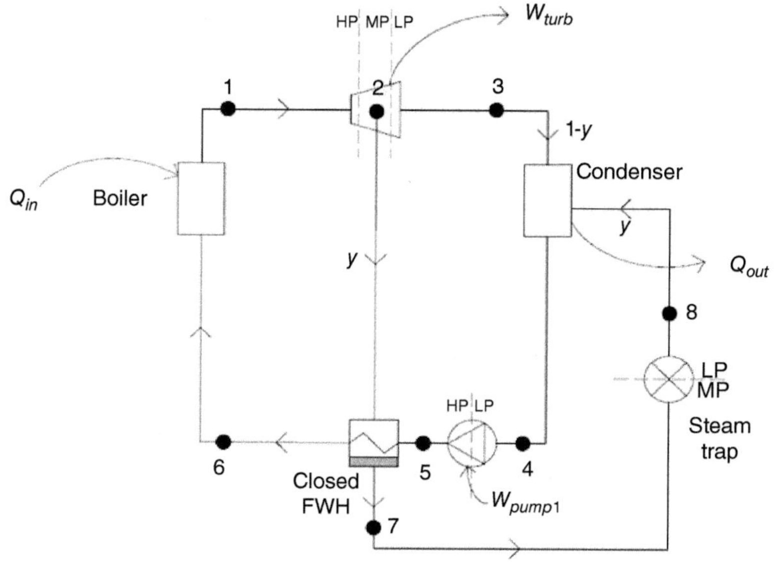

Figure 1.25 Schematic of a Rankine cycle using closed FWH and drain cascaded backward technology.

Figure 1.26 *T*–*s* diagram of a Rankine cycle using closed FWH and drain cascaded backward technology.

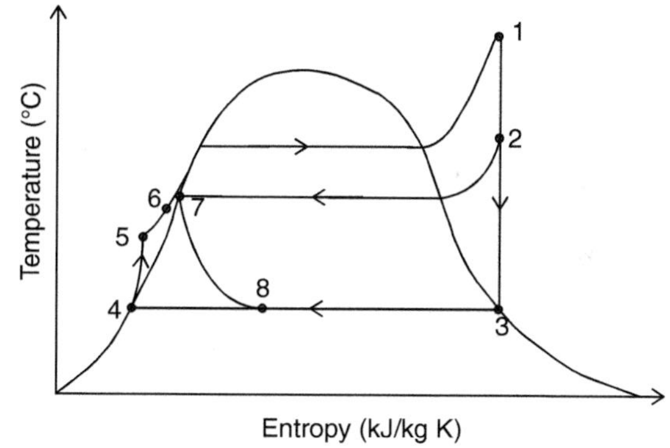

pressure from MP to HP, and then the condensate is mixed with the feed water in a mixing chamber, as shown in Figure 1.27. However, this process does produce higher energetic and exergetic efficiencies because the pump creates less irreversibility than a steam trap.

In a closed FWH (using the same notations as Figure 1.27), the energy balance leads to:

$$y \cdot (h_2 - h_7) = (1 - y) \cdot (h_6 - h_5) \tag{1.28}$$

This expression for the efficiency of the closed FWH cycle is the same as the expression for the efficiency of an open FWH. However (in Figure 1.27 for example), the benefit of the closed FWH is that the specific energy given by the boiler (q_{in}) is equal

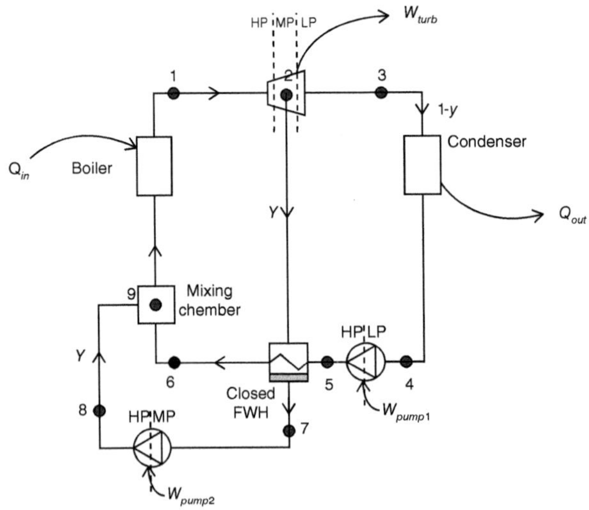

Figure 1.27 Schematic of a Rankine cycle using closed FWH and drain pumped forward technology.

to $h_1 - h_9$. Without it, the specific energy would have been $h_1 - h_5$, and since $h_5 < h_9$, q_{in} is reduced with a closed FWH.

Regenerative Rankine cycles are widely used in electrical power plants – the number of FWH systems varying from 6 to 10 and, in most of the cases, both open and closed FWH configurations can be used in the same plant.

1.2.4.6 Cogeneration

Cogeneration is the generation of two types of energy using only one source of primary energy. In most cases, cogeneration processes produce both electrical and thermal energies and so cogeneration is also known as *combined heat and power* (CHP).

The aim of a Rankine cycle-based power plant is to create electrical energy through the rotation of a steam turbine. However, in the process we also waste a lot of heat. Cogeneration offers a solution to utilise the thermal energy of this waste heat, thereby increasing the overall efficiency of the whole process. In Europe in 2011, 11% of heat and electricity requirements were produced with cogeneration plants [9].

Here is an example to better understand the benefits of cogeneration. To produce 350 kW electrical (kW_e) and 530 kW thermal (kW_{th}), a cogeneration power plant with an electrical efficiency (Eq. (1.29)) of 35% and a thermal efficiency (Eq. (1.30)) of 53% would use 1000 kW of primary energy. By comparison, a high efficiency power plant, such as a combined cycle gas turbine (CCGT) power plant, with an efficiency of 55% combined with a high efficiency boiler at an efficiency of 90% would require 1225 kW of primary energy [10]. Hence, in this example, the CHP cogeneration power plant saves 18% of the primary energy in comparison to the separate production of heat and electricity. The benefits of cogeneration are therefore: energetic, economic, and environmental.

$$\eta_e = \frac{P_{elec}}{P_{in}} \tag{1.29}$$

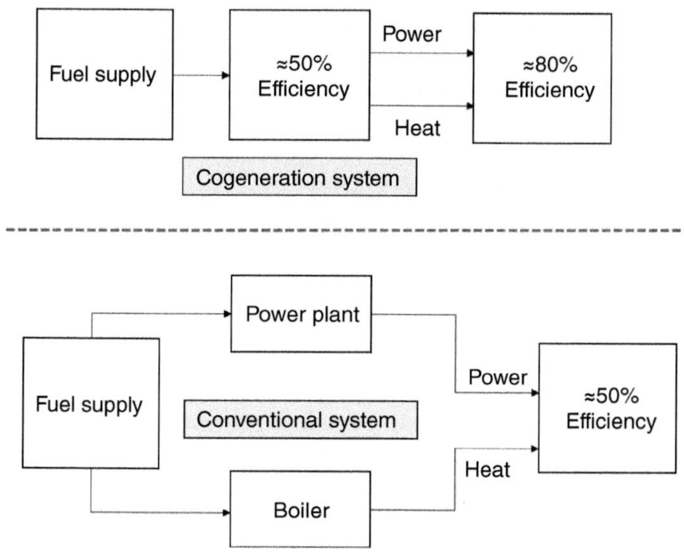

Figure 1.28 Cogeneration vs. conventional system.

$$\eta_{th} = \frac{P_{thermal}}{P_{in}} \tag{1.30}$$

where, P_{elec} is the electrical power output, $P_{thermal}$ is the thermal power output, and P_{in} is the total power input (kW). In this example, the overall efficiency of the cogeneration plant is 88% whereas the overall efficiency of the conventional system is 72%. Furthermore, the overall system efficiency is further reduced during the delivery of this conventional electrical power owing to approximately 10% (or more) power loss in transmission.

In its general application therefore, Figure 1.28 illustrates why cogeneration is a better solution than conventional systems for producing both heat and electricity. Notwithstanding other advantages, the average overall efficiency of a cogeneration plant is higher than the average overall efficiency of a conventional system.

The efficiency of a cogeneration power plant is given by:

$$\eta_{cogen} = \frac{P_{elec} + P_{thermal}}{P_{in}} \tag{1.31}$$

Although cogeneration power plants can use fossil fuel or uranium as their combustible, it is also possible to use renewable forms of energy such as solar, biomass, or geothermal.

The main disadvantages of a cogeneration power plant are that: (i) the production of electrical energy involves the production of a large quantity of thermal energy – which is sometimes difficult to use – and (ii) electrical energy is much more versatile in comparison with thermal energy. Common ways to harness the thermal energy produced by cogeneration plant is to use it for industrial applications or for district heating, e.g. through a heat network [11], but other applications are also practical such as using the heat to warm the water of a swimming pool. It is also possible to use a portion of the thermal heat output to produce cooling energy through a *sorption* chiller, this process being called trigeneration [12].

A cogeneration Rankine cycle requires a heat exchanger to recover heat from the fluid and in one of several possible cogeneration configurations, heat is recovered directly from the condenser – in which case, the cogeneration plant is operating as a back pressure plant. The temperature of the heating stream is directly linked to the pressure of the fluid within the condenser, this pressure being controlled by a valve. The higher the pressure of the condenser, the higher the temperature of the heating stream. However, we must be careful because the higher the pressure within the condenser, the lower the efficiency of the thermodynamic cycle.

So, to avoid influencing the LP of the cycle, it is also possible to extract a fraction (y) of the steam coming from the turbine to add heat to the heating stream. In this case, the heat recovery does not influence the LP of the system. Unfortunately, as the mass flow rate in the turbine is reduced, the work delivered by the fluid to the turbine is also reduced. As the fraction extracted is controlled with a valve, different heating needs can be met by varying this. Figure 1.29 is a schematic of this kind of cycle. Point 6 in the figure is the mixing chamber.

Figure 1.30 represents the T–s diagram of this cogeneration cycle, which is very similar to the T–s diagram for the regenerative Rankine cycle with an open FWH. The difference between those two cycles being that, with the open FWH configuration, the latent heat from the extracted steam is entirely given to the feed water for the regenerative cycle, whereas with the cogeneration cycle, a fraction of this latent heat is given to the heating stream. The value of this fraction is a function of the temperatures and flows within the heat exchanger – which can vary with the time. Hence, the position of point 2′ on the T–s diagram is variable. For example, if the process heater recovers 100% of the latent heat from the extracted steam, then this point will be on the left side of the vapour dome. However, if it is less than 100%, the point will be under the vapour dome, in which case, the rest of the latent heat is given to the feed water and the mixing chamber acts as an open FWH.

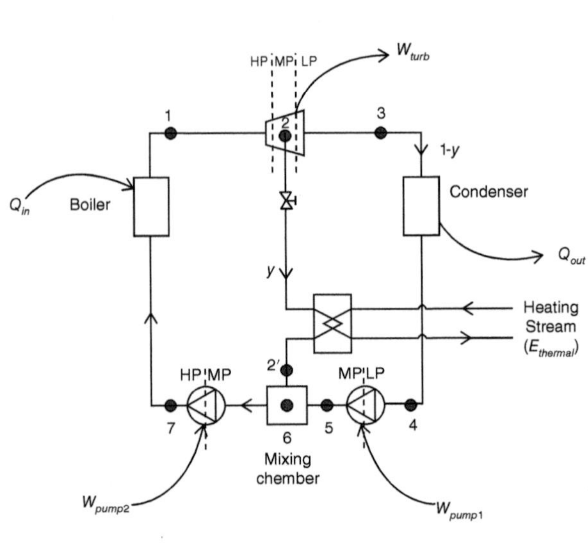

Figure 1.29 Schematic of a cogeneration Rankine cycle.

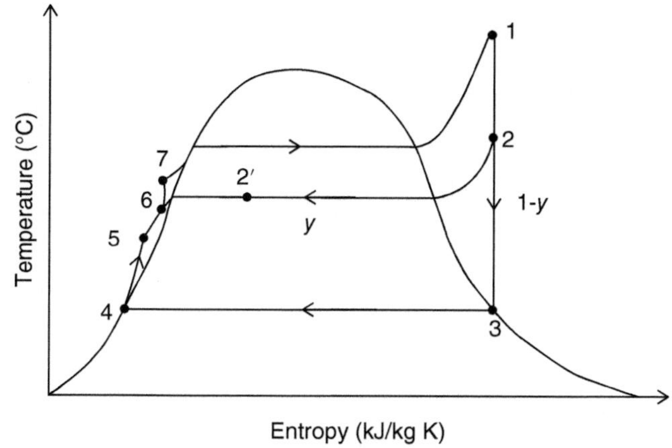

Figure 1.30 *T–s* diagram of the cogeneration Rankine cycle.

The overall efficiency of this thermodynamic cycle is given by:

$$\eta = \frac{|W_{turb}| + |E_{thermal}| - W_{pump1} - W_{pump2}}{Q_{in}} \tag{1.32}$$

$$\eta = \frac{h_1 - h_2 + (1-y)(h_2 - h_3) + y(h_2 - h_2') - (1-y)(h_5 - h_4) - (h_7 - h_6)}{h_1 - h_7}$$

where, h is the enthalpy at a certain point denoted by the number, W is the work done, $E_{thermal}$ is the thermal energy, and Q_{in} is the energy input.

Please note that the configuration shown in Figures 1.29 and 1.30 is just one possible configuration of the cogeneration Rankine cycle. Other configurations exist and each one is specific to a certain type of application. For example, Figure 1.31 is a cogeneration cycle used for hot temperature applications, where instead of extracting steam from the turbine at an intermediate pressure, the steam is extracted just after the boiler. Hence, this steam is hotter (because its pressure is higher) and for that reason the configuration is suitable for hot temperature applications. It is also controllable by means of an expansion valve positioned just before the heat exchanger which controls the pressure (and therefore the temperature) inside the exchanger allowing the system to adapt to different heating demands.

1.2.5 Other Configurations of the Rankine Cycle

So far, only the classical Rankine cycle and its possible improvements have been presented. This section now describes other thermodynamic cycles, derived from the Rankine cycle, which also use steam as a working fluid.

1.2.5.1 Supercritical Rankine Cycles

As we have noted previously, increasing the HP of the system results in a more work given to the turbine. However, because of the vapour dome, it also results in a lower steam quality at the end of the expansion phase. The supercritical Rankine cycle

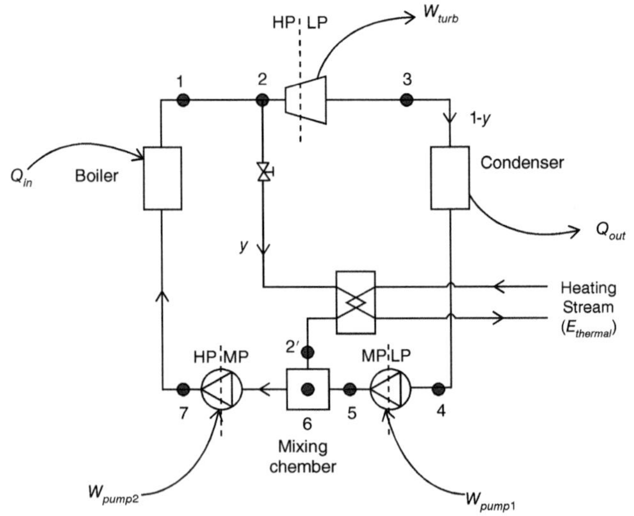

Figure 1.31 Schematic of a cogeneration Rankine cycle for hot temperature applications.

(SRC) bypasses this issue, because the working fluid is brought to such a condition that its pressure is above its critical pressure – a condition where the fluid is in a *supercritical* state and where there is now no longer a link between its vapour pressure and its temperature.

Advances in metallurgy and manufacturing techniques have made this possible, to the point where we can realistically increase the HP to a pressure greater than the critical pressure (for water) of 22.064 MPa (220.64 bar).

Hence, in an SRC, as the name implies, the fluid is in a supercritical state, i.e. it is behaving as a liquid and a gas at the same time. Working as a SRC increases the area of the cycle on the *T*–*s* diagram (see Figure 1.32), with a resultant increase in the work given by the fluid to the turbine.

As their working fluid, SRCs can use steam or organic fluids, but steam is still the most common fluid for both nuclear and fossil fuel power plants.

1.2.5.2 Reverse Rankine Cycles

The reverse Rankine cycle is another Rankine cycle widely used in the refrigeration sector, where instead of cycling clockwise on the *T*–*s* diagram, the cycle operates in a counterclockwise direction (as shown in Figure 1.33).

As we might expect, the components involved in this cycle are different and comprise:

- A compressor, used to compress the gas
- A condenser, used to cool and liquefy the fluid
- An expansion valve, used to drop the pressure of the liquid and so expand it back to a gas
- An evaporator, used to take heat from the environment

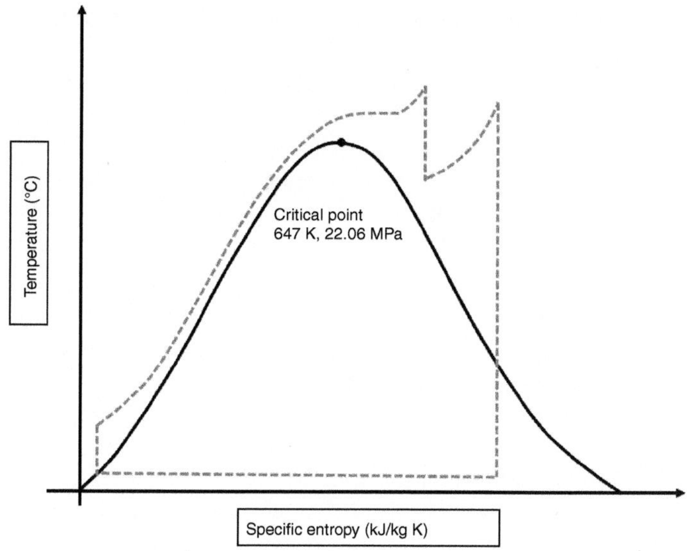

Figure 1.32 *T–s* diagram of a supercritical Rankine cycle using water as a working fluid.

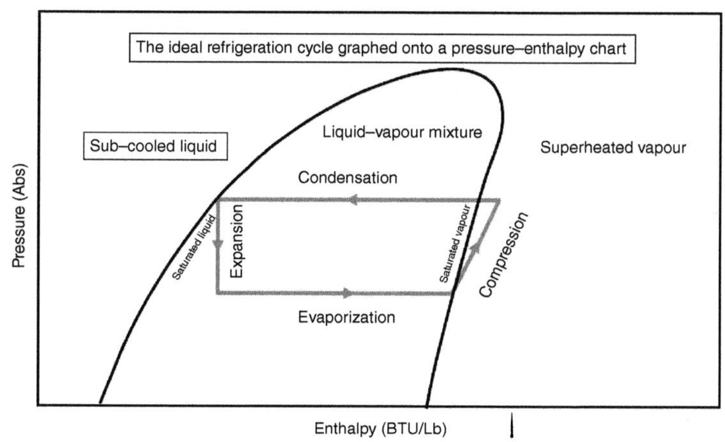

Figure 1.33 *P–V* diagram of the reverse Rankine cycle.

1.2.6 Rankine Cycles in Power Plants

The main application of the Rankine cycle is in converting thermal energy into mechanical energy in order to then obtain electrical energy. Hence, Rankine cycles are used in almost every high-temperature thermal power plant (fossil fuel, nuclear, geothermal, biomass, concentrated solar power, etc.). This section briefly showcases these different kinds of thermal power plant and their associated Rankine cycle.

1.2.6.1 Fossil Fuel Power Plants

Fossil fuel power plants consume either coal, fuel, or natural gas; the output power of such power plants typically range between 500 and 700 MW. The operation of such plant is based on complex Rankine cycles involving several turbines, pumps, reheaters, and regenerators, and sometimes, multiple thermodynamic cycles are used to improve the overall efficiency of the power plant. For example, the CCGT is a type of power plant using natural gas as fuel, which uses both the Rankine cycle and a gas turbine (Brayton) cycle to rotate two different electric generators.

In most cases, the cycle working fluid is still water because the temperature of the hot source is high enough to vapourise the fluid under HP conditions.

1.2.6.2 Nuclear Power Plants

Enriched uranium is a high-energy density fuel – producing 1000 kWh requires only 4 g of this material whereas it requires 350 kg of coal! The power output of a nuclear reactor is often between 0.9 and 1.5 GW [13] and different designs of nuclear reactor exist, the most widespread being light water reactors such as the EPR reactor or the AP1000 reactor. Heavy water, gas-cooled, and fast neutron reactors are less common but provide other advantages over light water reactors [14].

In light water reactors, three water circuits are used to avoid radiation contamination. The primary circuit is still a hot water loop and is the hot source for the second circuit, which then operates through a complex Rankine cycle. The third circuit is used in order to cool the fluid after the expansion phase and acts as a condenser.

1.2.6.3 Overall Efficiency of a Power Plant

To determine the best way to convert energy into electricity we can compare the overall efficiencies for different types of power plants. The overall efficiency of a power plant corresponding to the ratio between the thermal energy input and the electrical energy output over a given time being:

$$\eta_{overall} = \frac{\text{Electrical energy provided by the plant}}{\text{Thermal energy given by the combustible}} \tag{1.33}$$

For the fossil fuel power plants, the thermal energy given by the combustible corresponds to the heat of combustion for the mass of fuel used during the given time, and the current average value of the efficiency of a Rankine-based power plant using fossil fuel is 33% [15]; however, it can range up to 40% in some cases [16]. In comparison, CCGT power plants have an average efficiency of 45% but they can reach values in excess of 50% and the efficiencies of current nuclear power plants are similar to the efficiencies of the Rankine-based fossil fuel power plants, i.e. with an average value of 33%.

Figure 1.34 is an illustration of the losses occurring during the energy transformation process:

- Neither the combustion of the fluid nor the nuclear fission reaction are perfect processes. Hence it is necessary to introduce a boiler efficiency term (η_{boiler}) to quantify the amount of heat given to the system compared to the heat delivered by the combustible/reaction:

$$\eta_{boiler} = \frac{Q_{in}}{\text{Thermal energy given by the combustible}} \tag{1.34}$$

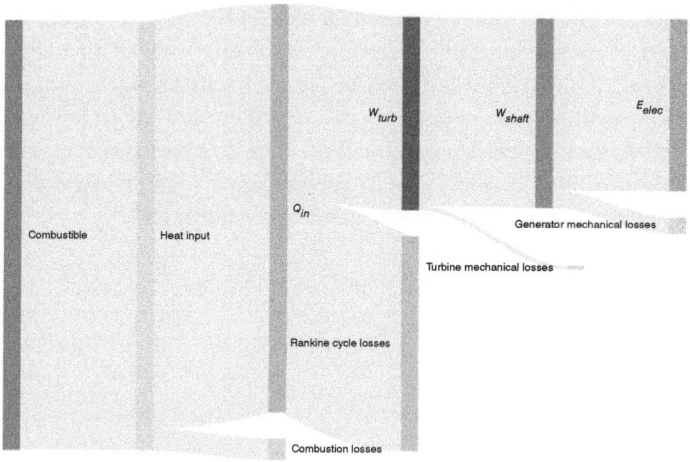

Figure 1.34 Sankey diagram of a Rankine-based power plant.

- The efficiency of the Rankine cycle (Eq. (1.14)) always induces energy losses. Indeed, even if the cycle was ideal and without irreversibilities, some losses would still be introduced by the cycle because the efficiency of this cycle cannot be higher than the Carnot's efficiency (Eq. (1.5)).
- Some of the work given by the fluid to the turbine is dissipated because of the turbine mechanical efficiency:

$$\eta_{turb} = \frac{W_{shaft}}{W_{turbine}} \tag{1.35}$$

- Finally, the generator does not convert 100% of the mechanical energy to electrical energy:

$$\eta_{generator} = \frac{E_{elec}}{W_{shaft}} \tag{1.36}$$

So, in the end and assuming the pumps of the system are powered by the plant itself, this leads to an overall efficiency equation for the power plant of [17]:

$$\eta_{overall} = \eta_{boiler} \cdot \eta_{actual\ cycle} \cdot \eta_{turb} \cdot \eta_{generator} \tag{1.37}$$

Of course, this 'overall' efficiency only takes into consideration the losses occurring during the production of the electrical power. It should also be noted that there are many other sources of loss that occur after this production phase, e.g. during the transport (transmission and distribution) phase and the consumption phase.

1.2.6.4 Case Studies

We need to appreciate that whilst real steam power plants are based on the Rankine cycle, they used very complex configurations of this thermodynamic cycle, involving a lot of components, in order to reach the maximum efficiency for a reasonable economic investment [18].

In these power plants, reheating and regeneration processes are widespread.

For instance, a combination of turbines are used as well as FWHs – all of them working under different pressure conditions. Among these only one is an open FWH, the others are closed FWHs using the drain cascaded backward technology. Pumps are not included on this diagram but, for this configuration, at least four pumps are necessary to increase the pressure of the fluid from one pressure stage to the next. Of course, to control this system, a lot of valves, sensors, and controllers are also required. So, when compared to the classical ideal Rankine cycle, this power plant cycle is very much more complex.

The overall efficiency of a steam power plant thermodynamic cycle is given by:

$$\eta = \frac{\Sigma|W_{turb}| + \Sigma|E_{thermal}| - \Sigma W_{pump}}{\Sigma Q_{in}} \tag{1.38}$$

1.3 Organic Rankine Cycle

The organic Rankine cycle (ORC) is very similar to the Rankine cycle, the prefix 'organic' being derived from the fact that ORC devices use organic compounds rather than water as their working fluid. Interest in ORC has accelerated in the past 10 years, particularly in the areas of waste heat recovery and renewable energy generation.

The typical parts found in an ORC are not dissimilar from those found in conventional Rankine cycle plant and these include:

1. Pump
2. Evaporator
3. Expansion device
4. Condenser

The devices are constructed to form a circuit in the order of the list provided earlier, however the configuration of an ORC system can vary depending on its application. Further stages such as preheating and regeneration may also be included to both improve the efficiency and also help by recovering energy from other sources. Furthermore, different working fluids also affect the performance of an ORC, with the selection of the working fluid being largely determined by its application, and where the temperature of the target heat source is highly influential. To analyse such systems, we need to note:

Mass conservation is governed by the following equation:

$$\sum \dot{m}_{out} = \sum \dot{m}_{in} \tag{1.39}$$

where $\dot{m}_{out}$ (kg/s) is the mass flow rate of working fluid out of the closed system, and $\dot{m}_{in}$ (kg/s) is the mass flow rate into the closed system.

The heat flow between any two points are defined using the following equation:

$$\dot{Q} = \dot{m} \cdot \Delta h \tag{1.40}$$

where $\dot{Q}$ (kW) is the heat flow, $\dot{m}$ (kg/s) is the mass flow rate, and Δh is (kJ/kg) the difference in enthalpy between two points in the ORC.

So, the net mechanical work is described using the following equations:

$$\dot{W}_{out} = \dot{W}_{gross} - \dot{W}_{in} \quad (1.41)$$

$$\dot{W}_{out} = \dot{m} \cdot \Delta h_{exp} \quad (1.42)$$

where $\dot{W}_{out}$ (kW) is the net mechanical power output from the ORC system, $\dot{W}_{gross}$ (kW) is the gross mechanical power output from the ORC, $\dot{W}_{in}$ (kW) is the gross mechanical power input into the ORC and Δh_{exp} (kJ/kg) is the enthalpy change across the expansion device.

Since the only input considered in the configurations is the pump, $\dot{W}_{in}$ can be calculated using Eq. (1.43).

$$\dot{W}_{in} = \frac{\dot{m}}{\rho} \cdot P \quad (1.43)$$

where $\dot{m}$ (kg/s) is the mass flow rate, ρ (kg/m) is the density of the working fluid under the working conditions, and P (N/m)] is the pressure.

1.3.1 Configurations of ORC

Like the Rankine cycle, the ORC exists in many different configurations each one of which has advantages and drawbacks and so is only suitable for a specific application.

1.3.1.1 Basic ORC Configuration

The basic configuration of an ORC is laid out in Figure 1.35; the system receiving and producing useful energy from one heat source, and Figure 1.36 shows the *T–S* graph for the same.

The evaporator between 2 and 3 heats the working fluid until it boils, and during boiling, although the temperature remains constant, the entropy increases. This boiling is represented by the flat line on the *T–S* graph. In the expander (between 3 and 4) the gaseous working fluid undergoes expansion where its temperature decreases. Finally, between 4 and 1 the working fluid passes through a condenser where the fluid returns to the liquid state ready for the process to begin again.

The equations governing the basic ORC are given subsequently.

$$\dot{Q}_{eva} = \dot{m} \cdot (h_3 - h_2) \quad (1.44)$$

$$\dot{W}_{exp} = \dot{m} \cdot (h_3 - h_4) \quad (1.45)$$

$$\dot{W}_{pump} = \dot{W}_{in} \quad (1.46)$$

$$\eta_{exp} = \frac{\dot{W}_{exp}}{\dot{Q}_{eva}} \quad (1.47)$$

$$\eta_{basic} = \frac{\dot{W}_{exp} - \dot{W}_{pump}}{\dot{Q}_{eva}} \quad (1.48)$$

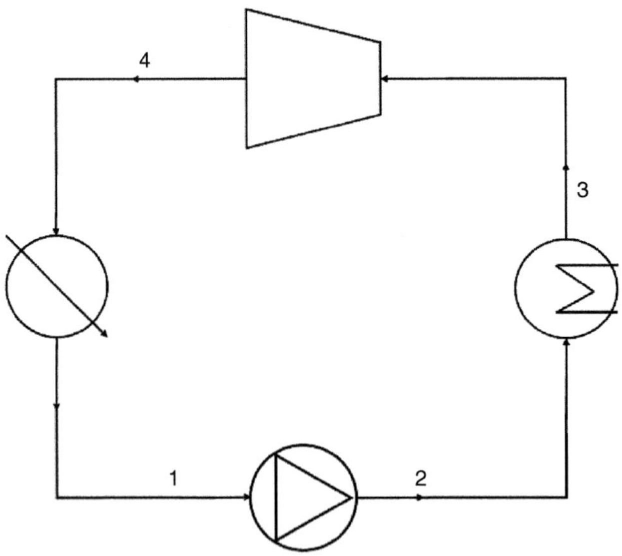

Figure 1.35 Schematic of a typical ORC device.

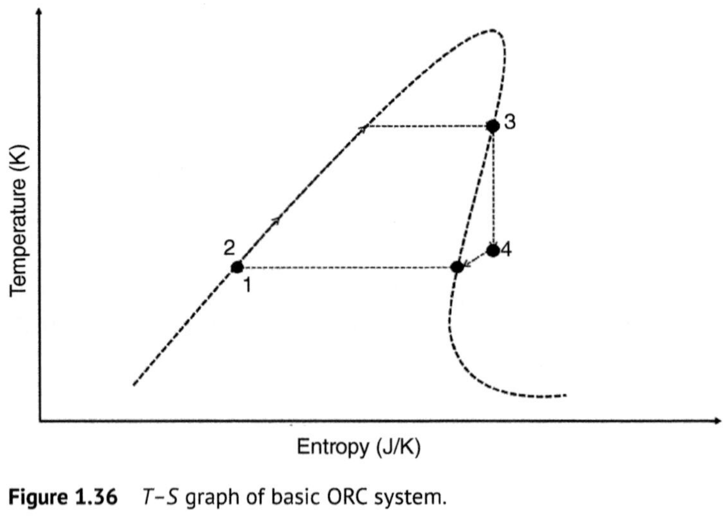

Figure 1.36 *T–S* graph of basic ORC system.

where η_{exp} is the efficiency of the expander in the recovery of energy from the evaporator (%) and η_{basic} is the thermal efficiency of the entire basic ORC configuration (%).

1.3.1.2 ORC with Preheating

Typically, configurations of ORC systems are more complicated where they are designed and constructed to recover energy from multiple heat sources.

In automotive applications, e.g. heat can be recovered from both the exhaust gases and the coolant liquid, so in this format the preheating system involves a heat

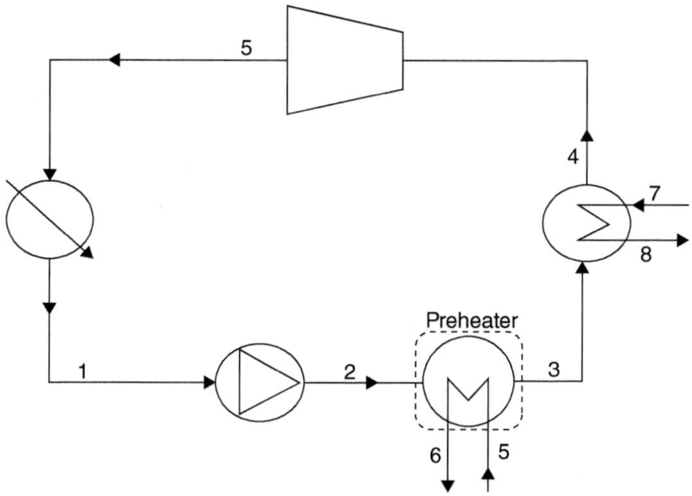

Figure 1.37 Schematic of an organic Rankine cycle with preheater.

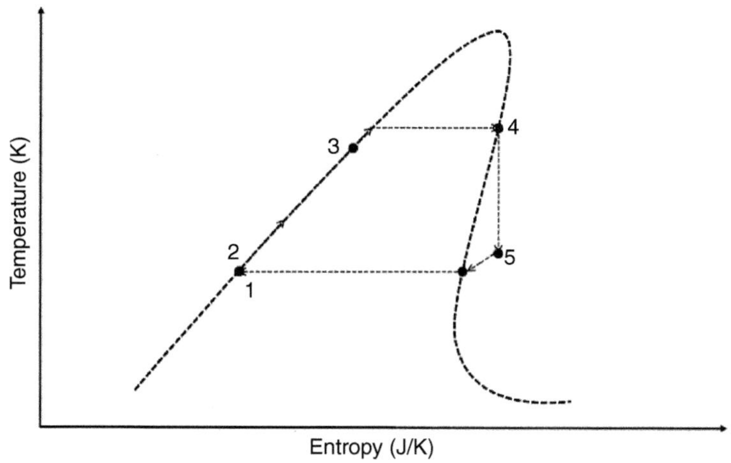

Figure 1.38 *T–S* graph of ORC system with preheater.

exchanger between the pump and the expansion device which then recovers heat energy from the engine coolant fluid and transfers it to the working fluid.

This is shown in Figure 1.37 – the preheater in this case increasing the energy content of the working fluid, so reducing the energy required for its evaporation in the evaporator. This process is represented in Figure 1.38 by the additional heating shown between 2 and 3, the preheating increasing the efficiency of the process by reducing the energy required in the evaporation for boiling to occur.

The equations governing the ORC with preheating are given subsequently.

$$\dot{Q}_{eva} = \dot{m}_{3-4} \cdot (h_4 - h_3) = \dot{m}_{7-8} \cdot (h_7 - h_8) \tag{1.49}$$

$$\dot{Q}_{pre} = \dot{m}_{2-3} \cdot (h_3 - h_2) = \dot{m}_{5-6} \cdot (h_5 - h_6) \tag{1.50}$$

$$\dot{W}_{exp} = \dot{m} \cdot (h_4 - h_5) \tag{1.51}$$

$$\dot{W}_{pump} = \dot{W}_{in} \tag{1.52}$$

$$\eta_{exp} = \frac{\dot{W}_{exp}}{\dot{Q}_{eva} + \dot{Q}_{pre}} \tag{1.53}$$

$$\eta_{pre} = \frac{\dot{W}_{exp} - \dot{W}_{pump}}{\dot{Q}_{eva} + \dot{Q}_{pre}} \tag{1.54}$$

1.3.1.3 Recuperative ORC

In order to improve the overall thermal performances of the system, preheating of the working fluid can also be achieved by recuperation as shown in Figure 1.39. As with normal preheating, this process impacts the working fluid both before the evaporator but unlike normal preheating, it also affects the working fluid after the evaporator, and these effects are shown between points 2 and 3 and 5 and 6 on Figure 1.40. As with preheating, the process reduces the energy needed for boiling to occur in the evaporator.

The equations governing the recuperative ORC are given subsequently.

$$\dot{Q}_{eva} = \dot{m} \cdot (h_4 - h_3) \tag{1.55}$$

$$\dot{Q}_{rec} = \dot{m} \cdot (h_3 - h_2) \tag{1.56}$$

$$\dot{W}_{exp} = \dot{m} \cdot (h_4 - h_5) \tag{1.57}$$

$$\dot{W}_{pump} = \dot{W}_{in} \tag{1.58}$$

$$\eta_{expansion} = \frac{\dot{W}_{exp}}{\dot{Q}_{eva} + \dot{Q}_{rec}} \tag{1.59}$$

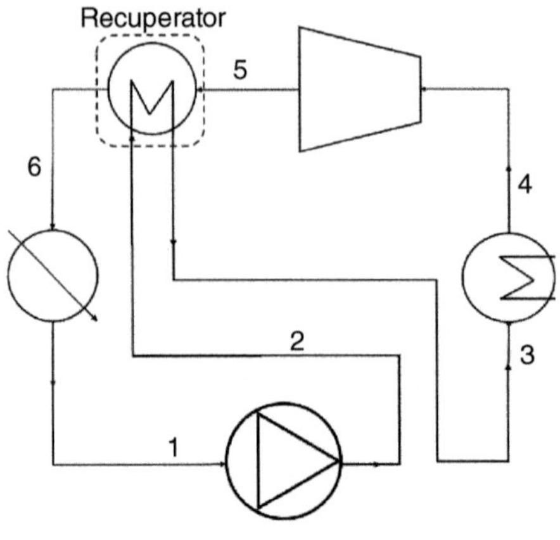

Figure 1.39 Schematic of an organic Rankine Cycle with recuperator.

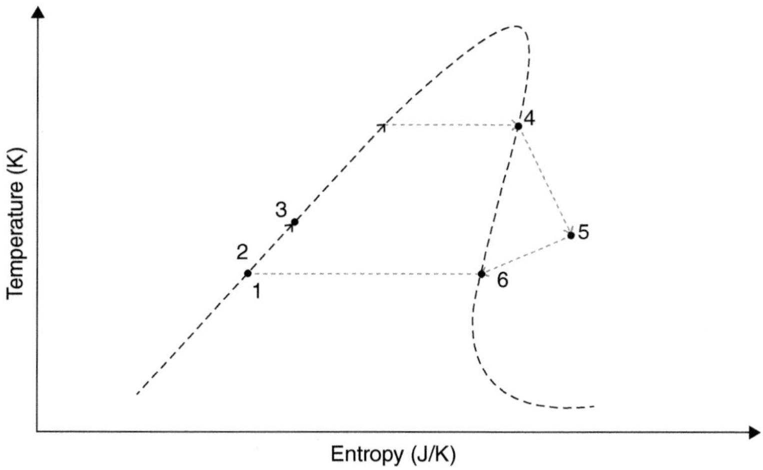

Figure 1.40 *T–S* graph of ORC system with recuperator.

$$\eta_{net} = \frac{\dot{W}_{exp} - \dot{W}_{pump}}{\dot{Q}_{eva} + \dot{Q}_{rec}} \tag{1.60}$$

1.3.1.4 Recuperative ORC with Preheating

In fact, both preheating and recuperation can be combined and operated within the same system and this is shown in Figure 1.41. This is a combination of both the processes that were shown in Figures 1.38 and 1.40; the entire process being described by the *T–S* graph in Figure 1.42 where recuperation and preheating reduce the energy required for boiling the fluid within the evaporator and also assist in the condensation process of the working fluid after the expansion device.

The equations governing the recuperative ORC with preheating are given subsequently.

$$\dot{Q}_{eva} = \dot{m} \cdot (h_5 - h_4) \tag{1.61}$$

$$\dot{Q}_{rec} = \dot{m}_{2-3} \cdot (h_3 - h_2) \tag{1.62}$$

$$\dot{Q}_{pre} = \dot{m} \cdot (h_4 - h_3) \tag{1.63}$$

$$\dot{W}_{exp} = \dot{m} \cdot (h_4 - h_5) \tag{1.64}$$

$$\dot{W}_{pump} = \dot{W}_{in} \tag{1.65}$$

$$\eta_{exp} = \frac{\dot{W}_{exp}}{\dot{Q}_{eva} + \dot{Q}_{rec} + \dot{Q}_{pre}} \tag{1.66}$$

$$\eta_{pre} = \frac{\dot{W}_{exp} - \dot{W}_{pump}}{\dot{Q}_{eva} + \dot{Q}_{rec} + \dot{Q}_{pre}} \tag{1.67}$$

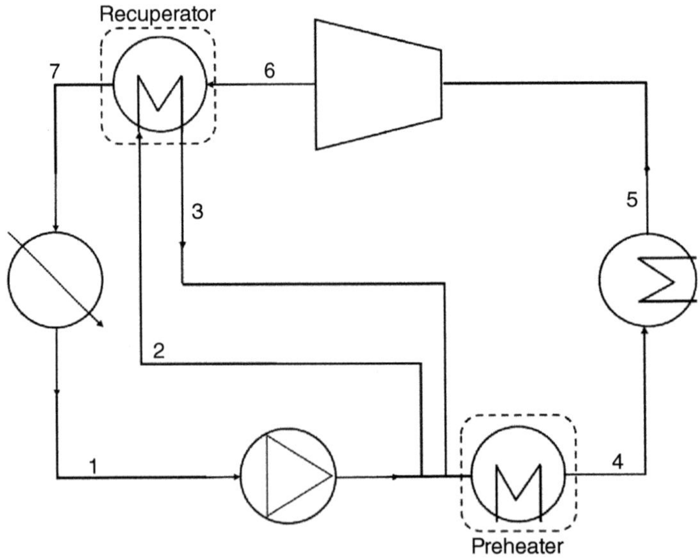

Figure 1.41 Schematic of an organic Rankine cycle with preheater and recuperator.

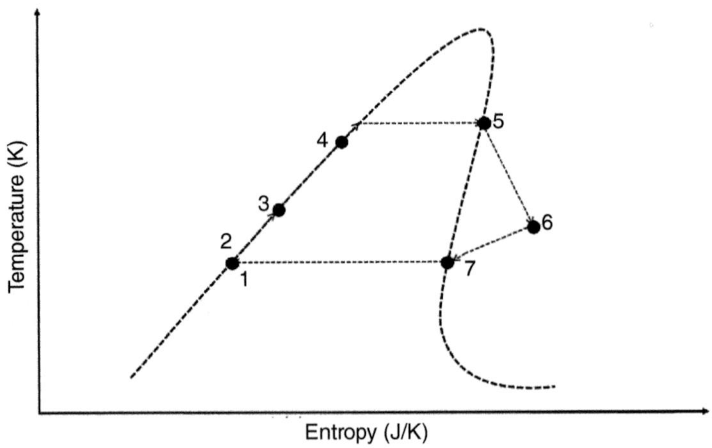

Figure 1.42 *T–S* graph of ORC system with recuperator and preheater.

1.3.2 Organic Working Fluids

Many different working fluids are used in ORC systems and the choice is largely determined by the conditions under which the system is designed to operate. Organic working fluids not only behave differently to water they each also have different characteristics, but these can be split into three categories known as: dry-, isentropic-, and wet-type working fluids. Wet, isentropic, and dry working fluids have $\frac{dT}{dS}$ values of <0, ∞, and >0 respectively, and the *T–S* graphs are displayed in Figures 1.43–1.45, respectively.

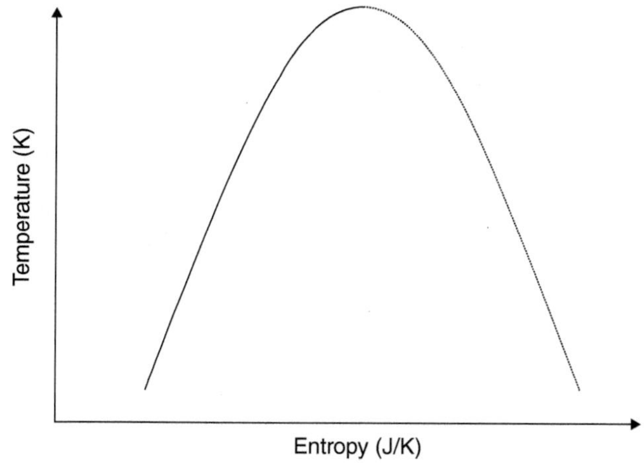

Figure 1.43 *T–S* diagram of a wet-type working fluid.

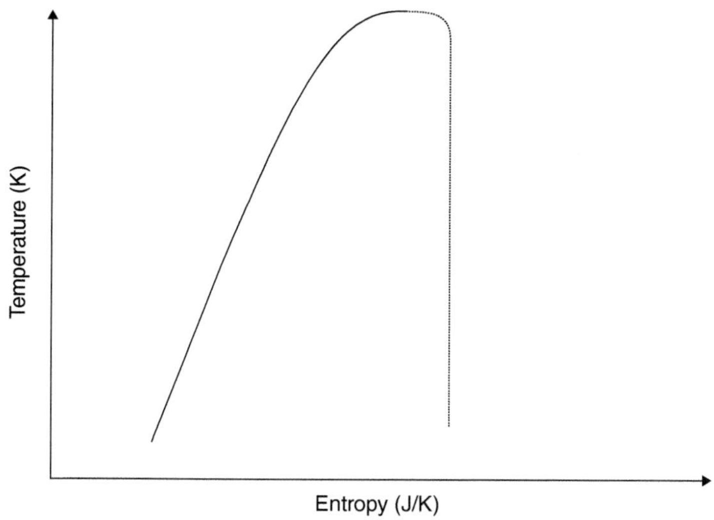

Figure 1.44 *T–S* diagram of an isentropic-type working fluid.

There are four major benefits of using organic working fluids over water in Rankine cycles:

1. *Less heat is required for evaporation*: This can be seen as the width of the *T–S* curve is narrower for dry and isentropic working fluids than for wet working fluids.
2. *Evaporation occurs at lower temperatures and pressures than water*: The reduced temperature of evaporation means that organic working fluids can be used to recover energy from lower temperature heat sources than conventional Rankine cycles.
3. *Outlet from expansion devices is a vapour at lower temperatures*: This reduces the likelihood of damage to the expansion device, because condensation should only

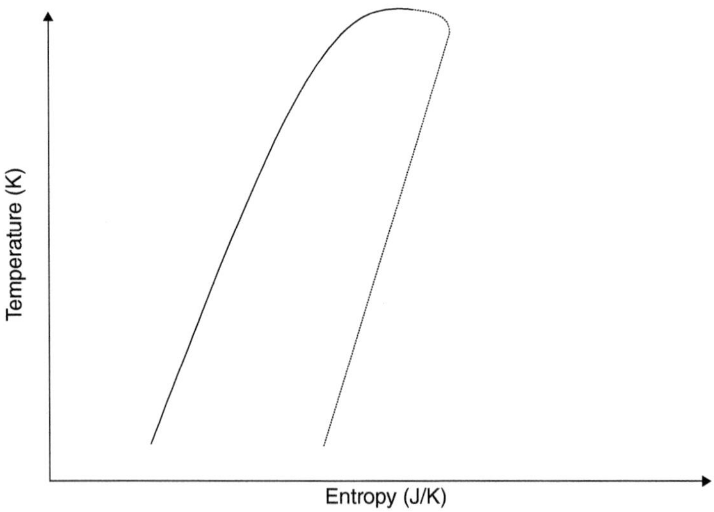

Figure 1.45 *T–S* diagram of a dry working fluid type.

occur after expansion. However, when the expansion inlet temperature is low and the inlet pressure is high, because wet organic working fluids have lower efficiencies than other working fluids they may start to condense within the expansion device potentially causing damage. This is one reason why dry and isentropic working fluids are more popular in ORC systems, where after isentropic expansion, they exist in the superheated state, reducing the risk of condensation and thus minimising potential damage to the expansion device [19].

4. *Smaller temperature difference between evaporation and condensation*: This results in reduced pressure drop across the condenser.

1.3.3 Organic Working Fluid Selection

The selection of the correct working fluid is critical to optimising the efficiency of an ORC system. As such, it is one if not *the* most important aspect of an ORC design. Working fluids are chosen based on their environmental sustainability, ozone depletion, global warming potential (GWP), safety including flammability, toxicity, and corrosivity, boiler vapour pressure, critical temperature, and thermal stability as shown in Table 1.1 [20].

These characteristics affect the behaviour of the fluids upon heating – as seen in the *T–s* graph for different working fluids in Figure 1.46 – and so under specified conditions, there should be a specific working fluid that will result in optimum ORC energy recovery. However, we should also note that certain conditions such as high temperatures and pressures make the use of some organic working fluids unsafe.

As discussed already, if the temperature in the evaporator is low, then the working fluid may condense within the expansion device resulting in decreased efficiency and possible damage. Additionally, any working fluid that requires more entropy to

Table 1.1 Basic properties of working fluids.

Fluid	Critical temperature (°C)	Critical pressure (kPa)	Density at 25 °C (kg/m^3)	Heat of vapourisation at 1 bar (kJ/kg)
R134a	101	4059	4.258	217
R227ea	102.8	2999	7.148	131.7
R245fa	154	3651	5.718	196
R123	183.68	3668	1464	170.6
R600	151.98	3796	2.441	358
Toluene	318.6	4126	862.2	361.3
iso-Butane	134.7	3640	2.44	165.5
iso-Pentane	187.2	3370	614.5	342.5
n-pentane	196.5	3364	620.8	358

Source: Table from Darvish et al. [20].

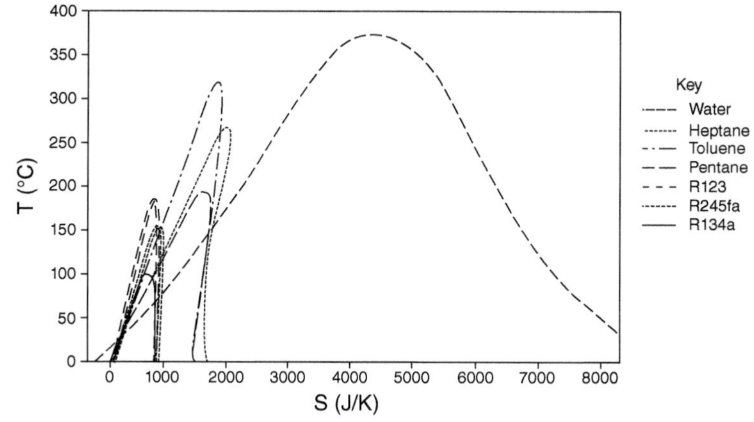

Figure 1.46 *T–S* diagram for different working fluids.

reach the superheated phase then also requires longer exposure within the evaporator, and this would result in a lower mass flow rate than selecting a different working fluid that requires less entropy to reach the superheated phase.

Different temperature–pressure charts are shown in Figure 1.47 for various common working fluids. This figure shows that different working fluids exist at different pressures under the same temperature; these pressure differences also increase with temperature. Choosing an incorrect working fluid could therefore present a safety issue – where pressures increase with temperature, then they exceed the safe limits of the system, resulting in its failure. Under such circumstances, a higher flow rate would be required in order to reduce the risk of this safety issues and such a pump

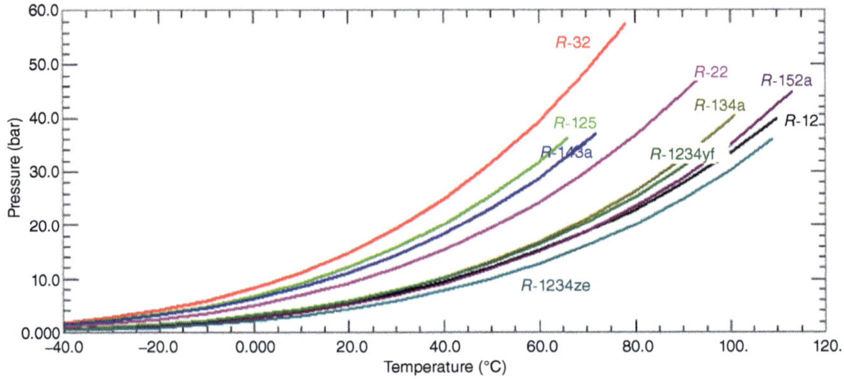

Figure 1.47 Temperature–pressure charts for different working fluids. Source: Abas et al. [21].

would require a higher energy input – potentially resulting in decreased system efficiency. The selection of the correct working fluid therefore is essential to ensure the safety and high efficiency of an ORC system.

These examples – extreme as they may seem – indicate in terms of both efficiency and safety the importance of selecting the correct organic working fluid for an ORC waste heat recovery system.

Although not always the case historically, in addition to system safety issues, working fluids are now also selected on their sustainability. In the past, organic working fluids included chlorofluorocarbons (CFCs), which due to their high ozone depletion potentials (ODPs) were subsequently banned, and working fluids were developed that had significantly smaller ozone reduction potentials. Nonetheless, despite the benefits of reduced ozone depletion capacity, these improved fluids still possessed very high GWPs and so, in turn, this then has led to the development of working fluids with reduced GWPs such as R123 and R600. The ozone depletion and GWPs of some common working fluids are shown in Table 1.2.

Note: ODP is defined as the relative amount of degradation to the ozone layer a substance can cause, with trichlorofluoromethane (R-11 or CFC-11) being fixed at 1.0.

Table 1.2 Sustainability aspects of some common working fluids.

Fluid	Ozone depletion potential	Global warming potential (equivalent CO_2)	Sources
R134a	0	1300	[22]
R227ea	0	3220	[23]
R245fa	0	1030	[24]
R123	0.02	0.02	[22]
R600	0	3	[25]

Table 1.3 Table showing typical applications of ORC in the industry.

Description/ working fluid	Focus	Industry	Temperature of source (°C)	Sources
Sewage heat recovery TY-1	Sustainability of system	Petrochemical industry	110	[26]
Heat recovery from feeder ship engine	Reducing NO_x and SO gases in exhaust emissions	Shipping	450	[27]
Heat recovery from car exhaust gas under different conditions	Energy recovery under three different ambient conditions, and different engine operating conditions	Automotive	500	[28]

1.3.4 Applications of the ORC

There are two main energy applications for the ORC: one is waste heat recovery (where ORC is used to improve the performance of an energy production process or recover energy from an exhaust stream) and the other being in the renewable energy sector. Here ORCs can be used to recover energy from geothermal energy, solar energy, and biomass combustion.

1.3.4.1 Waste Heat Recovery

The ORC is typically used for power output range between 100 kW and 2 MW, and Table 1.3 shows some example applications of ORC, typically in recovery of waste heat from exhaust gases in the industries identified. Configurations of the ORC can also vary depending on the application.

Renewable Energy

Biomass: Biomass energy generation is becoming more popular as it is now quite possible to produce energy from food waste and other organic materials that historically would be seen as waste materials. Furthermore, biomass fuels can be reproduced more quickly than fossil fuels – the length of the carbon cycle being much shorter for biomass than for fossil fuels. Unfortunately for this application, biomass fuels typically burn at lower temperatures than fossil fuels. Which means that conventional Rankine cycles are exposed to a higher risk of condensation within the expanders when used for power generation and biomass boilers. Consequently, conventional Rankine cycles may require superheating to allow their use in biomass energy recovery, and for this reason, ORCs are a popular energy recovery solution for use with biomass fuels [29].

Solar: ORCs make possible the recovery of thermal energy from solar radiation particularly in warmer countries. The low boiling temperatures of the organic working fluids used allows for the capture of energy from the Sun in such devices as flat panel collectors. Phase change materials can also be incorporated into such systems to store heat when energy is not required or when excess energy is being produced, this stored energy then being released later [30].

Geothermal: Ground source heat pumps allow for the cooling or heating of a property making use of the soil and rock as an energy store. This is practical because the year-round temperature of soil and rock at depths of only a few metres is assumed to be constant. Therefore, in summer, heat pumps can be used in reverse to take thermal energy from the air within a building and store it in the ground; whilst in winter, the process can be reversed, and energy taken from the ground can be used to heat the air inside the building. These systems can be improved by the use of ORC systems that help solve the 'cold accumulation problem' of such systems, by storing energy in seasons that require less heating [31].

1.4 Kalina Cycle

The Kalina cycles are a family of thermodynamic power cycles that utilise a mixture of fluids as their working fluid. The first Kalina cycle (also sometimes called the *reverse absorption* cycle) was developed in 1984 by Alexander Kalina, a Russian engineer [32], who then founded Exergy Inc., a company specialising in energetic efficiency and waste heat recovery.

The family of cycles was still inspired by the family of Rankine cycles. Their main goal being to further improve both the energetic and exegetic efficiencies of the thermodynamic power cycle by making use of the fact that in Rankine cycles a lot of energy is lost during the boiling phase. Moreover, as the temperature profiles of the working fluid and the heating fluid do not match the thermal behaviours of the source and working fluid diverge, thus irreversibilities are created and exergy is destroyed. Kalina cycles therefore offer a solution to this energetic issue and represent a big step forward for power cycles and their use in electricity generation [33].

In this section, the theoretical operation of Kalina cycles will be explained and illustrated. The advantages and drawbacks of this new family of power cycle are also discussed and compared to both classical steam cycles and ORCs. Next, applications for these thermodynamic cycles are presented and some case studies are used to clarify explanations. Lastly, current researches on Kalina cycles are briefly reviewed to give the reader an indication of the current state of research on this subject.

1.4.1 Cycle Fundamentals

As just stated, the Kalina cycle is a power cycle based on the Rankine cycle. The difference arises in using a mixture of fluids as the working fluid.

Using a mixture of fluids (instead of a pure substance) as the working fluid allows vapourisation and condensation to occur under a constant pressure through a non-isothermal process. This kind of mixture being called a non-azeotropic mixture (one with components that have different boiling points), as this way there is a better match between the temperature profiles of the working fluid and the heating fluid during the boiling phase. Helpfully, this is also the case between the working fluid and the cooling fluid in the condensation phase.

Currently, the most commonly used mixture for a Kalina cycle is an ammonia–water solution, and although relatively new, this mixture has proven its efficiency through various experiments and pilot projects [33].

It is successful because in a Kalina cycle, the mass fraction of ammonia has a new degree of freedom, allowing more flexibility and so reaching higher efficiencies. The main feature being that the mass fraction of ammonia is now directly linked to the saturation pressure of the mixture. So if, e.g. the pressure is held constant whilst the mass fraction of ammonia is decreasing, then vapourisation of the fluid will start at a higher temperature, as shown in Figure 1.48.

The ammonia vaporises first because it is more volatile than water and hence the mixture concentration of ammonia decreases, and this leads to an increase of the saturation temperature of the resultant mixture. This way, the boiling phase occurs over a range of temperatures and not just at one given temperature as would be the case for a pure substance.

Being able to control the mass fraction of ammonia in each part of the circuit is therefore an important factor and so, in comparison to more conventional steam cycles, Kalina cycles require new components. For example, a separator is certainly required to control the proportion of ammonia in the mixture. This control must occur at every stage of the circuit, so within the separator are two different phases: a liquid phase which is a lean ammonia–water solution (as ammonia is more volatile than water) and a gas phase which is rich in ammonia. Mixing one of these phases with the working ammonia–water solution is a way to either increase or decrease the mass fraction of the ammonia depending on which phase is being mixed with the solution.

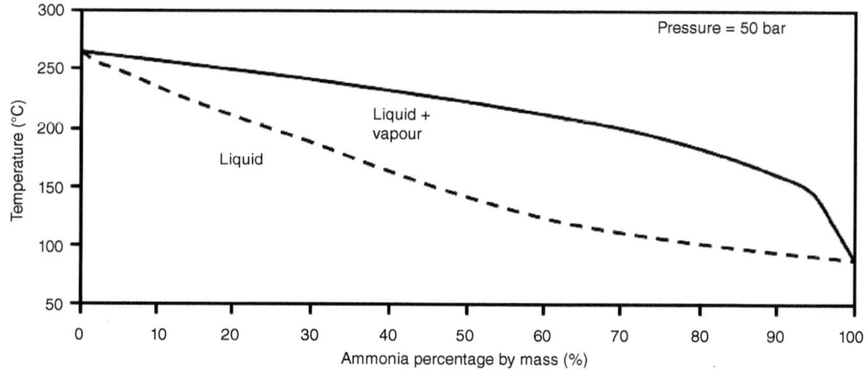

Figure 1.48 Saturation temperature as a function of ammonia concentration for an ammonia–water solution at a pressure of 50 bar.

To operate correctly, a high proportion of ammonia is required in the boiler to enhance the heat transfer. However, keeping this high proportion of ammonia during the condensation phase implies a very low condensation temperature or pressure. So, to overcome this issue, the concentration of ammonia in the ammonia–water solution is reduced before the condensation phase. It was used to demonstrate the feasibility of the Kalina cycles in Canoga, New York, in 1992 on a 6.5 MW power plant [34].

In this configuration, the boiler of the system is also called the *heat recovery vapour generator* (HRVG). Moreover, the subsystem operating the condensation and distillation phase is called *distillation and condensation subsystem* (DCSS). Those two acronyms are widely used for Kalina cycles. Throughout the cycle, the ammonia concentration of the mixture is varying. It is like different fluids were being used, i.e. the fluid in the boiler section is very different from the fluid in the condenser section. Many other configurations also exist, each tailored for a specific application.

1.4.1.1 Why Use Ammonia–Water Solution in Kalina Cycle?

At this time, Kalina cycles predominantly use the ammonia–water solution with hardly any other variants being reported. The mixture displays interesting physical, chemical, economic, and ecological properties, and by varying the percentage mix it can be made to behave like different fluids compared to a pure compound. The main reason therefore for choosing this solution is its flexibility, since by varying the ammonia concentration using distillation, we can alter either the temperature or the pressure of saturation.

Ammonia–water solution also has an excellent heat transfer coefficient and a low freezing point. In addition, because ammonia and water have similar molecular weights, it is possible to use standard components (turbine, pump, etc.) that have already proven their efficiencies in steam power plants. However, some materials such as copper, aluminium, and mild steel cannot be used with ammonia, but otherwise standard materials such as carbon steel and standard high-temperature alloys are fine.

In addition, from an economic point of view, the two components of the mixture are relatively inexpensive and widely available. Water is abundant and ammonia is produced on a world-scale for industrial and domestic uses.

The main issue with ammonia is its toxicity. If the human brain is exposed to high levels of this substance, it can lead to drowsiness and coma [35]. However, as ammonia is a gas under atmospheric pressure and temperature, it is relatively easy to vent off and so with the relevant safety procedures, already proven in industrial plants that use large quantities of ammonia, its use here is considered acceptable. Moreover, as ammonia has a strong smell, it is a self-alarming product which makes its detection easier.

However, the usage of ammonia can also lead to different types of corrosion, and these will be discussed in Section 1.4.2.

Contrary to the fluids used in ORCs, ammonia and water are ecological-friendly. Indeed, the GWP of each are zero, which means that the accidental release of those fluids into the atmosphere has no effect on the global warming. Conversely, the fluids used in ORCs have a proven effect on the global warming. So, in case of leakage from an ORC power plant, large quantities of fluid, harmful to the environment, could be released into the atmosphere. This is not the case for Kalina cycle-based power plants and furthermore, as with the modern organic fluids used in ORCs, the ODP of ammonia is also equal to 0.

1.4.2 Advantages and Drawbacks

To evaluate the potential of this technology, it is important to compare it with the current and most commonly used power cycles of steam systems and ORCs.

1.4.2.1 Advantages

Generally, Kalina cycles have better energetic and exergetic performances than Rankine cycles or ORCs.

As a reference, the Carnot's efficiency is given by:

$$\eta = 1 - \frac{T_C}{T_H} \tag{1.68}$$

In this formula, T_H (K) is the average hot source temperature and T_C (K) is the average cold source temperature. So if we can increase the average hot temperature and at the same time lower the average cold temperature, then the efficiency of the thermodynamic cycle will be improved. Kalina cycles achieve this, compared to not only the theoretical Carnot cycle, but also the very practical Rankine cycles. So for that reason alone, Kalina cycles have a better energetic efficiency.

Moreover, during the boiling and condensing phases, because the temperature profile of the mixture has a better match to the temperature profile of the heating/cooling fluid, heat exchanges are enhanced for Kalina cycles compared to Rankine cycles. Less exergy is destroyed during the heat exchange phases for Kalina cycles so the irreversibilities of the process are reduced. Simulations have showed that the best work output given by a Kalina cycle is 20% to 24% higher than the equivalent from a dual pressure Rankine cycle [33].

Also, when using a geothermal spring, Kalina cycles can generate from 10% to 50% more power than classical steam systems (ORCs), particularly when the temperature of the source is below 200 °C [36]. When the hot source is at medium temperature, e.g. with waste heat recovery, the efficiency increase provided by the Kalina cycle against steam cycles is between 10% and 40% [37].

The gains are not just in simulations. During the Global CempPower conference in 2012 [38], M.D. Mirolli (formerly from Wasabi Energy) and Kevin Happ (FLSmidth) reported that the use of Kalina cycles for waste heat recovery in the cement industry can lead to a 15–25% improvement of the recovery compared to current technologies (ORCs).

As we have already noted, because the properties of the working fluid can be changed during the operation of the plant, Kalina cycle-based plant can adapt easily to hot and cold days by varying the proportion of ammonia – this flexibility then allows Kalina cycles to harness energy from a wider range of low to medium–high-temperature heat sources.

The fact also that standard components can be used to operate a Kalina cycle is an advantage as it increases the reliability of the potential new power plants, because even if they are using a relatively new technology, their operation is based on tried and tested efficiency-proven components.

1.4.2.2 Drawbacks

Ammonia is a toxic fluid, so even if it is easy to vent off under atmospheric pressure and self-alarming due to its smell, it is relatively dangerous to use on a large scale. However, some safety procedures exist, and for the moment no health issue linked to Kalina cycle-based power plants have been recorded. Compared to the hydrocarbons used in ORCs, ammonia is non-flammable and does not represent an explosion hazard.

General corrosion can also occur with this type of power plant due to the presence of water. However, as ammonia raises the pH level of the solution, the phenomenon is less important than in classical steam cycles. Indeed, a high pH and a low oxygen proportion can result in the growth of protective magnetite layers (Fe_3O_4) on tube surfaces, but generally, a good control of the water chemistry is needed to avoid issues linked to corrosion.

The CO_2 level must also be controlled to avoid the formation of ammonium carbamate. This can occur if ammonia and carbon dioxide react under conditions of HP and high temperature. Furthermore, if the temperature of the mixture exceeds 400 °C, then ammonia becomes unstable and leading to nitride corrosion [39].

Currently, the physical properties of an ammonia–water solution are not yet fully known, particularly in the supercritical region where very little data is currently available. Some correlations exist and these are leading to smaller and smaller differences between experimental tests and simulations.

Kalina cycles are sometimes said to be more complex than steam cycles. This assumption is questionable, but it is undeniable that steam cycles have proved their efficiency and reliability across time. So even though pilot projects exist for the Kalina cycle it is relatively understandable why some industries are quite hostile to any additional investment in this technology at the present time.

1.4.3 Applications of the Kalina Cycle

Because the boiling temperature of the mixture varies, it can start to evaporate at a lower temperature than for the Rankine cycle at a given pressure. This is why Kalina cycles are mostly used in geothermal power plants (100–120 °C) and waste heat recovery (300–400 °C) [40]. For waste heat recovery applications, the most suitable applications are the cement industry, glass industry, petrochemical industry, steel industry, and potentially, for thermal power plants [39]. Using Kalina cycles as bottoming cycles in thermal power plant is possible, but for now it is not yet economically viable because the architecture of such a cycle is quite complex. Kalina cycles can also be used with alternative renewable energies other than geothermal, and some configurations of the cycle are made for solar thermal and ocean thermal [41] operations (Figure 1.49).

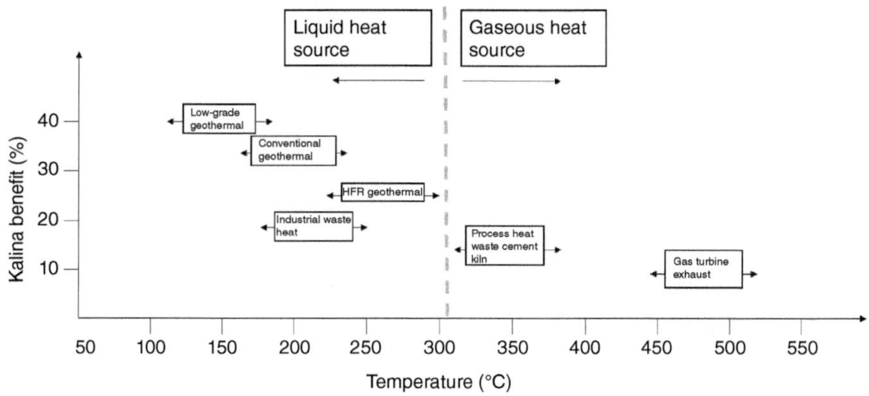

Figure 1.49 Kalina benefit as a function of heat temperature and types of applications.

1.4.3.1 The Different Configurations of the Cycle

Kalina cycles are part of a thermodynamic family of cycles. These vary by the number of each component, how they are arranged, and how they are linked together. To date, a lot of interesting Kalina cycle configurations have been designed and studied; every configuration being suitable for a particular type of application and so designated by a code. For example, KCS-1 was the very first cycle that used a mixture of fluids and was proposed by Kalina himself in 1984 [42].

All these thermodynamic cycles are split into two groups. The oldest ones are called *first-generation* Kalina cycles – the Kalina cycle being a trademark involving all the first-generation Kalina cycles [43].

Figure 1.50 shows three configurations of the Kalina cycle system [44], all suitable for geothermal applications. KCS-11 was designed for the high-temperature geothermal resources and KCS-34 for lower temperature geothermal resources. Those two cycles are the most common Kalina cycles used for geothermal application. KSG-1 is very similar to KCS-34, the KSG-1 patent is held by Siemens.

A second generation of Kalina cycle was developed by Dr. Alexander Kalina from 2008 to 2014. The patents for those new cycles are all owned by a company called Kalex LLC, founded by Kalina.

Those new cycles are more efficient and provide an improved power output. For example, the SG-2a cycle offers a better energetic and exergetic efficiency than either KCS-11 or KCS-34 when using geothermal heat sources in temperature range from 125 to 150 °C [45].

So even though these new cycles have a more complex architecture than their first-generation counterparts and so are consequently more expensive, they still offer a shorter payback period [46]. Second-generation Kalina cycles therefore seem most promising, and configurations now exist for biomass, geothermal, solar thermal, cement waste heat, bottoming cycle, and ocean solar thermal applications [47].

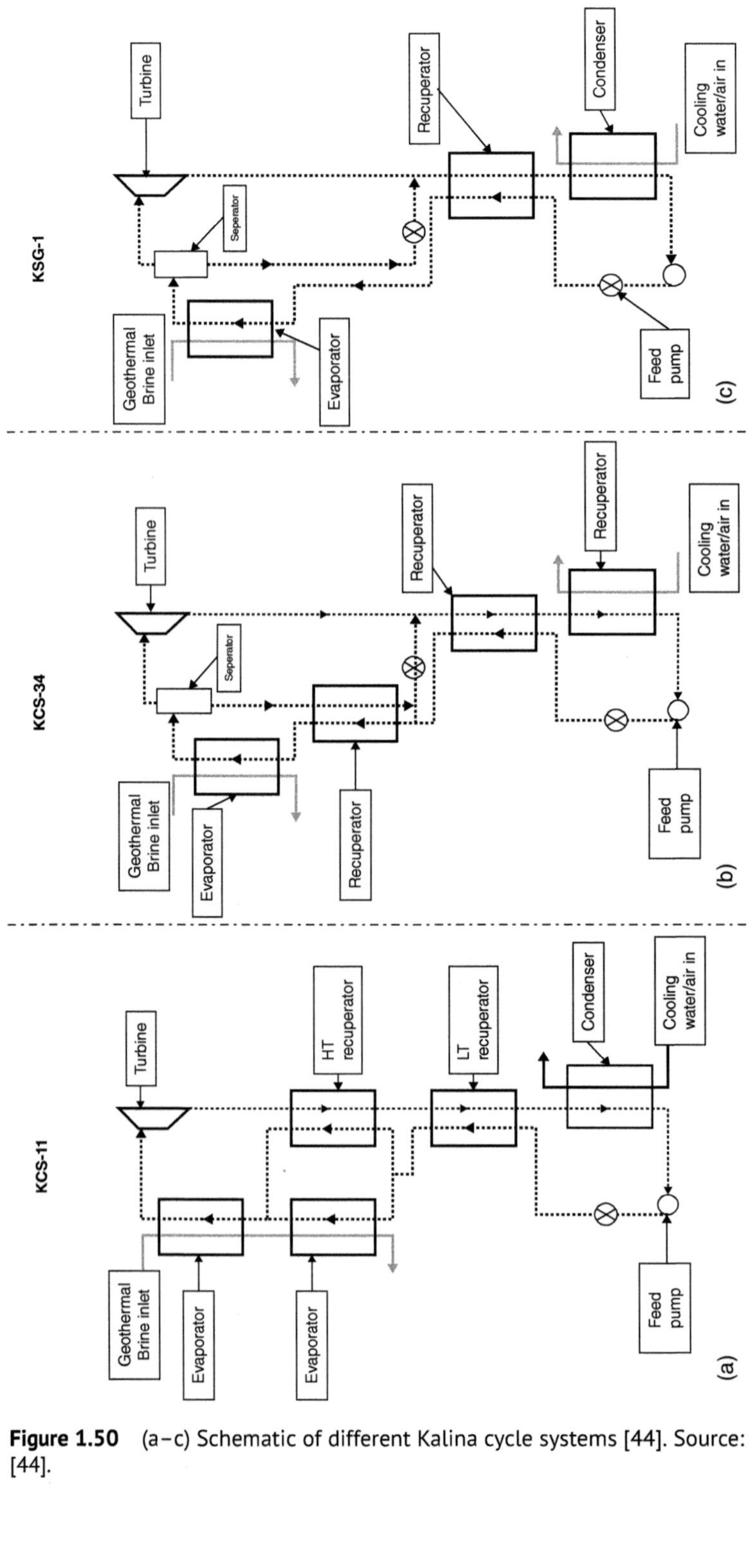

Figure 1.50 (a–c) Schematic of different Kalina cycle systems [44]. Source: Wang and Yu [44].

Table 1.4 Kalina cycle case studies from around the world [39].

Name	Country	Commissioned	Output (MW)	Heat sources
Canoga Park	USA	1992	6.5	Nuclear waste heat
Fukuoka	Japan	1998	4	Waste incineration
Sumitomo Metals	Japan	1999	3.5	Waste heat
Husavik	Iceland	2000	2	Geothermal
Fuji Oil	Japan	2005	3.9	Waste heat
Bruschal	Germany	2009	0.6	Geothermal
Unterhaching	Germany	2009	3.5	Geothermal
Shanghai Expo	China	2010	0.05	Solar hot water
Quingshui	Taiwan	2011	0.05	Geothermal

Source: Global cement [39].

1.4.4 Case Studies

Some pilot power plants exist and are currently producing energy using Kalina cycles. Table 1.4 lists case studies from around the world as of 2012.

The first large-scale Kalina cycle-based power plant was built in the United States in 1992 and operated for five years recovering nuclear waste heat and was able to produce 6.5 MW of electricity [39].

A famous Kalina-based plant (KCS-34) was built in Húsavík, Iceland. It is a 2 MW geothermal power plant and harnesses the energy of 125 °C geothermal brine, providing 80% of the electrical demand of the local town's 2500 inhabitants [48]. Another geothermal power plant (KCS34), also harnessing geothermal hot brine through a Kalina cycle, is located at Unterhachting in Germany. This power plant is generating 3.4 MW of electricity. The geothermal brine is also providing 38 MW of thermal energy to a district heating system. A third geothermal power plant (KCS-34) using the Kalina technology is the Bruchsal power plant [49], also in Germany.

Further Kalina cycle projects had been designed since 2012. For example, the Star Cement plant in Dubai is a 4.75 MW plant also using this technology [3], and it is anticipated, as we move forward, that more and more projects utilising Kalina cycles technology will evolve [50].

1.5 Brayton Cycle

The Brayton cycle is also known as the 'gas turbine cycle' or the *Joule* cycle. It is an open cycle, whereby the intake and exhaust points are both open to the environment [51].

Air enters the engine and is compressed by the compressor (resulting in HP air). Energy is then added by spraying fuel into the air and igniting it within the combustion chamber, and now this high-temperature pressurised gas is used to drive a turbine (mounted on the same shaft as the compressor) to produce work output. Finally, the combusted gases and any unused energy leaves through the exhaust and in doing so some of this unused energy can be recovered.

First though, the Brayton cycle itself. This consists of a compression stage (process 1–2) which takes place adiabatically followed by a second stage (process 2–3) – the heat addition process – which occurs isobarically. The third stage (process 3–4) is an expansion process which occurs adiabatically and the final stage (process 4–1), heat rejection, occurs isobarically, all as shown in Figure 1.51.

The Brayton cycle is different from the Otto and Diesel cycles in that the processes occur within an open system. Therefore, in order to determine the heat transfer and work for the process an open system and steady flow analysis are used.

The schematic of the open Brayton process is shown in Figure 1.52, however, to further define the process as a cycle we need to include the heat output as shown in Figure 1.53.

The thermal efficiency of the Brayton η_{th} cycle is given by the following equation:

$$\eta_{th} = \frac{W_{net}}{Q_{in}} = 1 - \frac{Q_{out}}{Q_{in}} \tag{1.69}$$

where W_{net} (kJ) is the net work output of the system, Q_{in} (kJ) is the heat input, and Q_{out} (kJ) is the heat output.

The conservation of mass then gives:

$$\dot{m}_{in} = \dot{m}_{out} \tag{1.70}$$

$$\dot{m}_2 = \dot{m}_3 = \dot{m} \tag{1.71}$$

where m_2 and m_3 are the mass flow rate of fluid leaving the compressor and the mass flow rate entering the turbine (kg/s), respectively.

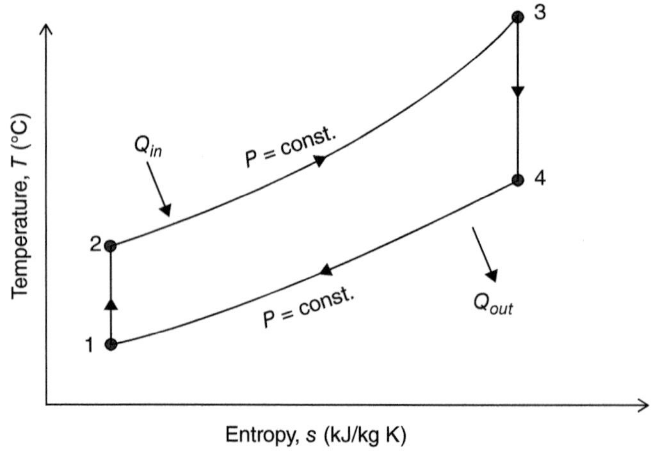

Figure 1.51 Temperature entropy graph for a Brayton cycle.

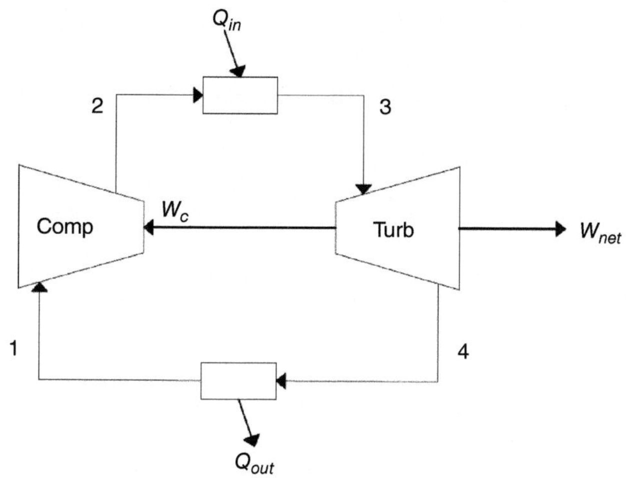

Figure 1.52 Gas turbine engine.

Figure 1.53 Closed Brayton cycle.

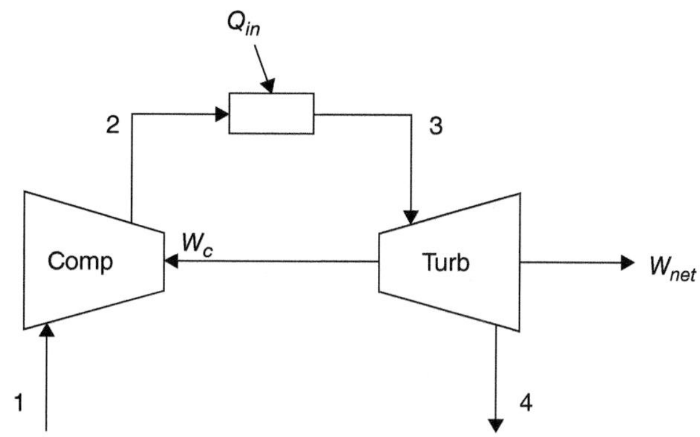

Therefore,

$$\eta_{th} = 1 - \frac{q_{out}}{q_{in}} = 1 - \frac{C_p(T_4 - T_1)}{C_p(T_3 - T_2)} \tag{1.72}$$

where $q_{out}(\text{kJ/kg}) = \frac{Q_{out}}{m}$, C_p (kJ/kg K) is the specific heat capacity of the working fluid (depending on the temperature), and T_x (K) is the temperature of the fluid given point.

Assuming the specific heat is constant (cold air standard analysis) gives:

$$\eta_{th} = 1 - \frac{T_4 - T_1}{T_3 - T_2} = 1 - \frac{T_1\left(\frac{T_4}{T_1} - 1\right)}{T_2\left(\frac{T_3}{T_2} - 1\right)} \tag{1.73}$$

And as processes 1–2 and 3–4 are isentropic:

$$\frac{T_2}{T_1} = \left(\frac{P_2}{P_1}\right)^{\frac{k-1}{k}} \tag{1.74}$$

$$\frac{T_3}{T_4} = \left(\frac{P_3}{P_4}\right)^{\frac{k-1}{k}} \tag{1.75}$$

In these last two equations, k is the heat capacity ratio, also known as the adiabatic index.

Moreover, as processes 2–3 and 4–1 are isobaric:

$$P_2 = P_3 \tag{1.76}$$

$$P_4 = P_1 \tag{1.77}$$

Thus,

$$\frac{T_2}{T_1} = \frac{T_3}{T_4} \tag{1.78}$$

Which can also be written this way:

$$\frac{T_4}{T_1} = \frac{T_3}{T_2} \tag{1.79}$$

So that finally, the efficiency of the ideal Brayton cycle is given by:

$$\eta_{th} = 1 - \frac{T_1}{T_2} \tag{1.80}$$

Now as the pressure ratio is defined as:

$$r_p = \frac{P_2}{P_1} \tag{1.81}$$

Figure 1.54 shows a typical pressure ratio curve for gas turbine engine.

Then the efficiency of the cycle can also be written using this coefficient as:

$$\eta_{th} = 1 - \frac{1}{r_p^{\frac{k-1}{k}}} \tag{1.82}$$

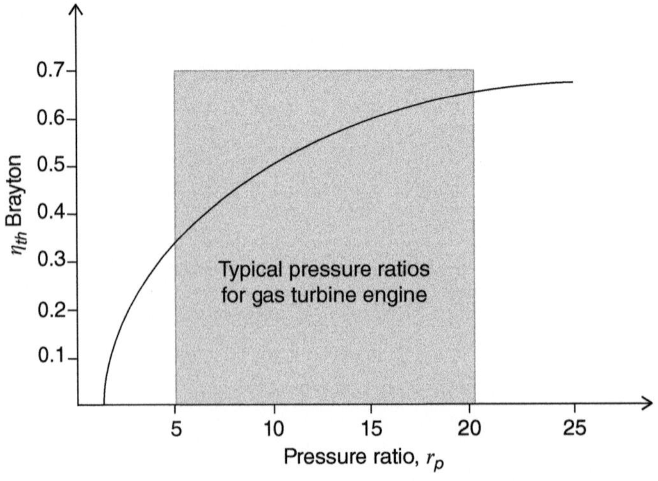

Figure 1.54 Typical pressure ratios for a gas turbine engine.

However, this is not the whole of the story because some of work produced by the turbine is also used by the compressor, the amount given back being defined by the back-work ratio (BWR) from:

$$BWR = \frac{W_{in}}{W_{out}} = \frac{W_{comp}}{W_{turb}} \tag{1.83}$$

1.5.1 Regenerative Brayton Cycle (Regenerator)

Because the temperature of the turbine exhaust is higher than the exit stream of the compressor, it presents an ideal opportunity to install a heat exchanger located between the hot exhaust of the turbine and the cooler gas leaving the compressor. Such a heat exchanger is also known as a regenerator or a recuperator. Figure 1.55 shows a schematic of the regenerative Brayton cycle.

And so in describing the efficiency of this regenerator, a *regenerator efficiency* term needs to be introduced, its calculation involving $q_{regen,max}$ and $q_{regen,act}$,and which can be obtained through enthalpy differences expressed in kJ/kg (see Figure 1.56).

$$\varepsilon_{regen} = \frac{q_{regen,max}\ (\text{kJ/kg})}{q_{regen,act}\ (\text{kJ/kg})} \tag{1.84}$$

$$q_{regen,max} = h_5' - h_2 = h_4 - h_2 \tag{1.85}$$

$$q_{regen,act} = h_5 - h_2 \tag{1.86}$$

Hence,

$$\varepsilon_{regen} = \frac{h_5 - h_2}{h_4 - h_2} \tag{1.87}$$

For ideal gases using the cold air standard analysis with constant specific heat, the regenerator effectiveness becomes:

$$\varepsilon_{regen} = \frac{T_5 - T_2}{T_4 - T_2} \tag{1.88}$$

So, for the efficiency of the cycle (using the closed cycle analysis) this leads to:

$$\eta_{th,regen} = 1 - \frac{q_{out}}{q_{in}} = 1 - \frac{h_6 - h_1}{h_3 - h_5} \tag{1.89}$$

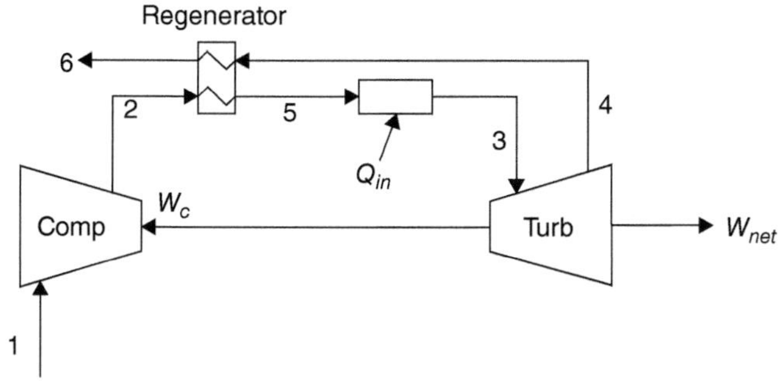

Figure 1.55 Schematic of the regenerative Brayton cycle.

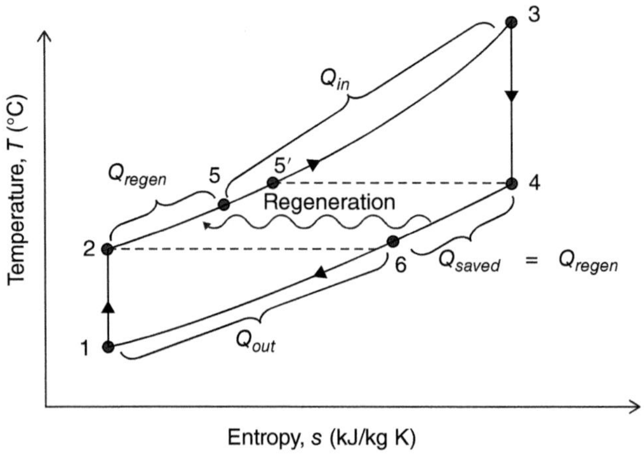

Figure 1.56 Temperature entropy graph for a regenerative Brayton cycle.

And assuming the efficiency of the regenerator is equal to 1, this then gives:

$$\eta_{th,regen} = 1 - \frac{T_1}{T_3} r_p^{\frac{k-1}{k}} \tag{1.90}$$

Figure 1.56 shows the temperature entropy profile for a regenerative Brayton cycle.

1.5.1.1 Compressor Analysis

The work (kW) used by the compressor is given by the following equation:

$$\dot{W}_{act,comp} = \dot{m}(h_{2a} - h_1) \tag{1.91}$$

where $\dot{m}$ is the mass flow rate of working fluid (in kg/s) and h_x is the enthalpy at the given point (kJ/kg).

Hence the adiabatic efficiency of the compressor is given by:

$$\eta_{comp} = \frac{\dot{W}_{isen,comp}}{\dot{W}_{act,comp}} = \frac{h_{2s} - h_1}{h_{2a} - h_1} \cong \frac{T_{2s} - T_1}{T_{2a} - T_1} \tag{1.92}$$

And the isentropic temperature at the outlet of the compressor is given by:

$$T_{2s} = T_1 \left(\frac{P_2}{P_1}\right)^{\frac{k-1}{k}} \tag{1.93}$$

1.5.1.2 Turbine Analysis

The work provided by the turbine (kW) is given by the following equation:

$$\dot{W}_{act,turb} = \dot{m}(h_3 - h_{4a}) \tag{1.94}$$

From which, the adiabatic efficiency of the turbine is given by:

$$\eta_{turb} = \frac{\dot{W}_{act,turb}}{\dot{W}_{isen,turb}} = \frac{h_3 - h_{4a}}{h_3 - h_{4s}} \cong \frac{T_3 - T_{4a}}{T_3 - T_{4s}} \tag{1.95}$$

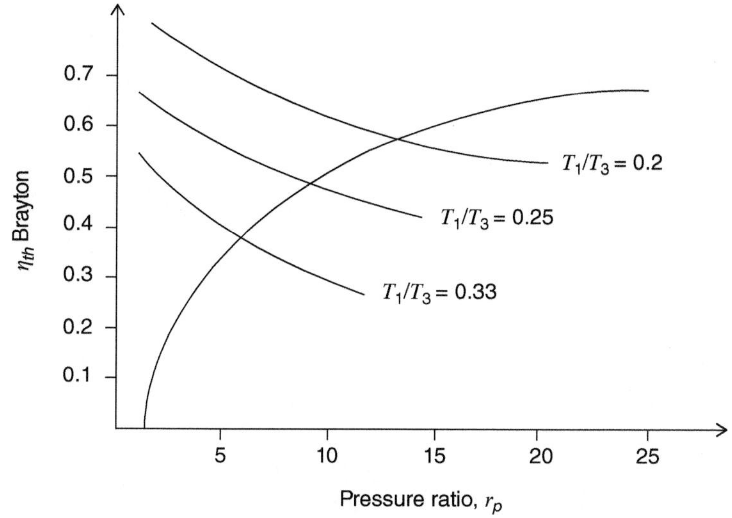

Figure 1.57 Regenerative Brayton cycle efficiency as a function of the pressure ratio and minimum and maximum temperature ratio.

And so, the isentropic temperature at the outlet of the turbine is given by:

$$T_{4s} = T_3\left(\frac{P_4}{P_3}\right)^{\frac{k-1}{k}} \tag{1.96}$$

1.5.1.3 Heat Supplied to the Cycle

The heat supplied to the cycle (kJ/kg) for process 5 to 3 can be calculated as follows:

$$q_{in} = h_3 - h_5 = C_p(T_3 - T_5) \tag{1.97}$$

The plot in Figure 1.57 shows the regenerative Brayton cycle efficiency as a function of the pressure ratio and minimum and maximum temperature ratio, ${}^{T_1}/_{T_3}$.

Thus, when the efficiency of the regenerative cycle is equal to the standard Brayton cycle then:

$$r_p = \left(\frac{T_3}{T_1}\right)^{\frac{k}{2(k-1)}} \tag{1.98}$$

1.5.2 Regenerative Brayton Cycle (Reheater and Intercooler)

The performance of a Brayton cycle can be further improved by incorporating an intercooler and a reheat process, the process schematic in Figure 1.58 shows a potential layout for a Brayton cycle with such additions. The concept is simple. During multi-stage compression, cooling the working fluid between each stage ultimately reduces the amount of work that is required by each compressor. This is because if we reduce the specific volume of the working fluid by cooling, then less work is required to achieve a set pressure increase.

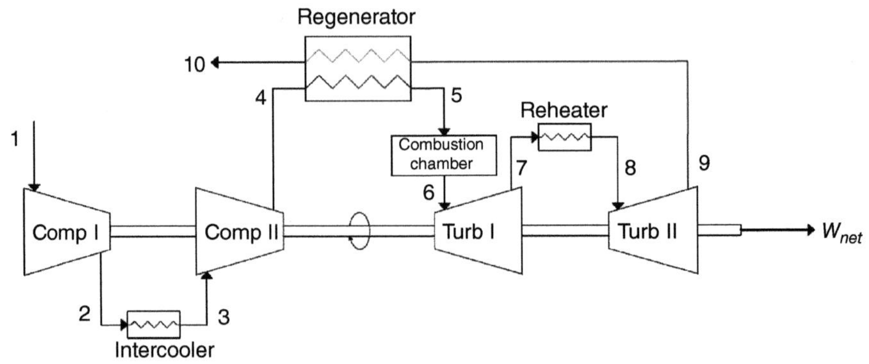

Figure 1.58 Schematic of the regenerative Brayton cycle with intercooler and reheat.

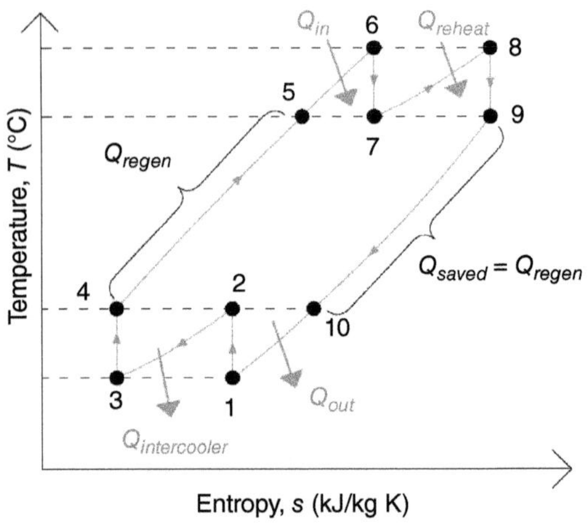

Figure 1.59 Temperature entropy graph for a regenerative Brayton cycle with a reheater and an intercooler.

So, incorporating a reheating process between two or more turbines of a multi-stage expansion ultimately results in an increase in the net work done and therefore a more efficient cycle. The temperature against entropy for this thermodynamic cycle is shown in Figure 1.59.

1.5.2.1 Intercooling

To minimise the work used by the compressors, P_2 (Pa) should be set as:

$$P_2 = \sqrt{P_1 P_4} \tag{1.99}$$

As the process 2–3 is isobaric, it can also be written as follows:

$$\frac{P_2}{P_1} = \frac{P_4}{P_2} = \frac{P_4}{P_3} \tag{1.100}$$

1.5.2.2 Reheating

To maximise the turbine work, P_7 should be set as:

$$P_7 = \sqrt{P_6 P_9} \tag{1.101}$$

As the process 7–8 is isobaric, it can also be written this way:

$$\frac{P_6}{P_7} = \frac{P_7}{P_9} = \frac{P_8}{P_9} \tag{1.102}$$

1.6 Chapter Summary

To conclude, the Rankine cycle can be considered (alongside the Carnot cycle) as the basis for all the external combustion thermodynamic cycles that aim to convert thermal energy into mechanical energy. In most cases the mechanical energy is then converted into electrical energy because of its greater versatility and adaptability as a form of energy.

Even though the ideal Rankine cycle is quite simple, a significant number of improvements can still be made to it in order to enhance the performance of the system and increase its energetic efficiency – reheating, regeneration, increasing or decreasing the pressure can all be considered, in order to make the most of a power plant.

In most cases, water is used as the working fluid because it is non-toxic, non-reactive, abundant, cost-effective, and displays good thermodynamic properties. However, using an organic fluid is also an option to help further increase the efficiency of the Rankine cycle or to harness medium temperature hot sources such as geothermal, biomass solar energy, or for waste heat recovery. In which case, this new cycle is known as an ORC, but even this can be further improved in many ways using reheaters and/or recuperators for instance. The use of new fluids also allows for greater flexibility, in that different fluids may be more suitable for specific applications, thus allowing them to reach their greatest possible efficiency. Hence, as discussed earlier, the selection of the right working fluid is crucial to efficient operation. To assist in this choice, organic fluids are divided into three families: wet-type fluids, dry-type fluids, and isentropic fluids, and currently a huge number of studies are focusing on the integration of ORC with existing systems so as to improve their overall efficiency.

Recent technological advances also allow the working fluid (water or organic fluid) to be used in supercritical conditions in order to improve the amount of work provided by the turbine(s) of a power plant.

In this respect, the Kalina cycle can be considered as the biggest step forward since the development of the Rankine steam cycles and ORCs. The technology presents a better energetic efficiency than steam cycles particularly for low-grade temperature applications such as geothermal and waste heat recovery.

The technology is relatively new; hence the capital expenditure necessary for a Kalina cycle-based power plant is, in most cases, higher than for a conventional system. However, as the Kalina cycle systems have a higher efficiency, their payback period is reduced, particularly for the latest second-generation cycles. It is also important to note that the Kalina cycles are a family of thermodynamic cycles and so a vast number of configurations exist, each one being suitable for a specific application. In most cases, a water–ammonia mixture is used as the working fluid in those employing binary cycles. Indeed, this fluid offers interesting thermal performance; is compatible with standard materials and components; is ecological-friendly, abundant, and relatively inexpensive.

The reverse Kalina cycle is also a widely known thermodynamic cycle in that it is the absorption refrigeration cycle. This cycle uses an absorber and a generator instead of a compressor to carry out the suction and compression processes, meaning that almost no electricity is required to run the cycle. The future of Kalina cycles seems promising. An ever-increasing number of scientific studies are addressing the integration of this cycle in various systems: geothermal power plants, cement plants, concentrated solar power plants, and biomass power plants.

Now the Brayton cycle, although different to the Rankine cycle and its derivatives in that it is an open thermodynamic cycle, still has the same role to play as the closed power cycle – converting heat into mechanical energy – and the cycle efficiency can also be improved by using different technologies such as regeneration, reheating, and intercooling. The Brayton cycle can also use organic fluids as their working fluid (CO_2 in most cases) and sometimes these fluids too can operate under supercritical conditions with a further increase in efficiency.

But we are not just thinking about systems in isolation. Even if Rankine cycles and Brayton cycles are quite different, they can still be used together in combined cycles. In the CCGT plant, those two cycles are used together to reach a relatively high overall thermal efficiency (from 50% to 63%). So high, that the association of the Rankine cycle with the Brayton cycle is currently the most efficient way of producing electricity on a large scale. For example, the Chubu Electric Nishi-Nagoya power plant Block-1 in Japan is achieving a gross efficiency of 63.08%, and so is currently recognised as the world's most efficient combined cycle power plant.

References

1 Zhou, F., Joshi, S.N., Rhote-Vaney, R., and Dede, E.M. (2017). A review and future application of Rankine cycle to passenger vehicles for waste heat recovery. *Renewable Sustainable Energy Rev.* 75: 1008–1021. https://doi.org/10.1016/j.rser.2016.11.080.

2 Sattari, A. (2017). A novel design of a low Cost CSP using turbocharger as an expander. *J. Sol. Energy Res.* 2: 20–24.

3 Janie Ling-Chin, H.B.Z.M., Taylor, W., and Roskilly, A.P. (2019). Chapter 4: State-of-the-art technologies on low-grade heat recovery and utilization in industry. In: *Energy Conversion. Current Technologies and Future Trends*. IntertechOpen https://doi.org/10.5772/intechopen.78701.

4 Sethi, A., Vera Becerra, E., and Yana Motta, S. (2016). Low GWP R134a replacements for small refrigeration (plug-in) applications. *Int. J. Refrig.* 66: 64–72. https://doi.org/10.1016/j.ijrefrig.2016.02.005.

5 Shang, R., Zhang, Y., Shi, W. et al. (2014). Fresh look and understanding on carnot cycle. *Energy Procedia* 61: 2898–2901. https://doi.org/10.1016/j.egypro.2014.12.213.

6 Xu, W., Deng, S., Su, W. et al. (2018). How to approach Carnot cycle via zeotropic working fluid: Research methodology and case study. *Energy* 144: 576–586. https://doi.org/10.1016/j.energy.2017.12.041.

7 Greitzer, E.M. (n.d.). 8.6 Enhancements of Rankine cycles.ZSSI.

8 Weston, K.C. (2000). *Fundamentals of Steam Power 2.1 Introduction*, 1e. University of Tulsa.

9 Riddoch, F. and Craenen, S. (n.d.). Cogeneration at the foundation of Europe's energy policy.

10 IBGE. (n.d.). Guide Cogénération.

11 Mat, N., Chee, I., Tan, W., and Yatim, A.H.M. (2018). A comprehensive review of cogeneration system in a microgrid: a perspective from architecture and operating system. *Renewable and Sustainable Energy Rev.* 81 (Part 2): 2236–2263. https://doi.org/10.1016/j.rser.2017.06.034.

12 Dabwan, Y.N. and Pei, G. (2020). A novel integrated solar gas turbine trigeneration system for production of power, heat and cooling: thermodynamic-economic-environmental analysis. *Renewable Energy* 152: 925–941. https://doi.org/10.1016/j.renene.2020.01.088.

13 Goncalv, E., Silva, D., Thibault, J. (2008). Cycles thermodynamiques des machines thermiques. Institut Polytechnique de Grenoble, 153.

14 Association WN. (n.d.). Advanced Nuclear Power Reactors | Generation III+ Nuclear Reactors.

15 Administration USEI. (n.d.). SAS Output.

16 Quoilin, S., Van Den, B.M., Declaye, S. et al. (2013). Techno-economic survey of organic rankine cycle (ORC) systems. *Renewable Sustainable Energy Rev.* 22: 168–186. https://doi.org/10.1016/j.rser.2013.01.028.

17 Al-Taha WH, Osman HA. Performance analysis of a steam power plant: a case study 2018. MATEC Web of Conferences, Volume 225, UPT-UMP-VIT Symposium on Energy Systems doi:https://doi.org/10.1051/matecconf/201822505023.

18 Inc AP. (n.d.). Economics and feasibility of rankine cycle improvements for coal fired power plants.

19 Minea, V. (2014). Power generation with ORC machines using low-grade waste heat or renewable energy. *Appl. Therm. Eng.* 69: 143–154. https://doi.org/10.1016/j.applthermaleng.2014.04.054.

20 Darvish, K., Aliehyaei, M., Atabi, F., and Rosen, M. (2015). Selection of optimum working fluid for organic Rankine cycles by exergy and exergy-economic analyses. *Sustainability* 7: 15362–15383. https://doi.org/10.3390/su71115362.

21 Abas, N., Kalair, A.R., Khan, N. et al. (2018). Natural and synthetic refrigerants, global warming: a review. *Renewable Sustainable Energy Rev.* 90: 557–569. https://doi.org/10.1016/j.rser.2018.03.099.

22 Nair, V. (2021). HFO refrigerants: a review of present status and future prospects. *Int. J. Refrig.* 122: 156–170. https://doi.org/10.1016/j.ijrefrig.2020.10.039.

23 Linde Group (2013). Safety Data Sheet – Heptafluorpropane (R227).

24 Linde Group (2014). Safety Data Sheet – Pentafluoropropane (R245fa).

25 Linde Group (2013). Safety Data Sheet – Isobutane (R600a).

26 Yang, H., Xu, C., Yang, B. et al. (2020). Performance analysis of an organic Rankine cycle system using evaporative condenser for sewage heat recovery in the petrochemical industry. *Energy Convers. Manage.* 205: 112402. https://doi.org/10.1016/j.enconman.2019.112402.

27 Baldasso, E., Andreasen, J.G., Mondejar, M.E. et al. (2019). Technical and economic feasibility of organic Rankine cycle-based waste heat recovery systems on feeder ships: impact of nitrogen oxides emission abatement technologies. *Energy Convers. Manage.* 183: 577–589. https://doi.org/10.1016/j.enconman.2018.12.114.

28 Yue, C., Tong, L., and Zhang, S. (2019). Thermal and economic analysis on vehicle energy supplying system based on waste heat recovery organic Rankine cycle. *Appl. Energy* 248: 241–255. https://doi.org/10.1016/j.apenergy.2019.04.081.

29 Sikarwar, S.S., Surywanshi, G.D., Patnaikuni, V.S. et al. (2020). Chemical looping combustion integrated organic Rankine cycled biomass-fired power plant – Energy and exergy analyses. *Renewable Energy* 155: 931–949. https://doi.org/10.1016/j.renene.2020.03.114.

30 Alvi, J.Z., Feng, Y., Wang, Q. et al. (2020). Modelling, simulation and comparison of phase change material storage based direct and indirect solar organic Rankine cycle systems. *Appl. Therm. Eng.* 170: 114780. https://doi.org/10.1016/j.applthermaleng.2019.114780.

31 Li, W., Lin, X., Cao, C. et al. (2018). Organic Rankine cycle-assisted ground source heat pump combisystem for space heating in cold regions. *Energy Convers. Manage.* 165: 195–205. https://doi.org/10.1016/j.enconman.2018.03.062.

32 Saghafifar, M., Omar, A., Mohammadi, K. et al. (2018). A review of unconventional bottoming cycles for waste heat recovery: Part I-Analysis, design, and optimization. *Energy Convers. Manag.* 198 https://doi.org/10.1016/j.enconman.2018.10.047.

33 Zhang, X., He, M., and Zhang, Y. (n.d.). A review of research on the Kalina cycle. *Renewable Sustainable Energy Rev.* 16 (7) https://doi.org/10.1016/j.rser.2012.05.040.

34 Globalgeothermal. (n.d.) Global Geothermal-Kalina Cycle application.

35 Schenker, S., Mccandless, D.W., Brophy, E., and Lewis, M.S. (1967). Studies on the intracerebral toxicity of ammonia. *J. Clin. Investig.* 46: 838–848.

36 Ambriz-Díaz, V.M., Rubio-Maya, C., Chávez, O. et al. (2021). Thermodynamic performance and economic feasibility of Kalina, Goswami and Organic Rankine Cycles coupled to a polygeneration plant using geothermal energy of low-grade temperature. *Energy Convers. Manag.* 243: 114362. https://doi.org/10.1016/j.enconman.2021.114362.

37 Global cement. (n.d.). Kalina bottoming cycle vs Rankine (ORC) in thermal power plant design.

38 Mirolli, M., Happ, K. (2012). Global CempPower Conference. FLSmidth & Co A/S.

39 Global cement. (2012). Kalina Cycle power systems in waste heat recovery applications.

40 Omar, A., Saghafifar, M., Mohammadi, K. et al. (2018). A review of unconventional bottoming cycles for waste heat recovery: Part II-Applications. *Energy Convers. Manage.* https://doi.org/10.1016/j.enconman.2018.10.088.
41 Kalinapower. (n.d.). Kalina – Electricity from heat.
42 Blanco, M. and Lourdes, R.S. (2017). *Advances in Concentrating Solar Thermal Research and Technology*. Google Livres.
43 Nasruddin, U.R., Rifaldi, M., and Noor, A. (2009). Energy and exergy analysis of kalina cycle system (KCS) 34 with mass fraction ammonia-water mixture variation. *J. Mech. Sci. Technol.* 23: 1871–1876. https://doi.org/10.1007/s12206-009-0617-8.
44 Wang, E. and Yu, Z. (2016). A numerical analysis of a composition-adjustable Kalina cycle power plant for power generation from low-temperature geothermal sources. *Appl. Energy* 180: 834–848. https://doi.org/10.1016/j.apenergy.2016.08.032.
45 Kalex Systems LLC. (2009). Kalex Kalina Cycle Power Systems for Geothermal Applications.
46 Renz, M., Engelhard, M., Zander, M. (2006). The New Generation Kalina Cycle Contribution to the conference "*Electricity Generation from Enhanced Geothermal Systems*."
47 Kalex Systems LLC. (2009). Kalex Kalina Cycle Power Systems for Geothermal Applications.
48 Lcak HM, Ark M, Irolli M, Hjartarson H, Húsavíkur O. (2002). Notes from the North: A Report on the Debut Year of the 2 MW Kalina Cycle ® Geothermal Power Plant in Húsavík, Iceland.
49 G-u M. (2010). *The Geothermal Power Plant Bruchsal.*
50 Kalina Power. (2016). Introduction to Kalina Cycle®. Kalina Power Limited.
51 Huang, K., Marthinsen, K., Zhao, Q., and Logé, R. (2018). The double-edge effect of second-phase particles on the recrystallization behaviour and associated mechanical properties of metallic materials. *Prog. Mater. Sci.* 92: 284–359. https://doi.org/10.1016/j.pmatsci.2017.10.004.

2

Waste Heat Recovery

2.1 Burner and Air Preheaters

Air preheating is a common application for recovered waste heat; the energy content within a hot fluid being transferred to the cooler air stream and a selection of the technologies used for this purpose are described within this section. The main technologies are recuperators, regenerators, and run-around coils with each technology having multiple variants. The differences, advantages, and disadvantages are also highlighted.

2.1.1 Recuperators

The simplest form a recuperator can take is a radiative recuperator (Figure 2.1). This consists of two concentric metal tubes where hot gases, such as exhaust gases, pass through the inner tube and a cooler fluid, usually a gas, passes through the outer tube. Heat is extracted primarily by the radiation mechanism. This design exhibits very low-pressure losses in comparison to other technologies [1], but it has been noted that the technology only has between 40% and 60% heat transfer efficiencies – which falls short of other available technologies. Additionally, they require routine maintenance as fouling from condensable vapours and particulates can be an issue [2]. To apply this technology to higher temperature streams, dilution can be used to lower the temperature before passing through the recuperator so as to avoid overheating the structure material and to prevent thermal stresses. The simplicity of the design allows these to be incorporated directly into stacks where the exhaust gases energy can be recovered close to source.

Convection recuperators (Figure 2.2) consist of a tube bundle around which a hot gas passes. A shell, sometimes baffled, encloses the tubes in which the cooler fluid flows. They are suitable for low-temperature applications and those with a high mass flow rate [3]. Metallic units can be used at gas temperatures <1000 °C; however, exposure to higher temperatures decreases their lifespan. Ceramic recuperators are often applied for higher temperature applications up to 1550 °C [4].

Better heat transfer effectiveness can be achieved by combining multiple energy transfer mechanisms into one recuperator. These are known as hybrid recuperators, as shown in Figure 2.3. These tend to have two sections that recover by radiation and

Waste Heat Recovery in Process Industries, First Edition. Hussam Jouhara.

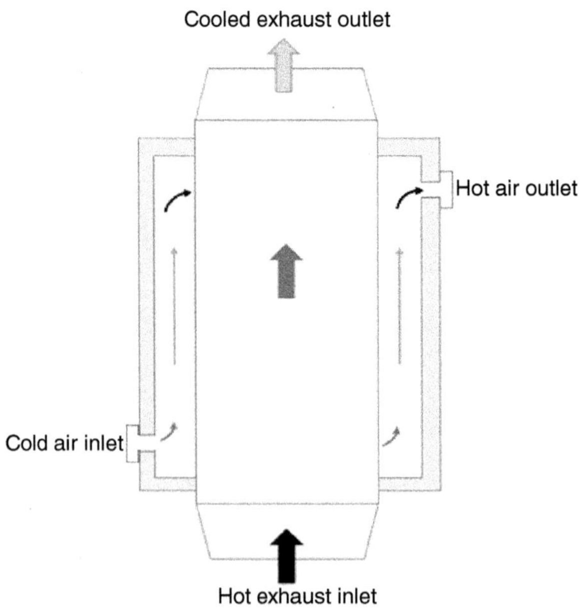

Figure 2.1 A radiative recuperator.

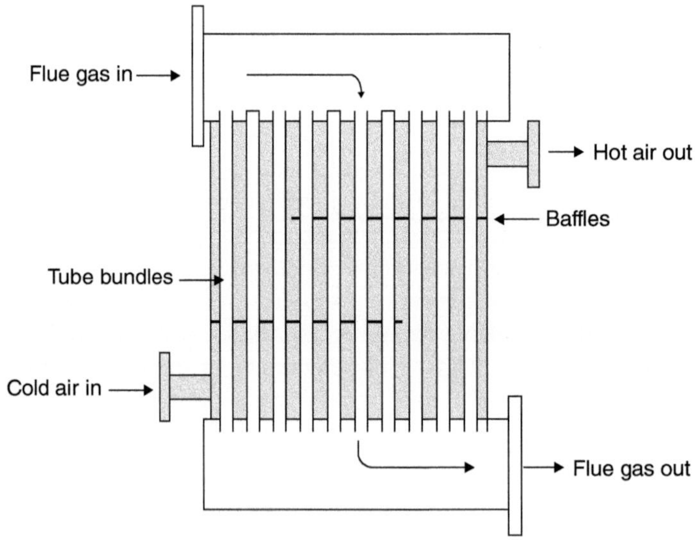

Figure 2.2 A convective recuperator.

convection. The first section cools a hot fluid using radiation. In the second section, the hot fluid is passed through tubes and the heat transfer cooling fluid is passed through the shell and recovers energy with a convective recuperator.

2.1.1.1 Recuperative Burners

Many industrial processes use combustion to heat or melt a product by using burners which control the combustion reaction by providing a uniform and controlled

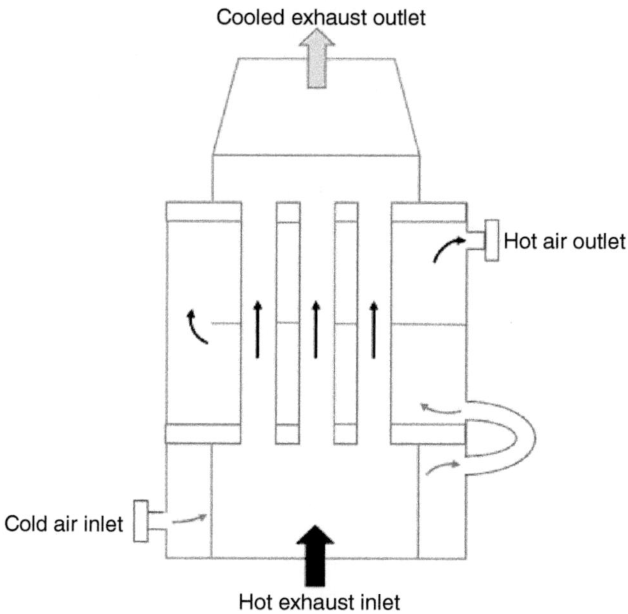

Figure 2.3 A hybrid recuperator.

heat input to that process. For example, furnaces to melt aluminium, kilns to fire ceramic products, or spray dryers to evaporate moisture and generate a powdered product. A burner is an energy transducer which converts one form of energy to another – chemical energy in the form of fuel into heat energy – the energy content provided to the process being controlled in this way. Full combustion is shown by:

$$\text{Fuel} + O_2 \rightarrow CO_2 + H_2O + \Delta H \tag{2.1}$$

The oxygen for the reaction can be supplied by the roughly 21% oxygen content of air (known as combustion air), or supplied by oxygen tanks by either injecting pure oxygen or enriching the combustion air with a higher proportion of oxygen. ΔH is the enthalpy and shows the energy released, the majority of which is in the form of heat. The combustion products are usually vented by a flue to be treated and then released to atmosphere. However, a proportion of their energy still remains that hasn't been transferred to the process. Three opportunities therefore exist for recovering heat from a combustion processes: from the product, the exhaust gases or from shell or cooling water losses. To recover the heat from combustion processes, recuperative and regenerative burners have been developed to harness this waste heat by using the energy to preheat the combustion air for the combustion process. Using a heat recovery burner is of paramount importance, if not the most important factor to ensure fuel use is reduced as much as possible whilst ensuring an efficient and cost-effective process.

The reason that fuel consumption is reduced by the increase in combustion air temperature is threefold. Firstly, the enthalpy within the waste gas is reintroduced into the furnace, a hotter flame is achieved, and the activation energy of a reaction is more readily reached, and as such there is increased re-radiation leading to more efficient combustion so greater energy release. Initially, a flame is ignited on a burner which will continue to burn as long as the combustion precursors are present. The heat is fed back from the reaction to continue the combustion by the mechanisms of re-radiation and recirculation [5]. Re-radiation is a mechanism that contributes to flame stability. Energy returns from the high-temperature burning zone, the product, and the combustion chamber walls. Modern burners rely on recirculation, shown by Figure 2.4. The presence of recirculating hot combustion products close to the burner nozzle raises the fuel temperature and aids ignition. The recirculation happens either outside the flame (external) or within the axis of the burner nozzle due to entrainment by a turbulent jet (internal). Internal recirculation can be promoted by flow deviation using bluff bodies or by creating swirl within the combustion air.

Recuperative, also known as self-recuperative, burners specifically work by transporting the exhaust gases past the combustion air separated by a thermally conductive material [6]. The exhaust gases are forced back through the burner by an educator. The exhaust gas transfers energy to the material by convection, the energy transfers across the material by conduction and then the cooler combustion air receives this energy and increases in temperature. For this reason they tend to be gas burners but they can also be configured for liquid burners, and for a recuperative burner, this conductive layer is integral to the burner design. Independent heat exchangers (HEXs) can be installed within the system to preheat combustion air (Figure 2.5) but this section specifically deals with recuperative burners because it tends to be simpler to integrate the heat exchange mechanism into the burner design which is then more cost-effective. The recuperative burners continuously transfer heat between two fluids and so tend to operate in a steady state and are more suited to lower temperatures, but recuperative burners have a lower heat recovery efficiency than other technologies and cannot be used with particulate-loaded or condensable vapour-containing exhaust gases [2].

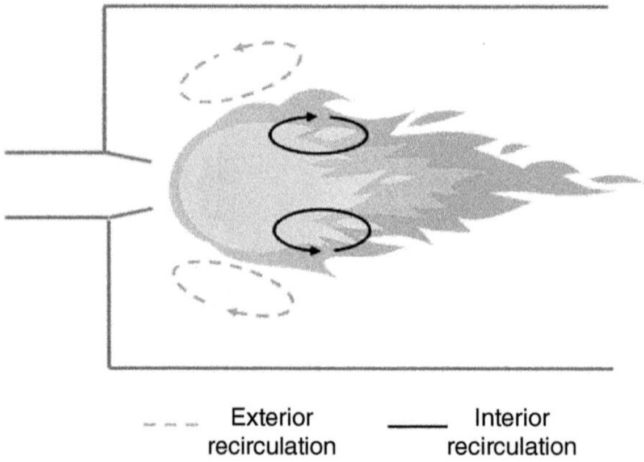

Figure 2.4 Examples of burner recirculation.

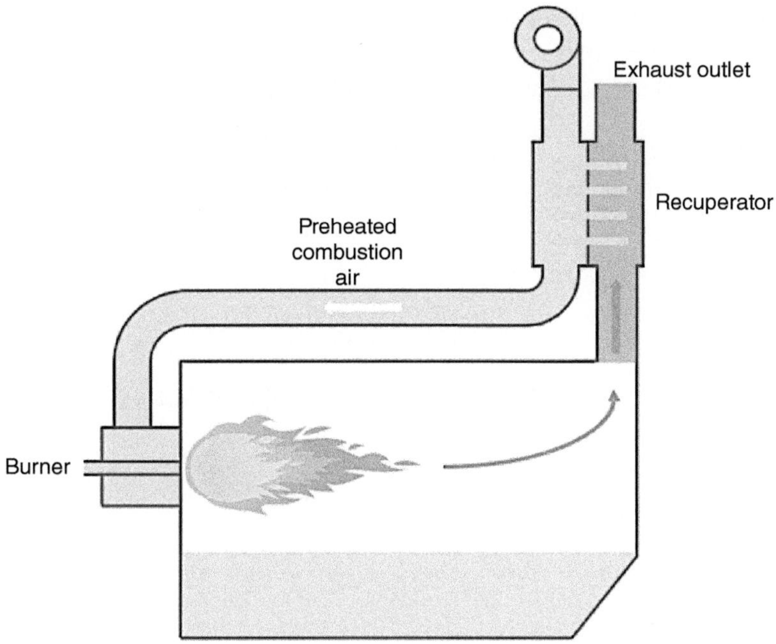

Figure 2.5 Schematic of a recuperator being used to preheat combustion air.

2.1.1.2 Classifying Recuperative Burners

Recuperative burners are generally classified by fuel type, design of the flow (Table 2.1), the flame profile, and size. Counter-flow is most effective for heat transfer as the heat sink can reach similar temperatures to the exhaust gas inlet but parallel flow is limited to equivalent source and sink outlet temperatures. In terms of fuel type, methane, natural gas, liquefied petroleum gas (LPG), and propane are typical examples of gaseous fuels. Heavy and light fuel oil (HFO or LFO), as well as diesel, are examples of liquid fuels used in recuperative burners. Gas fuel systems can be either premixed or turbulent jet diffusion burners, and are used when a small but high-intensity flame is required. The premixing can be achieved using a venturi or a mixing tube, the air and gas being supplied at a high pressure to prevent the flame returning into the burner and to ensure sufficient mixing. This leads to very high combustion efficiency. Partial premix burners can be used by having secondary airports which control flame temperature and profile and minimise NO_x formation. Jet diffusion is the predominant industrial burner technology. A fuel jet is introduced separately and coaxially surrounding the combustion air. Within the combustion chamber these streams mix via entrainment and ignite. [5]

When sizing a burner, over- or underheating can cause major issues with a process. Consequently, the burner used must correspond to the demands of the system. If overheating occurs burners will cycle, leading to greater fluctuations in temperature and unnecessary overuse of system controls. Conversely, underperformance can lead to longer process times and inadequate final products.

Table 2.1 Recuperative burner flow design.

Parallel flow	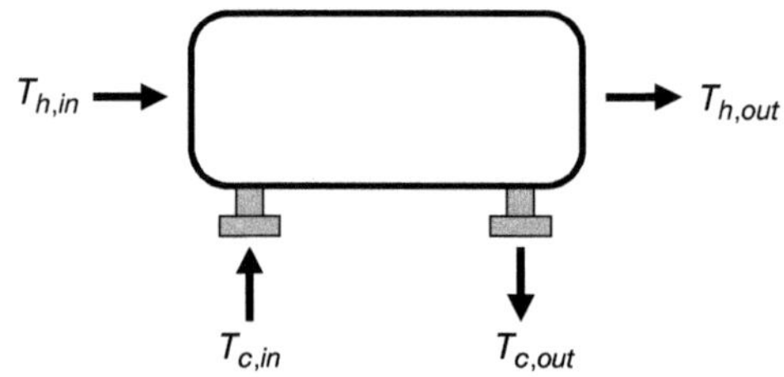
Counter-flow	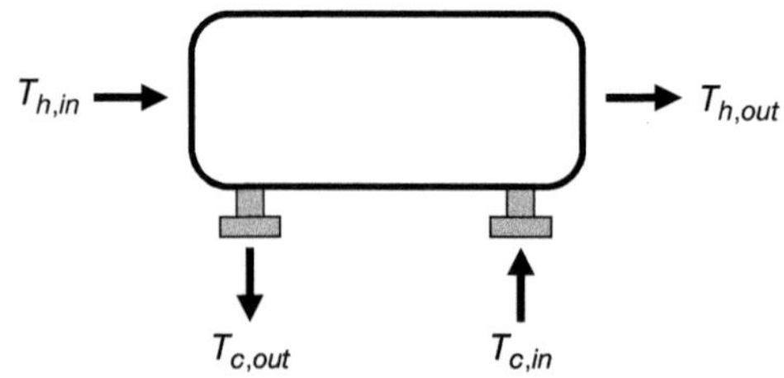
Single-pass cross flow	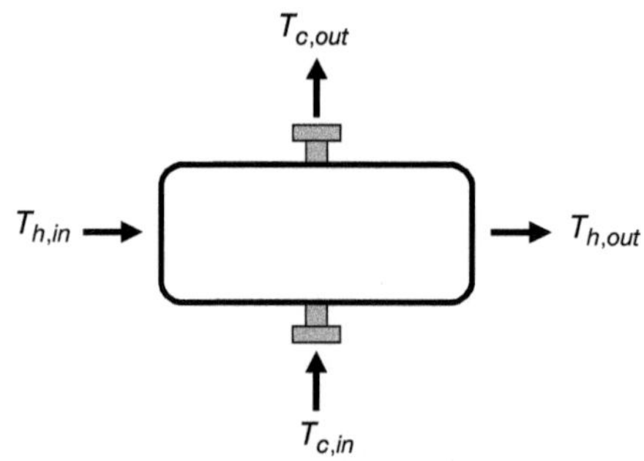
Multi-pass cross flow	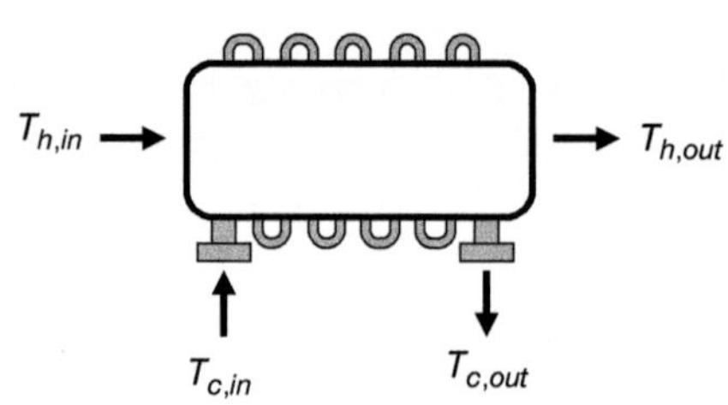

2.1.1.3 Efficiency Improvement and Fuel Savings

It has been reported that preheating combustion air can increase furnace efficiency by up to 50%. Table 2.2 reflects the furnace efficiency as a result of introducing preheated combustion air.

The capital expenditure for the installation of these burners can be high. Therefore, reduced energy costs, improvements in process efficiency, and lower emissions of NO_x are important factors to consider. Table 2.3 provides an overall comparison of the advantages and disadvantages for a recuperative burner.

Table 2.2 Furnace efficiency improvement by preheating combustion air.

Furnace outlet temperature (°C)	Combustion air preheat temperature				
	204 °C	316 °C	427 °C	538 °C	649 °C
1427	22%	30%	37%	43%	48%
1316	18%	26%	33%	38%	43%
1204	16%	23%	29%	34%	39%
1093	14%	20%	26%	31%	36%
982	13%	19%	24%	29%	33%
871	11%	17%	22%	26%	30%
760	10%	16%	20%	25%	28%

Source: Adapted from Johnson et al. [7].

Table 2.3 Recuperative burner advantages and disadvantages.

Advantages	Disadvantages
Only one burner and flue required	Can suffer from expansion and contractions issues such as stress fractures
No cycling valves or change of flow direction	Potential risk of corrosion issues if metallic
Available in high temperature materials such as ceramics	Risk of cross-contamination
Simple controls	Issues can lead to a short lifespan
Reduced fuel usage	More effective technology available
Reduced exhaust volume and NO_x	Single heat input point
Can produce higher flame temperatures and so increase the heat transfer rate to the product	
Faster heat up times	

When considering a heat recovery burner over a standard burner, its thermal efficiency can be compared in two ways – either the fuel reduction for providing an equivalent heat to the load or the increased thermal output for the same fuel usage. However, since even updated processes normally still require the same operating temperatures, the first method of calculating the costs savings from fuel reduction is predominantly used.

The thermal efficiency for a recuperative burner can be measured using three equations [8]:

$$\text{Net available heat (\%)} = \left(\frac{\text{Firing rate} - \text{Heat in exhaust gas}}{\text{Firing rate}}\right) \times 100 \qquad (2.2)$$

$$\text{Heat to load (\%)} = \left(\frac{\text{Net available heat} - \text{Heat loss}}{\text{Firing rate}} \right) \times 100 \tag{2.3}$$

$$\text{Effectiveness, } \varepsilon = \frac{\text{Heated air temperature}}{\text{Exhaust temperature entering heat exchanger}} \tag{2.4}$$

Burner performance is measured by the effectiveness, ε, shown by Eq. (2.4) which describes the transfer of heat from the exhaust gases to the combustion air, and typically recuperators have an efficiency between 0.3 and 0.6 [2, 8]. Of course, in addition to the advantages of using less fuel, heat recovery burners also release less CO_2 into the atmosphere.

2.1.2 Regenerators

A HEX is designated as a regenerator if it includes a thermal storage (TES) media that extracts and stores energy within the cooling a fluid, and which is then released to another fluid. As these units have a high surface area to volume ratio, they are more economical compared to recuperators.

2.1.2.1 Rotary Regenerators

Rotary regenerators (also known as rotary air preheaters) include heat wheels, Rothemuhle, Munter, or Ljungstrom wheels. Typical uses are for low- to medium-temperature ranges but high temperatures are possible whilst still providing a high efficiency of heat transfer [9, 10]. Traditionally, for waste heat recovery (WHR), they have been applied to the combustion processes in power plants, but applications for heating, ventilation, and air conditioning (HVAC) and other industrial processes are now available – wheels having been built with diameters as large as 21 m and a throughput of 1130 m^3/min of air [4].

Rotary regenerators span across two isolated ducts: one containing the hot stream and the other the cool stream, usually in counter-flow. The regenerator can be constructed from a woven mesh of knitted aluminium or stainless steel wire, or a honeycomb glass ceramic structure or silicon nitrides – amongst others – so as to create a larger heat transfer surface with a high heat capacity. The choice of material depends on the temperature of the streams, the contamination load, and the capital expense budget. This particular regenerator reduced 1600 °C exhaust gases to 140–310 °C by preheating combustion air for burners to 1400 °C from 20 °C. This provides a 62% fuel reduction – an attractive proposition [11, 12]. Overall efficiency has been reported as high as 85% [2]. The Rothemuhle regenerator is an adaption of this standard design, in that a static wheel is used with rotating hoods to distribute the airstreams across the matrix. This design is not commonly applied but has typically been utilised in power generating plants.

Rotary regenerators have issues with fouling and high pressure drop [2] due to matrix plugging, however a honeycomb structure can been used to counteract these issues ('plugging' is a fouling phenomenon in which particles adhere to surface elements and so gradually block or 'plug' the flow path). There are also the possibilities of cross-contamination between the streams and gas leakages. Also, since there is a continuous heating and cooling cycle, high thermal stresses occur within the regenerator and as such can lead to cracking or degradation of the matrix. Flexible seals act to minimise these leakages, but reliability is still a concern. Because of the risk, only applications where some cross-contamination is acceptable between streams can use these designs. Self-cleaning has been applied to reduce the impact of fouling and some designs have purge sections which remove contamination, such as particulates, before they enter the different stream. However, and notwithstanding these problems, the benefits of this system are the relatively high efficiencies.

The general equation for heat transferred from the fluid to heat sink matrix is:

$$\overline{\propto} A(T_{fluid} - T_{solid}) = Mc_s \frac{\delta T_s}{\delta t} \tag{2.5}$$

where $\overline{\propto}$ is the bulk heat transfer coefficient (W/m^2/K), M is the mass of solid heat sink (kg) behind the surface area (A) exposed to the heating fluid (m^2), T is the absolute temperature (K), δ is the thickness of heat sink (m), and t is time (seconds).

2.1.2.2 Static Regenerators

Static, or fixed bed, regenerators use two or more operating regenerators that operate out of phase. The regenerator heat sink can be enclosed within a duct or pipes and the valves switch the flow of the two streams resulting in one storing energy and the other extracting it. These tend to be for very high-temperature applications and require a refractory or ceramic heat sink. These need to withstand thermal stresses, corrosive environments, and particulate loading. An example of static regenerators is seen in Section 2.1.2.3.

2.1.2.3 Regenerative Burners

Like recuperative burners, regenerative burners work in a repeating cycle preheating the combustion air but typically using alternating burners operating in pairs. One burner fires, and the exhaust gases pass through a port in the second burner body to a HEX medium – predominantly, 2–3 cm ceramic alumina balls [13]. The medium extracts and temporarily stores the heat energy from the exhaust. Once the medium reaches a set temperature (or periodically), the fuel exhaust and air valves reverse the direction of their flow so that the combustion air of the second burner now passes through this hot medium and is preheated. As with recuperative burners, the burner combustion efficiency is increased because the flame is hotter and so a greater enthalpy content entering the combustion chamber – the repeating cycle leading to lower fuel consumption and lower CO_2 emissions per unit of heat. Literature reports that fuel consumption can be reduced by as much as 40–45% [14, 15].

Due to their design, regenerative burners are more suited to high-temperature recovery and can obtain higher efficiency as they recover more heat than recuperative burners. Preheating combustion air efficiency improvements are seen in Section 2.1.4. A fan is required to draw the exhaust gases through the unit as there is a higher resistance. There is a non-steady flow due to the cyclic nature of operation but also the preheat temperature of the combustion air decreases over each half-cycle as the energy content is removed from the thermal medium. Advantageously, these burners are very suitable for dirty exhaust gases as the burner is less prone to clogging if thermal media is used as the initial capture. However, this does mean that the burner medium requires maintenance and the regular removal of particulates.

If not present in the original design for a process, retrofitting these burners requires modifications to the system with the installation of increased pipework, ductwork, and controls. The majority of applications use a dual burner system, but self-regenerative burners exist as well as other minor system variants [16]. These are more compact, made so by the integration of two units into one and a burner of this type using a ceramic honeycomb heat transfer medium has been developed by NKK [17].

There are conflicting reports of regenerative burner success:

- Voyer and Caron [18] built an energy-efficiency model and determined that installation of regenerative burners was cost-effective for melting furnaces but not for holding furnaces.
- Schalles [19] reports short return on investment (ROI) periods, reduction in CO_2 emissions, and the maintenance being a minor consideration when taking fuel savings into account.
- Hassan and Al Kindi [14] report that regenerative burners are not profitable for saving energy when applied to aluminium holding furnaces and have a negative effect on the life of the furnace.

Table 2.4 provides a list of regenerator applications with technology, type of operation, and source and sink fluids.

Table 2.5 provides a comparison of advantages and disadvantages to installing regenerative burner technology.

2.1.3 Burner Technology Comparison

Table 2.6 provides an overview comparison between the two burner technologies and standard burners. Thermal efficiency, operating costs, emissions, and fuel usage are improved but there is an increased initial cost with more maintenance required.

Table 2.4 Applications of regenerators.

Type	Operation	Heat source	Heat sink	Applications
Brick checker-work	Cyclical	Exhaust gases	Air	Open hearth furnace
Pebble bed	Cyclical	Gas or liquid	Liquid or gas	Flameless combustion burners
Rotary wheel	Rotary	Exhaust gases	Air	Air preheater, water tube boiler, gas turbine
Heat pipes	Closed loop	Liquid, vapour, gas	Liquid, vapour, gas	Multiple small-scale

Source: Adapted from Mullinger and Jenkins [5].

Table 2.5 Advantages and disadvantages of regenerative burners.

Advantages	Disadvantages
No thermal expansion or corrosion issues	Requires multiple burners, valves, and flues
Compact per unit of heat recovered	Control system is more complex
High temperature achievable with ceramic construction	Thermal media requires refreshing or cleaning for optimal performance
High effectiveness	
Heat pattern can be spread	
Reduce fuel usage	
Reduce exhaust volume and NO_x	
Can produce higher flame temperature and more heat transfer rate to the product	
Faster heat up times	
Smaller furnaces in new installations	

Table 2.6 A high (H), medium (M), and low (L) comparison of burner technology.

	Burner type		
Consideration	Cold Air	Recuperative	Regenerative
Thermal efficiency	L	M	H
Fuel usage	H	M	L
CO_2 emissions	H	M	L
Maintenance	L	H	H
Initial cost	L	M	H
Operating cost	H	M	L

Source: Adapted from Schlesinger [13].

Figure 2.6 shows a comparison between regenerative and recuperative burner efficiency, the theoretical maximum efficiency, and the efficiency if there is no WHR.

2.1.4 NO_x Formation

Though not directly related to heat recovery, the formation of NO_x compounds goes hand in hand with burner technology. The majority of the relevant discussions are out of scope for this chapter, but a small section has been added here in view of the importance and the benefits of controlling the release of NO_x emissions.

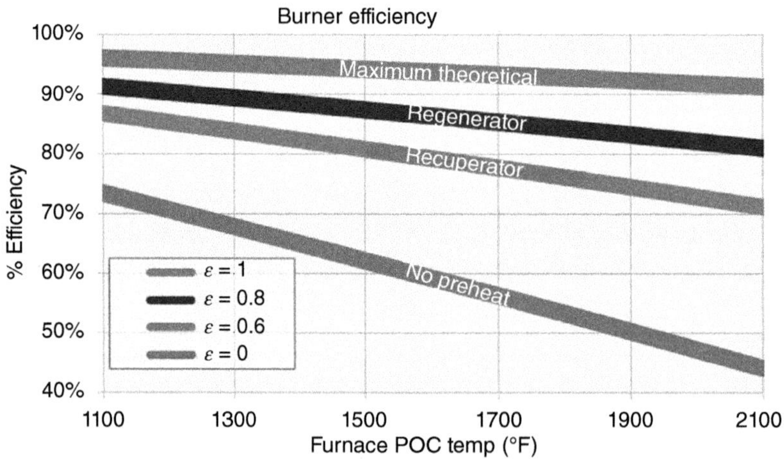

Figure 2.6 An efficiency comparison between regenerative and recuperative burners.

The reaction mechanisms are extremely complex with many pathways, so the equations presented subsequently only show the overall reactions without the intermediate pathways. Nitrogen in air is relatively stable; however, at high temperatures it can react with oxygen to form NO_x molecules. The subscript, x, depends on the amount of oxygen that reacts with the nitrogen. The majority is NO (95–99%) with NO_2 forming 1–5% [20].

$$N_2 + O_2 \rightarrow NO + N \tag{2.6}$$

$$N + O_2 \rightarrow NO + O \tag{2.7}$$

$$N + OH \rightarrow NO + H \tag{2.8}$$

These molecules are detrimental to the environment as they can form smog and ozone, and as such, the concentrations that can be released to the atmosphere are heavily regulated. They are thought to contribute to 30% of acid rain formation [20]. Increasing the flame temperature can cause an increase in NO_x formation as the activation energy for the reaction is more readily reached. Significant work has been done on the design of burners to reduce the formation of NO_x to develop more environmentally friendly burner technology.

2.1.5 Run-Around Coil

A run-around coil (RAC) can be used for liquid or air preheating purposes but tends to be used to recover heat from the exhaust of an air handling system to preheat the supply air. They are also named run-around loops or pump-around coils. They are a sensible HEX as they do not change moisture content, only the temperature of the airstreams.

The system preheats air by pumping a heat transfer fluid between two HEXs [21]. A hot stream is passed through a HEX which warms the heat transfer fluid. The heat transfer fluid is circulated to a further HEX where cooler air is passed through, extracting the energy from the heat transfer fluid. The now cool fluid continues to be pumped to the first HEX and the cycle repeats. A RAC is chosen if the heat streams are at separate locations or at a distance from each other for practical or safety reasons. Energy is required to pump the fluid so there are losses in the pipework and low effectiveness has been reported [9], 40–60% being typical [9].

The HEXs used for the loop are multi-row tube coils with fins to direct the fluid and increase the heat transfer surface. The heat transfer fluid is normally water due to the temperature range targeted but can contain glycol to prevent freezing. This can have an adverse effect by increasing the resistance within the system. However, RACs do have the advantage that no cross-contamination is possible and the cost of pipework is low in standard sizes.

The working fluid is the main consideration when looking at this application as higher temperatures require the use of organic or other fluids which can thermally degrade or change phase. Higher temperature applications require design changes such as the removal of fins in favour of a spiral around the ductwork, and higher temperature applications only show a 10–15% effectiveness [9] – much lower than using water as a working fluid.

2.2 Heat Exchangers

A HEX, as the name suggests, is simply a device used to transfer heat from one fluid (typically a liquid or a gas) to another fluid, but (in most cases) without the two fluids having to mix or come into contact with each other. For example, consider the exhaust gases from a power plant. Rather than venting these hot fumes to the atmosphere, we might be able to recover some of their heat by using a HEX.

Heat transfer is a fundamental discipline of engineering and of particularly relevance for example, in the design of refrigeration systems; in establishing the overall energy balance for a building and hence its heating or cooling load; in the use of electrical equipment such as generators, motors, and transformers where the removal of the heat produced as a by-product of their operation is essential to help limit their operating temperatures, and many other applications. They come in many sizes and configurations but typically they all share a common principle of operation – a solid wall (usually metallic for high thermal conductivity) separating the two moving fluids, one gaining heat (from heat flowing through the wall) whilst the other loses its heat.

As mentioned, HEXs come in many forms and can be classified in a number of ways. For example, by transfer process, number of fluids, surface compactness, flow arrangement, and the like. Figure 2.7 shows a common method of classification – by construction, a few examples of which we will consider in greater detail within this chapter.

2.2.1 Shell and Tube HEXs

A common class of HEX is the shell and tube exchangers, typically used in the petrochemical industry and the nuclear power generation industry and suited to

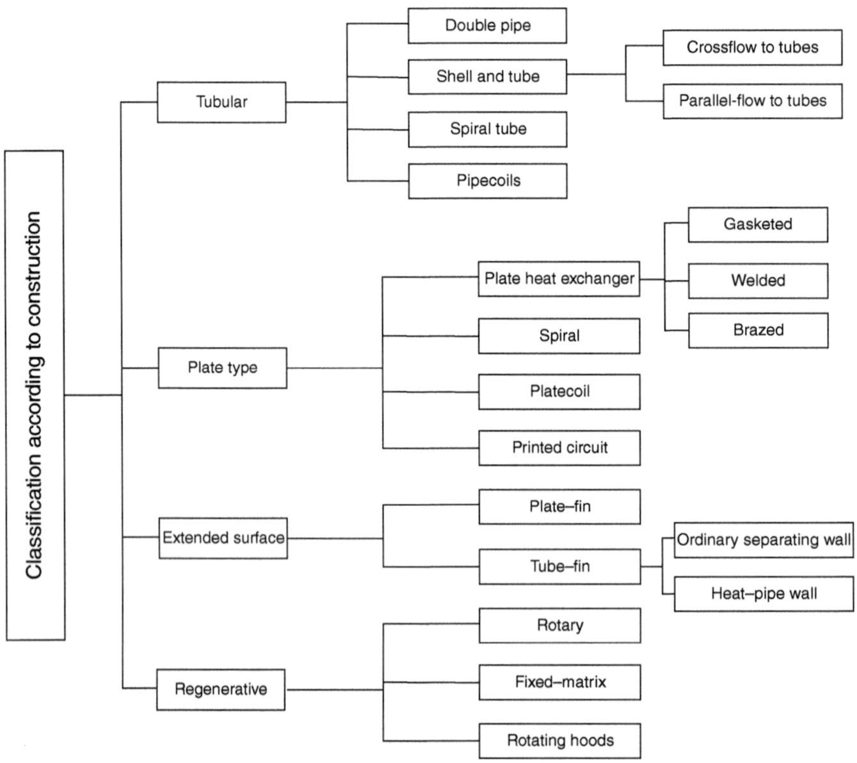

Figure 2.7 Heat exchanger classification by construction.

high-pressure applications. As the name implies it essentially comprises a large pressure vessel (the shell) with a large number or 'bundle' of tubes passing through it, these can be either plain or finned. One fluid flows through the tubes, whilst the second fluid flows through the shell surrounding the tubes and so heat exchange takes place between them.

2.2.1.1 Construction

The various possible contact patterns between the shell and tube bundle are denoted by a pair of numbers (1–2, 1–4, 2–4, etc.), the first number denoting the number of shell passes and the second the number of tube passes (for each tube of the bundle) – the term 'pass' representing flow from one end of the HEX to the other – and there are two methods of mounting the tubes: (i) straight tubes arrangements and (ii) U-tube arrangements. However, irrespective of the tube mounting method, an odd number of tube passes are rarely used due to expansion as a result of temperature change, an odd number making the associated stress problems are harder to manage [22].

Fixed tubesheet and floating head are both examples of straight tube exchangers and both have plenums at each end to allow the tube fluids to make another pass. The fixed tube version has its straight tubes fixed at each end making it the simplest and cheapest form of construction but it means that the outside of the tubes cannot

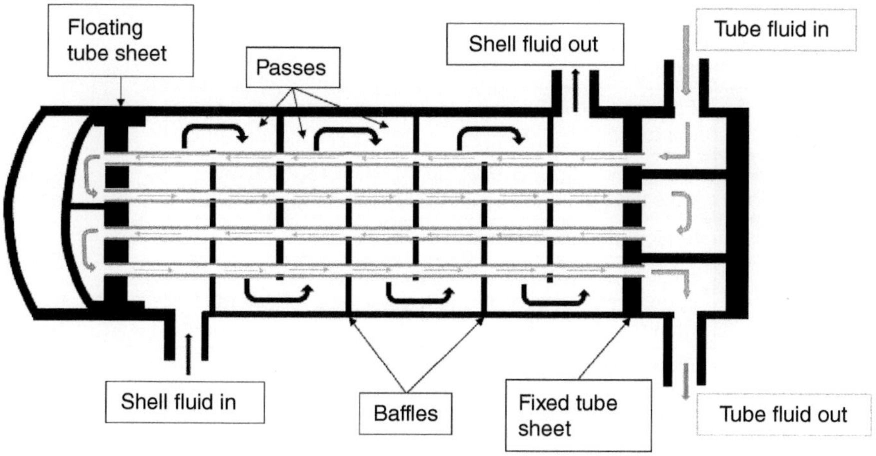

Figure 2.8 Floating tubesheet example of a shell and tube heat exchanger.

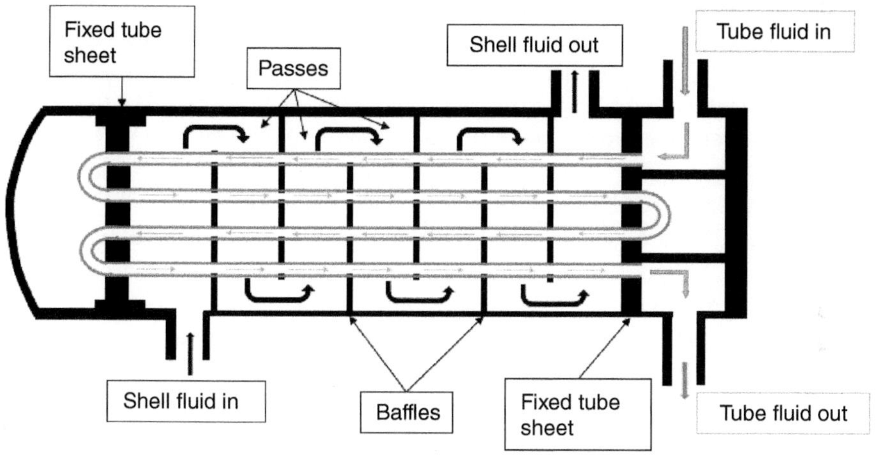

Figure 2.9 *U*-tube example of a shell and tube heat exchanger.

be mechanically cleaned. The floating head version is more versatile in that one of the tube-sheets is free to float within the shell to allow for tube expansions as shown in (Figure 2.8). This also ensures cleaning of both the inside and outside of the tubes is possible.

U-tube HEXs, as the name suggests, are where the tubes are bent into a 'U'-shape to make the necessary number of tube passes and are intermediate in cost (Figure 2.9). The tube bundle can be removed but only the tube exterior can be mechanically cleaned.

In addition to their tube arrangements, shell and tube HEXs also possess other essential features such as their baffles. These are used to support the tube bundle, reduce vibration, and prevent tube sagging, but importantly they also establish a turbulent flow regime which helps enhance convective heat transfer. The baffles

Table 2.7 Shell and tube, typical fluid allocation.

Shell-side	Tube-side
Viscous fluids	Fluids prone to fouling – easier to clean the inside of the tubes
Lower flow rates	Corrosive fluids – avoids the expense of special and expensive materials for both the tubes and the shell
Cleaner fluids	Toxic fluids – less chance of leakage
	High-pressure flows – more cost-effective to strengthen the tubes for high-pressure applications than the shell
	Flows requiring low-pressure drop – tube-side pressure drop is usually greater than shell-side, so less viscous fluid should pass through the tubes
	Cooling water

can also be arranged in a number of ways, for example when used for sensible heat transfer (heating or cooling) the baffle 'cut-away' should be horizontal, whilst for shell – side condensation a vertical cut-away is better.

The choice of which fluid flows through the tubes is largely governed by the practicalities of maintenance and operation. Table 2.7 highlights typical fluid allocations.

2.2.1.2 Applications and Limitations

Shell and tube HEXs are typically used for high pressure (>30 bar) and high temperature (>260 °C) applications, but as with all HEX construction, materials may also play a major role in the selection process especially if thermal and/or structural stress effects need to be considered. Usually these do not pose any serious problems at pressures below 15 bar and temperatures below 150 °C, but at pressures above 70 bar and temperatures above 550 °C material choices can be very limited.

Temperature differences between the tubes and shell may also cause problems especially where the temperature difference is greater than 50 °C. A large temperature difference can cause thermal expansion problems. Also, as you might expect in the case of corrosive fluids, there will be a need to select corrosion-resistance materials such as stainless steel or titanium to provide longer in-service component lifetimes; or a leak-proof construction as required where toxic or very expensive fluids are involved. Finally, despite their extensive use, shell and tube HEXs are not suitable for use in the transport industries (automotive and aircraft) due to their large size and weight.

2.2.2 Plate Heat Exchanger

The plate HEX (PHE) is a widely used, more compact form of HEX and consists of a number of thin rectangular corrugated or embossed metal plates, sealed around the edges, and held together in a frame. They are used to transfer heat between two (or more as shown in Figure 2.10) fluids and have the advantage that the fluids are

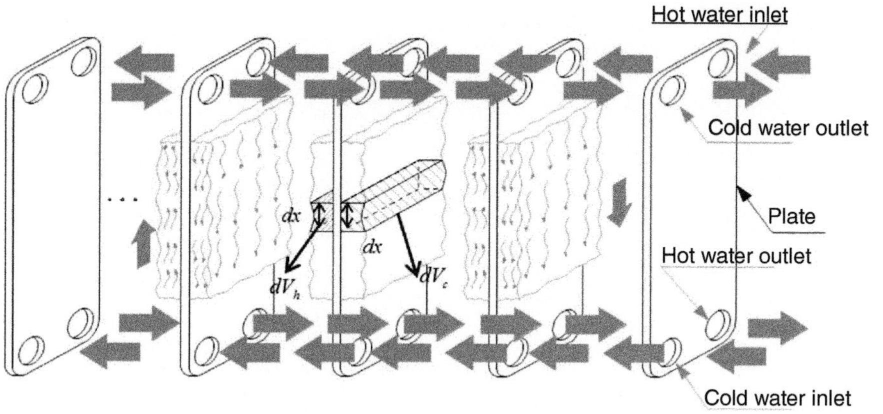

Figure 2.10 Plate heat exchanger example. Source: Miao et al. [23].

exposed to a large surface area which increases the heat transfer rate. They are suitable for medium- and low-pressure applications and are available in a wide range of sizes, for instance, very small versions are being used in the hot water section of domestic combination boilers to ensure a good flow rates [24].

Plate surface formations have evolved over the years to enhance heat transfer coefficients and reduce unit size, and are typically manufactured from stainless steel or titanium to produce units with operating limits between −35 and 220 °C with operating pressures up to 25 bar and flow rates up to 5000 m^3/h. By using brazed plates, the operating limits can be extended from −195 to 200 °C and pressures up to 30 bar, but the plates can then only be cleaned by applying chemicals. Welded-plate exchangers allow even greater temperatures (dependant on the materials) and pressure limits of up to 60 bar.

2.2.2.1 Spiral Plate Heat Exchanger

One variation of the PHE is the spiral PHE. Formed from two long strips – typically of carbon steel, stainless steel, or titanium – the strips, as the name suggests, are wrapped to form concentric spirals with normal operating limits up to 400 °C and pressures up to 25 bar (Figure 2.11).

2.2.3 Heat Pipe Heat Exchanger

A heat pipe HEX (HPHE) is another heat transfer device but this time relying on both thermal conductivity and a phase change (change of state) between a liquid and a gas.

A typical heat pipe consists of three sections: a condenser, an evaporator, and an adiabatic section. The tube is hermetically sealed and partially filled with a liquid. When the evaporator section is placed within a hot stream, the liquid will vapourise, the vapour then travels along the pipe and condenses back to a liquid at the cooler end of the pipe (condenser section) releasing its latent heat to pass through the walls

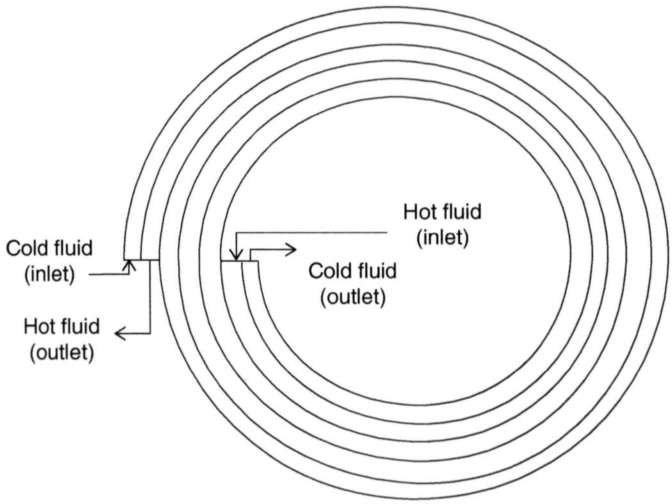

Figure 2.11 Spiral plate heat exchanger. Source: Picón-Núñez et al. [25].

Table 2.8 Heat pipe working fluids.

Working fluid	Temperature range (K)
Liquid helium	2–4
Ammonia	213–373
Ethanol	273–403
Methanol	283–403
Water	293–573
Mercury	523–923
Sodium	8731473
Indium	2000–3000

Source: Based on Jouhara et al. [26].

of the tube and so transferring heat. The condensed liquid then returns to the evaporator section of the tube though either capillary action, centrifugal force, or gravity and the cycle repeats.

Heat pipes are fast acting and very effective thermal conductors owing to the very high thermal heat transfer coefficients for vapourisation and condensation, but the working fluid must be chosen so that the pipe contains both liquid and vapour over its operating temperature range, examples of which are noted in Table 2.8.

Pipe materials are typically copper for water heat pipes or aluminium for ammonia heat pipes and although not essential, they invariably operate under vacuum or partial vacuum conditions. It is important to note that the pipe must contain both saturated liquid and vapour to transfer heat (Figure 2.12).

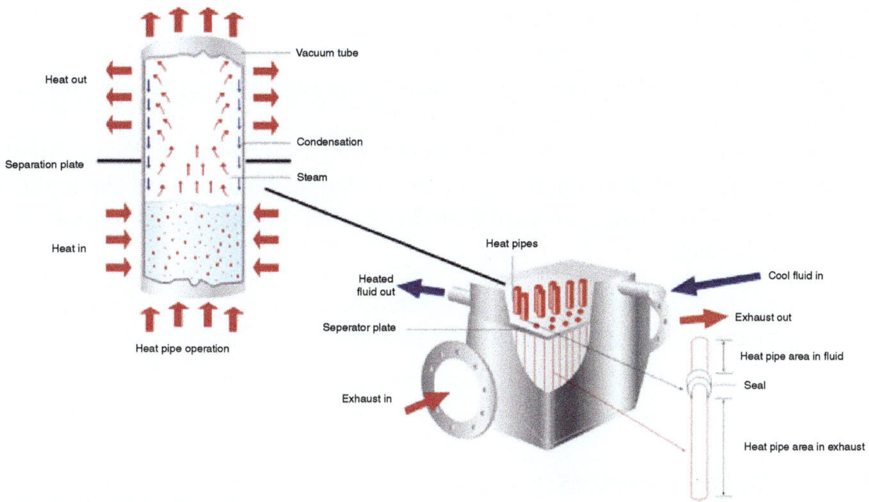

Figure 2.12 Example heat pipe module for waste heat recovery from an industrial exhaust stream. Source: Courtesy of Spirax Sarco Ltd.

A heat pipe contains no mechanically moving parts and so requires minimal maintenance, although over its life, its effectiveness may reduce due to breakdown of the working fluid, impurities absorbed from the tube wall, or scaling effects on the outside of the tube wall. However, their advantages over other forms of HEXs are their speed and high efficiency at transferring heat with only a comparatively small temperature differential being necessary for operation.

2.2.4 Compact HEX

HEXs classified as being 'compact' are those that are manufactured with a large ratio of heat transfer surface area to volume – a ratio known as the 'area density' or surface compactness (β); resulting in smaller and lighter HEX modules. They are commonly used for gas-to-gas and gas-to-liquid (or liquid-to-gas) HEXs in order to counteract the low heat transfer coefficients associated with gases.

Of course, 'compactness' is a matter of degree. Typical compact heat transfer surface area to unit volume ratios for compact liquid exchangers are those where the beta values are in excess of 400 m^2/m^3, whilst for gases, the beta values are those in excess of 700 m^2/m^3. However, that is not the whole story, for laminar flow or 'meso' HEXs beta values are greater than 3000 m^2/m^3, and the term 'micro' HEX is used where the surface area density is greater than 15 000 m^2/m^3. In contrast, a typical industrial shell-and-tube exchanger has a surface density of less than 100 m^2/m^3.

Basic flow arrangements for two-fluid stream compact HEXs are invariably: single-pass cross-flow, single-pass counter-flow, and multi-pass; the latter two leading to highly effective units or for situations with a small temperature differential between the fluid streams. However, there are a number of important design and operating considerations relating to extended surface compact HEXs that need to be considered:

- Usually at least one of the 'fluids' is a gas
- Fluids should be 'clean' and non-corrosive
- Operating pressures, temperature, and relative required pump power are limited compared to shell-and-tube exchangers
- The use of a highly compact surface results in a design which typically has a large facing area and short flow length, hence the design of the header for the liquid fluid is critical to ensuring a uniform flow distribution through a large number of small flow channels.

Fouling or scale formation can also be a potential problem as many fin geometries cannot be mechanically cleaned adding additional operating costs for chemical cleaning.

2.3 Waste Heat Boilers

WHR is a method of thermal absorption, that is, the reuse of heat energy that would be either disposed of or actually emitted into the atmosphere. Implementing recuperation of waste heat, plants will minimise energy costs and reduce CO_2 emissions, whist, at the same time, also increasing energy efficiency [27]. WHR techniques such as waste heat boilers (WHBs) implement the collection of the waste heat so as to ensure the optimal performance of the WHR process [28]

WHBs on industrial plants recover heat produced in furnaces or exothermic chemical reactions and are employed because these are locations that can contain a considerable amount of thermal energy that should not be released into the atmosphere. For example, in a standard coal power plant the hot exhaust gas produced from the combustion process after leaving the combustion chamber has a temperature of up to 1000 °C [29].This heat could then be collected in a WHB to produce low-to-medium pressure steam. WHBs may also be utilised to extract heat from the process fluids which require cooling prior to transport and storage.

The steam produced in WHBs may be used for heating purposes, for turbines generating electricity, compressing vapours, or for pumping liquids. However, the steam from the WHBs may contain substantial moisture, therefore it is necessary to install a high-performance separator in combination with a ventilation system to ensure that the waste heat boiler provides maximum steam quality to the receiving process [30].

WHBs consist of multiple water tubes which are positioned parallel to one another and follow the direction of heat exiting the system as shown in Figure 2.13. WHBs are available in a number of capacities, allowing gas intakes from 1000 to 1 million ft^3/min (0.47–470 m^3/s). But also, in situations where the waste heat is not adequate to produce the required amount of steam, it is possible to install auxiliary burners or after-burners to achieve a higher steam performance. The steam produced can be used either for process heating or for power generation, but the generation of superheated steam would involve the installation of an additional superheater to the system [31].

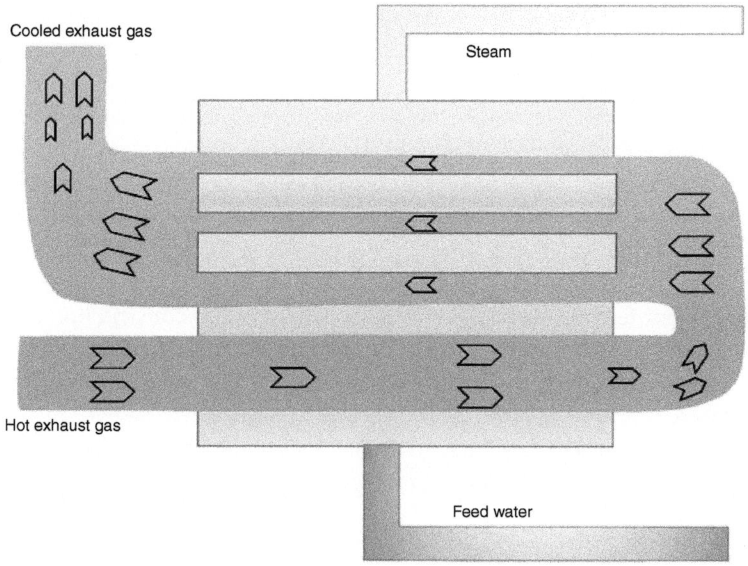

Figure 2.13 Waste heat boiler demonstration.

WHBs can also be combined with other WHR devices, such as after-burners, pre-heaters, and finned–tubed evaporators, to increase their performance by preheating the feed water and, if possible, to generate superheated steam [28]. Moreover, an economiser can be added to preheat the feed water before it enters the evaporator, that will boost the steam flow from the heat recovery boiler [32] – economisers being a type of HEX that transfers heat to boiler feed water whilst cooling gases entering intermediate and final absorption towers [33]. WHBs focus primarily on heat transfer convection; the gas velocity and the draught loss are both high for high transfer rates [28]. As a result, an induced fan is normally mounted with a WHB to collect and expel flue gas from combustion chambers by producing a region of negative air pressure [34].

2.3.1 Different WHB Designs

WHBs can be classified into several types based on their usage and applications such as:

1. *Water tube boilers*: for clean or dusty flue gas.
2. *Gas tube boiler*: for relatively clean flue gases.
3. *Bent tube boiler*: suitable for heavy dusty gases.
4. *Positive circulation boiler*: for clean and low-temperature gases.
5. *Supercharged boiler*: for gas turbine exhaust.
6. *Three-drum-low-head boiler*: suitable for light and dusty loading.
7. *Water wall bidrum-type boiler*: for gases that comprise sticky particles.

When selecting the correct and applicable WHBs type, the following criteria needs to be considered: heat load, temperature of the gases available for WHR for steam generation, chemical nature and the corrosiveness of the gases, available draught, dust load, chemical composition of the gases, pressure exerted, and the available footprint.

However in most instances, water tube boilers are more commonly applied for WHR due to their vessels' ability to withstand extremely high pressures without a reduction in efficiency. In addition, these boilers have the ability to endure thermal shock due to changes in gas temperature because the wall of the furnace can be efficiently cooled by adding water wall tubes. Moreover, since water is circulated in the tube side, and can be easily maintained, water tube WHBs are not susceptible to damage in cases of poor feed water quality. Other benefits of their construction are that issues of sluggishness and degradation can be reduced by adjusting the size and spacing of the tubes and maintenance more easily facilitated because dust particles may also be retrieved.

Approximately, 95% of the heat generated in the steam is wasted by the gases passing through the boiler and it can be deduced from Eq. (2.9) [28]:

$$\text{Heat lost by hot gases} = V * s\,(T_1 - T_2)\ \text{MJ/s} \tag{2.9}$$

where $V = \text{m}^3$ gas/s and s is the specific heat.

$$\text{Heat absorbed by water} = 0.95\ V * s(T_1 - T_2)\ \text{MJ/s} \tag{2.10}$$

where, T_1 and T_2 are the temperature of hot gases before and after the waste heat boiler in °C.

Overall, the total heat as steam can be deduced from the summation of the sensible heat in feed water and the heat absorbed by water in boiler [28].

2.3.2 WHB Methodologies

Waste heat boiler deployment has been playing a vital role in industry. Currently, WHBs are commonly used in the chemical, energy, medicine, and fuel industry to recover waste heat from flue gas [35, 36].

2.3.2.1 Feed Water Preheating Effect

Feedwater heating is important in some situations to prevent condensation of sulfuric acid in the boiler, sulfuric acid in the exhaust gas being formed by the sulfur content of the fuel. A study investigated the effect of feed water heating on a specific WHB. The feed water is fed into this boiler at the gas outlet and passes through the centre until it leaves the other end in the desired state (superheated).

The conditions of the steam exiting are regulated by the rate of flow for the feed water because low water flow rates yield high superheat, high minimum wall temperatures, and poor heat recovery. This control function was used, for this analysis, to achieve optimum heat recovery at an appropriate lower wall temperature [37], and the effect of WHB weight and volume on the outcome of a LM2500-based combined gas and steam turbine system (COGAS) was studied. The boiler was controlled to extract the maximum heat from the gas turbine exhaust whilst maintaining an acceptable minimum wall temperature. At a gas turbine power of 8.95 MW, the boiler – without feed water heating – produced sufficient steam to generate 2120 kW at a turbine efficiency of 80%. The boiler, including diffuser, weighed 8770 kg. The steam power was then improved to 2980 kW with a boiler weight of 17 400 kg when the feed water is heated. The respective 24% and 33% increase in power for these two systems converts into a 15% and 20% rise in fuel usage over the LM2500 gas turbine at higher COGAS power levels. At a lower power stage, all COGAS systems produced a 20% increase in fuel consumption over the standard gas turbine. Overall, it was observed that feed water heating allows more heat to be extracted from the gas turbine [37].

2.3.2.2 Optimising Thermodynamic Cycles

Industries use WHR boilers for the flue gas from diesel engine. The combustion of fuel in an engine normally resulting in 30–40% of heat energy being emitted by flue gas, increases the industrial application of heat recovery boilers for energy savings [38]. For example, the floating production storage and offloading Facility (FPSO) is a facility floating above sea level for the initial manufacture, storage, and discharge of crude oil. FPSO has the combined functions of production, accommodation, and management of petroleum products for its employees and many diesel engines are used to supply power to the FPSO needs.

For the purpose of recovering the released waste heat, waste heat recovery boilers (WHRBs) are regularly employed with diesel engines. However, due to the limited space on a FPSO, the design and installation of WHRBs are relatively challenging. Liu et al. [39] demonstrated a theoretical analysis and carried out onsite tests on the thermal performance of a WHR system for offshore oil production facilities, and these included the components of diesel engines, thermal boilers, and WHBs. A standard Brayton cycle was used to investigate the thermal performance, but compared to a conventional configuration, the fans of the WHRB engine outlet were removed due to the restricted capacity of the offshore platform. Cases with and without the addition of a fan were correlated with the system's output in terms of thermal stability, energy consumption, and thermoeconomic index.

As illustrated in Figure 2.14, the FPSO system contains diesel engines, WHRB and thermal oil boilers. For power generation, the crude oil was used as the fuel for the diesel engines. Generally, as in this case, the high-temperature gas in the range of 350–400 °C is normally discharged into the atmosphere.

But with the modification, the gas emitted from the diesel engine is transferred into the WHRB to heat the medium hot oil. The distance between the exhaust manifold of each engine and the WHRB can be more than 10 m depending on the local circumstances, even so, under standard operation conditions, the WHRB will still provide adequate heat to satisfy the thermal demand of FPSO, the thermal boilers only being started when the heat supply is inadequate. These two types of boilers are mounted in parallel and the WHRB installation is shown in Figure 2.15. Each WHRB being fitted with a single diesel engine to prevent inter-relationships between different diesel engines. After installation of the WHRB and the silencer, the flue gas from the diesel engine is supplied to the WHRB under standard operation. A bypass

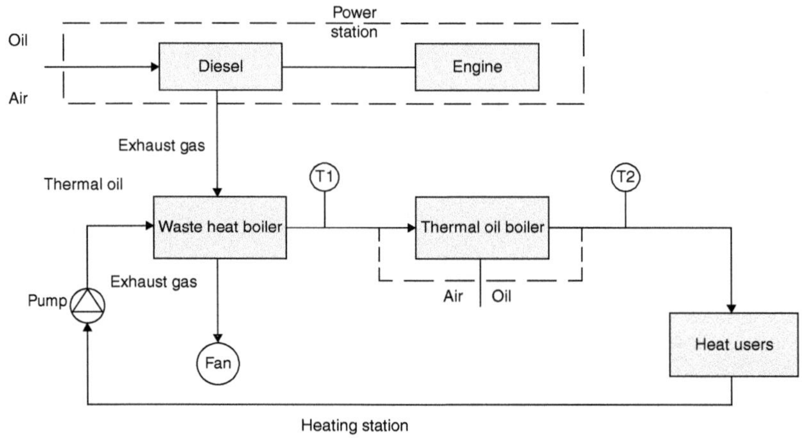

Figure 2.14 The heating station and power station diagram of FPSO. Source: Liu et al. [39].

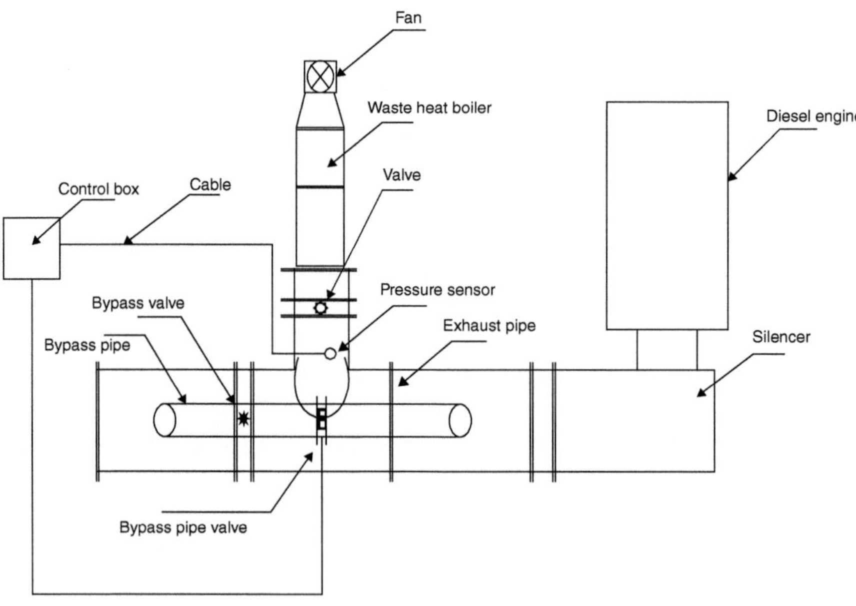

Figure 2.15 WHRB installation demonstration. Source: Liu et al. [39].

pipe was also attached for safety in the event the system is operated at much higher capacity than the designed value.

The findings indicate that the WHRB procedure improves the energy consumption of the whole system but raises the flow resistance inside the duct. As the WHRB takes the place of the thermal boiler, the energy consumption of the whole system without a fan is marginally decreased but the heat recovery performance is increased. So in all, this research offers valuable guidelines for optimising the recovery of waste heat for offshore oil production plants through the use of WHRB.

2.3.2.3 Heat Recovery Boiler with Water Spray Systems

The demand for high thermal performance coupled with the low emission of contaminants is growing for all thermal systems that use fossil fuels. Gaseous fuel systems have some benefits in terms of emitting filterable particulate matter (PM) as a solid or aerosol from the exhaust stack; nonetheless, they may still emit large quantities of NO_X, which will gradually be converted into atmospheric PM by secondary processes within the atmosphere [40]. NO_x emissions from industrial boilers are mostly regulated using after-treatment facilities such as selective catalytic reactors (SCRs) to maintain their stricter specifications. However, domestic and commercial NO_x boiler emissions can have a significant PM forming effect because small boilers are not ideal for SCRs [41]. Hence there is also a continuous need for advanced technologies to maximise thermal efficiency and at the same time reduce total NO_x emissions without excessive post-treatments. In a study, Kim et al. [42] aimed at exploring a thermodynamic analysis approach to assessing thermal efficiency and to provide information on the power design requirements for each device component. For the study, a chain-block structure analysis system was implemented using water spray heat recovery boilers (HR-B/WS) where water is sprayed into the inlet air stream and/or reaction field.

Design of Water Spray Heat Recovery Boilers Figure 2.16 indicates the proposed heat recovery boiler with water spray (HR-B/WS) considered in the research. The basic boiler subsystem (BBS) on the left comprises only the burner and primary HEX in the combustion chamber. The heat recovery subsystem (X) on the right recovers waste energy discharged from the BBS.

A counter-flow HEX (X) with air stream path (AX) and exhaust gas stream path (GX) was also used. The waste energy stored in emission gas (EG) is transmitted by a noncontact HEX into the inlet air system. To the bottom right, the water recirculation subsystem (WRS) consists of a reservoir (R), pumps, and spray nozzles. Water condensed at C and GX is held in the R and hot water is sprayed by recirculation into the inlet airflow and reaction zone. The air and water mixture are heated in the AX from hot exhaust gas. Inlet air is humidified by the evaporation of the small droplets from the water stream, which requires a greater portion of the extracted energy. Vapour in the preheated air reduces the temperature of the flame, reducing NO_x emission similar to exhaust gas recirculation (EGR). Water sprayed on the combustion chamber also lowers the temperature of the component, which reduces the production of NO_x in the post-flame zone. The actual flow path lengths, including air pipes, exhaust gas pipes, water feed lines, and the like, are ignored for the thermodynamic analysis [43].

Subsequently, the suggested method of research was tested and the boiler system was studied in operation. With the evidence of the study, HR-B/WSs are expected to achieve up to 95% performance with decreased NO_x emissions and additional water spray to the combustion chamber can be a useful tool to minimise NO_x emissions whilst maintaining high performance when inadequate water is supplied to the HEX. The current research therefore indicates that HR-B/WSs offer excellent thermal performa nce and reduced NO_x emissions.

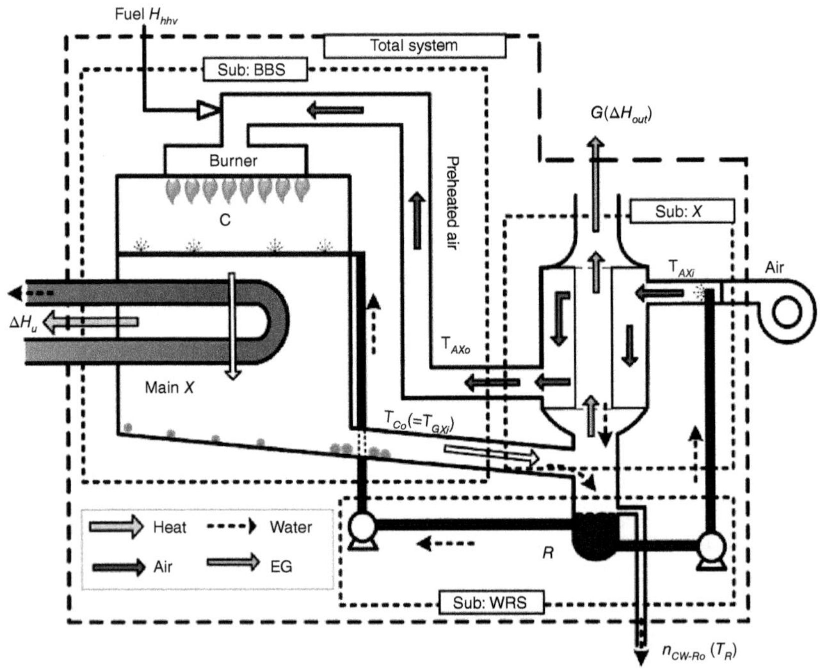

Figure 2.16 Typical heat recovery boiler with water spray. Source: Kim et al. [42]

2.3.3 Failure Modes

HEX tubes are a crucial component of any waste heat boiler, but they are often exposed to demanding operation settings, including high temperature, high pressure, corrosive flue gas, alkaline fly ash, and so on. Accordingly, HEX tubes of a waste heat boiler frequently encounter severe failures; the typical failure causes found in WHBs being categorised into the following classes: defects of the material [43], high-temperature corrosion [44, 45], stress corrosion cracking (SCC) [46, 47], and dew point corrosion [48, 49]

2.3.3.1 Failure Modes Analysis

A well-known German chemical corporation used a waste heat boiler to work with waste methyl methacrylate (MMA) liquid from upstream facilities, not just using exhaust gas, but also ensuring the pollutants were environmentally safe [50]. A schematic of the industrial procedure of the waste to energy conversation with this WHRB. As can be seen, the liquid waste collected by S-1930, atomised and mixed with natural gas, is sprayed into the thermal oxidiser where the gas mixture burns with preheated air producing high-temperature exhaust gas. The hot flue gas is then treated with aqueous ammonia to remove the harmful nitrogen oxides and so make it environmentally friendly. Afterwards, the gas flows through the waste heat boiler, which is composed of evaporator X-1981, superheater X-1984, evaporator X-1982, evaporator X-1983, and economiser X-1985, allowing the feed water to absorb latent enthalpy from the hot gas.

At the end of the process, the low-temperature flue gas can be release into atmosphere openly with minimal influences on the environment and the high-pressure steam is transferred to downstream plants to be reused.

Frequent shell-side corrosion occurred on the economiser tubes of a waste heat boiler and a series of analysis procedures were performed to investigate the root cause of the corrosion. Both the macroscopic and microscopic properties of the failed tubes were then closely studied followed by an investigation of the phase compositions of their deposits and then an investigation which included the accurate determination of the sulfur and nitrogen content. The findings showed that repeated failures were mainly due to sulfuric acid dew point corrosion and the sulfur was mainly exposed from MMA pollution [50].

2.4 Heat Recovery Steam Generators

As a result of the recent developments and technological advancements within the field of WHR, significant efficiency improvements can also be achieved for energy-intensive industries. Recovering waste heat from industrial processes is regarded as an ideal solution because it brings substantial financial or economic saving benefits and helps with reduction of greenhouse gas emissions. As described by Jouhara et al. [51], there are various technologies and applications which offer different techniques of WHR.

Unfortunately, and for all today's extensive efforts towards energy production from green and renewable sources, still nearly 70% of global power is produced from fossil fuels. For instance, as has already been investigated, fossil fuels involve a large portion of energy composition for many major industries, such as, food processing, pulp and paper, chemicals, and primary metals production. This clearly indicates that the employment of WHR technologies and applications show a promising solution to help reducing the dependency on fossil fuels in the mentioned industries [51].

WHR systems take advantage of harnessing the energy that is lost due to the inefficiencies of a process and converting this to useful energy. This, in return, improves the net useful energy that can be utilised on a system and hence reduces the overall amount of fuel consumption. Alternatively, the waste energy can also be absorbed and then used to generate other forms of energy or be used to create other products that can have a useful impact on the overall efficiency of the system [52–54]

A good example of such a technology is the waste heat recovery steam generator or (HRSGs) which has been shown as an ideal solution for converting wasted heat from the exhaust of a process to help produce steam. WHR steam generators use several different components and HEXs to convert these wasted hot gases from the stack to useful steam – the steam then being used for heating, processing, or even generating electricity [55].

Investigations have shown that waste HRSGs can increase the energy efficiency of an industrial process, which in turn brings significant economic and environmental benefits [56]. For instance, in a recent study conducted by Krishnan et al. [57],

Table 2.9 Temperature range and characteristics of different waste heats.

Waste heat source	Temperature range (°C)	Cleanliness
Furnace or heating system exhaust gases	316–1100	Varies
Gas (combustion) turbine exhaust gases	480–600	Clean
Jacket cooling water	90–100	Clean
Exhaust gases (for gas fuels)	480–600	Mostly clean
Hot surfaces	65–316	Clean
Compressor after or inter-cooler water	38–82	Clean
Hot products	100–1370	Mostly clean

Source: Thekdi [58].

nearly 371 MW of thermal energy was recovered from the exhaust of a factory flowing at almost 700 kg/s that was cooled down by almost 400 °C. This clearly indicates the huge advantage in employing waste heat steam generators for energy recovery, especially from high quality and clean waste heat. Table 2.9 below shows the sources, quality, and temperature ranges of different waste heat.

2.4.1 Construction of Waste HRSG

Waste HRSGs are mainly categorised into three different groups, namely single, dual, and triple-pressure types, depending on the configuration and number of drums and HEXs in the system. The cost of investment increases as the system gets more complex, with the triple-pressure system being the most expensive. Having noted that, the effectiveness of the system is also dependent on how many passes the waste heat can go through, so a triple system offers the greatest efficiency. Regardless of this, investigations have shown that a dual pressure system is the most widely used because it offers a good balance between efficiency and cost [59].

Others have shown that firstly the detailed full design of any selected WHR steam generator must be considered since the overall performance of the installation entirely depends on the configuration and number of the components available within it. And in order to produce an efficient and economically feasible system, the quality and quantity of the exhaust gas must also be fully studied since this is the parameter that guides the minimum heat exchange area of the system components. Hence indicating the correct design parameters are essential in ensuring the maximum utilisation and steam generation from the system. In this regard, many studies have been conducted which take into consideration different design parameters for adopting a waste heat steam generator system within a manufacturing processes. For instance, Nag and De [60] conducted investigation into the functionality and performance of a WHR steam generator that can be used in a combination of a gas and a steam power cycle and concluded that the entropy is decreased when the system is being fully operated. Franco et al. [61] performed researches into different designs of WHR steam generators and discovered that exergy losses can be reduced based on the irreversibility of the process. And it also further evolved that combining a WHR steam generator with a gas turbine can increase the efficiency of a system by nearly 60%. Furthermore, and in a cogeneration setting, by employing more advanced components, system efficiency can increase by up to 75%. Today, recent advances in WHR steam generator design comprising a so-called 'double-pressure' system allows the capability of maximum heat recovery – the system incorporating auxiliary heating unit and high-performance superheaters that can be used to efficiently generate steam [59].

The first step therefore in designing a WHR steam generator is to indicate the amount, quantity, and quality of steam required to be generated by the system. In this regard, the temperature and flow rate parameters of the generated steam need to be considered so that the sizing of the main system components such as the economiser, evaporator, and superheater can be performed.

Additionally, WHR steam generators can be rather large in size and in this respect, the method of construction and transportation of the system also needs to be taken into consideration. For example, the size of a unit can be so significant that it could potentially be made in several sections or modular parts to be assembled at a later stage on site at the facility. Whereas smaller units can be transported to the site as whole and in a fully assembled configuration [62].

There are many differing designs of WHR steam generators, but all follow a similar operational methodology. The basic configuration includes a set of HEXs, placed at the outlet of exhaust gases and which allow water circulation through them, so that radiative and conductive heat transfer occurs from the exhaust outlet, through these HEXs, allowing the water to be transformed into steam at high pressure [51].

2.4.1.1 HRSG Design and Construction

A typical WHR steam generator can comprise several pressurising segments, such as low-, intermediate-, and high-pressure sections. Each of these typically include an evaporator and steam drum components which transform water into steam. The higher the number of WHR sections, the higher the efficiency of the system. After the steam is generated at the final stage, it will then pass through a superheater which ensures that the pressure and temperature of the generated steam exceeds its point of saturation [63]. If the WHR steam generator needs to be switched off, then a diverting valve can be placed at the inlet of the system to divert the outlet exhaust flow to the stack.

There are usually four components included in a WHR steam generator, namely an evaporator, steam drum, superheater, and economiser. This section will cover how each these components plays its role within a WHR steam generator system [64].

2.4.1.2 Evaporator

The evaporator is the section in which the waste heat from the exhaust is turned into steam. The segment contains a HEX that comprises several tubes in which water is passed, and could be defined as the heart of the WHR system since it dictates the overall steam generator configuration and performance. However, research has shown that the quantity of thermal energy removed by this segment is somewhat limited because the difference in the inlet and outlet temperatures of the fluid through this stage are not significantly different to the saturation temperature at the system pressure.

2.4.1.3 Superheater

The superheater component is typically placed at the first stage of a WHR steam generator and directly faces the incoming hot exhaust gas. The component can be described as the first layer of defence in terms of waste heat absorption and is used not as an evaporator but rather as a dryer of the saturated steam that is diverted from the steam drum. The component is connected directly to the steam drum and has the role of topping up the amount of thermal energy absorbed by the evaporator by increasing the temperature of the generated steam to reach the saturation point or just beyond. In this regard, the design and size of the evaporator is directly dependent on the type of the steam drum used.

2.4.1.4 Economiser

As apparent from the name of the component, the economiser is responsible for improving the efficiency of waste heat absorption so making the system more economical. With this in mind, the component is placed at the outlet of the waste heat steam generator and is configured to preheat the water that is being fed into the evaporator. Thus it acts as a last resort for absorbing any waste heat that could not be picked up by the evaporator and thereby improving the overall efficiency of the system. Like the superheater, the design of the economiser also depends on the type and size of evaporator being used.

2.4.1.5 Steam Drum

Steam drums can be designed in the form of either high or low-pressure systems. The component comprises several internal separators and channels which comprise various baffles and diverters. The channels within the drum allowing for the collection and segregation of water from steam before entering the superheater HEX.

2.4.1.6 Evaporator Types

Evaporators come in different shapes and designs. The following are the commonly used evaporator types which are available in different configurations and different operational sizes [65]:

D-Frame The D-frame evaporator is the most compact type of evaporator and is ideal for WHR steam generator infrastructures with smaller exhaust sizes as shown in Figure 2.17. Thanks also to their compact components, the size of the HEX is convenient for transportation and so can be installed as a whole unit with no assembly complications.

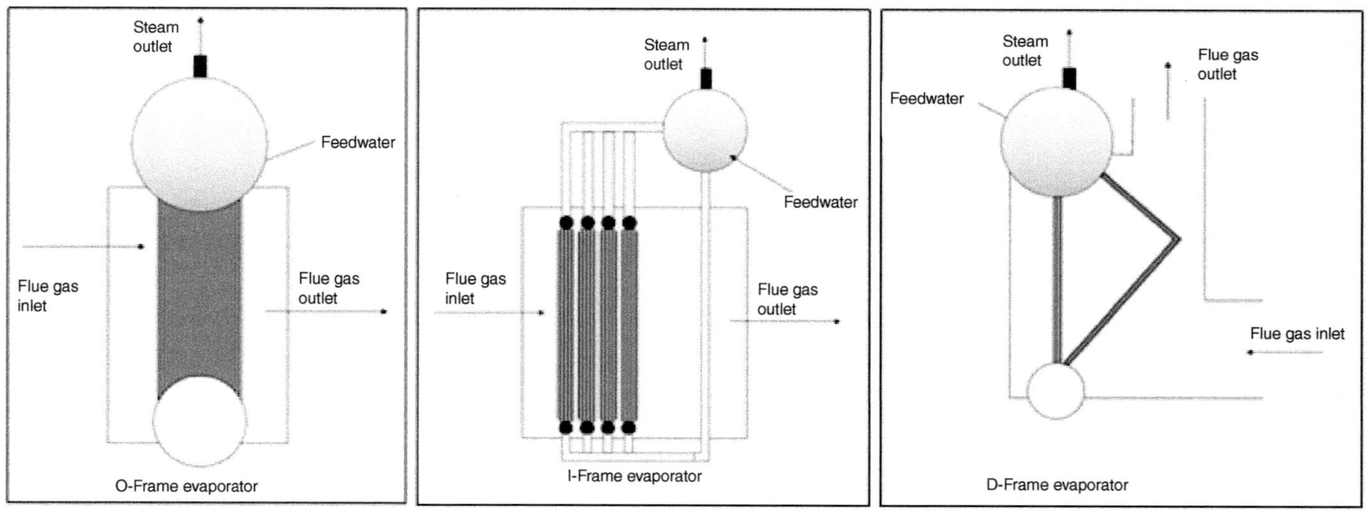

Figure 2.17 HRSG frame types.

I-Frame The I-frame evaporators are made of several axial or lateral sections and can be used to recover any flow of waste heat, and the design of this particular type of evaporator is very economical and practical to manufacture. The I-frame evaporator HEXs can be made in arrangements of up to three rows of tubes and headers in each bundle set as shown in Figure 2.17.

O-Frame The O-frame evaporator is the most conventional type of HEX used in WHR steam generators as shown in Figure 2.17. The specific configuration of the HEX allows the header of the unit to be used as a steam segregation device and thanks to this unique design feature, several evaporator HEXs can be connected to the same steam drum. The design makes this type of HEX suitable and simple to assemble for larger WHR steam generators.

2.4.1.7 Horizontal Tube HEXs

Horizontal tube HEXs are found to be the cheapest types of design to manufacture. The HEX employs several sets of large diameter horizontal tubes which are placed in a specific arrangement to absorb the waste heat from flue gases. Having noted that and the consequently large size of HEX, the assembly and transportation of a system employing this type of HEX may be more complex and higher in cost [66].

2.4.1.8 Natural Circulation HRSGs

Natural circulation HRSGs comprise several stacks of tube HEXs that are placed vertically perpendicular to the outlet of the exhaust path. This particular configuration allows for water within the HEX to heat and then rise naturally to the header drum through its change in density and unrestricted expansion, whilst at the same time, getting replaced by a counter-flow due to gravity. This natural circulation within the water/steam cycle brings several advantages to the overall performance of the system [67], some of which are:

- There is no requirement for the installation of a circulation pump. This could potentially make the system more reliable, cheaper and, in the longer term, save money since less energy is consumed.
- The inlet duct to the HEX can be installed at a lower height or on ground level. This allows for easier installation, and could also make for a more robust structure.
- If used for WHR from a furnace, the configuration create a naturally water-cooled zone for the system

However, having indicated the advantages of natural circulation WHR steam generators, there are also several disadvantages associated with the system as indicated below:

- The system has been shown to be more difficult and complex to disassemble and clean
- The HEXs generally require regular cleaning and moving the tube banks is not easy or, in some cases, not feasible.
- System start and heat-up may take much longer time to complete so lowering overall efficiency and also invoking a higher carbon footprint.
- Rapid processing with frequent stopping and restarts can result in the build-up of deposits, fatigue, and leakages within the tubes.

2.4.1.9 Assisted (or Forced) Circulation HRSGs

Assisted circulation WHR steam generators are configured so that the stack of HEX tubes are in a horizontal orientation but perpendicular to the incoming exhaust flow. The HEXs in this system comprise an inlet and an outlet drum header – where water is distributed inside and steam is generated. The order of the drums is arranged so that steam can be generated by injecting or forcing water at the inlet drum that is connected to several tube banks. The structure of the HEXs is supported by individual metallic sheets that are held in place by a steel support and which are usually relatively tall. As apparent from the name of the system, there is also a need for recirculation pumps to move water from the lower level to the collector drums and direct the water–steam mixture within the evaporators [68].

There are certain advantages associated with the use of assisted circulation WHR steam generators which makes the technology a better option over naturally circulated system. These benefits can include:

- More rapid start-up from cold so faster to reach operational running
- Quicker response to conditional changes of the flow of steam
- The system generally requires a lower water content and so offers a smaller carbon footprint
- The system is compact in size and can be placed on a unit in cases where space is restricted.

Conversely, the following disadvantages are associated with forced circulation WHR steam generators:

- The need for circulation pumps to be integrated into the system, which ultimately results in higher active energy consumption and potentially more maintenance work
- The units are very tall in height and so may pose installation restrictions
- Because the HEXs are placed in a horizontal orientation, there is the potential for the water within the tubes to dry, with the possibility of overheating, fatigue, and eventually corrosion.

2.4.1.10 Tube Materials

HEX tubes can be made from various materials based on the temperature range at which the system is being operated. The majority of WHR steam generators operate in the relatively low- to medium-temperature range; however, the system can also be used in combination with supplementary firing at higher temperature ranges. The evaporator and economiser are usually made using carbon steel because these components only have to deal with the lower temperatures within the system. But having said that, higher alloy materials with a reduced wall thickness are often used in order to reduce the cost of the system. In order to increase the rate of heat transfer through the HEX the tubes are usually covered with helical fins. The superheaters are constructed using T11 alloy tubes so that they can deal with higher ranges of heat from the exhaust. We should also note however, that due to the occurrence of oxide and carbon deposits passing through the exhaust, the tubes require cleaning on a regular basis to reduce the reduction of heat transfer efficiency and performance [69].

2.4.1.11 The 'Pinch Point' and Other Effects

There are many variables that dictate the energy balance of WHR steam generators, but the pinch point analogy has been noted as a method that allows for the selection of an appropriate temperature for the operation of the system. To explain this further, e.g. if the pinch temperature for evaporation is not properly defined, then the quantity of the energy recovered from the evaporator section can be limited because both inlet and outlet temperatures to the HEX can be at their saturation temperature relative to the system pressure. Therefore, the pinch point or more simply the optimum operating temperature of one WHR system can be different to that of another, and entirely depend on the heat source. In addition, this also affects the entire design of the HEXs and the coils used within a system.

Investigators have therefore concluded that several different heat balance variables need to be taken into consideration when designing a WHR system and some of these important parameters include the exhaust outlet mass flow rate and temperature, the maximum and minimum steam generation flow and temperature requirements, exhaust thermal and pressure properties, and other limitations related to the noise and emissions produced by the system. Based on these, if the desired output from a system cannot be delivered, introducing an additional process – such as supplementary firing – might be required [59, 70].

Supplementary Firing Supplementary or in-duct firing is a terminology that is used to describe a configuration that can be integrated with the exhaust of a WHR steam generator to provide additional output to the system if and when required. This can increase the system output to a required rate and so can result in an increase in efficiency.

Additionally, a supplementary firing system can also help generate more power from a system, especially at peak times when higher demand is required The supplementary firing configuration is usually placed at the exhaust of a WHR steam generation system, and can be used to adjust the inlet heat and so throttle the amount of flow generated to the WHR system. One obvious disadvantage associated with the use of supplementary firing is that harmful emissions can be produced, which is not usually the case with a typical standalone system.

2.5 Heat Pumps

2.5.1 Fundamental Principles of Heat Pumps

Heat pumps are essentially devices that absorb *heat* from one space and release it to another. In theory, the process is the perfect reverse of the more commonly known term 'refrigeration' – where the function is to extract heat from a space that requires cooling, the function of the heat pump being to supply heat at an elevated temperature to a space that requires heating. That being said, *reversible* heat pumps can work in either direction to provide heating or cooling to an internal space.

Looking at Figure 2.18, the two modes of operation are as follows.

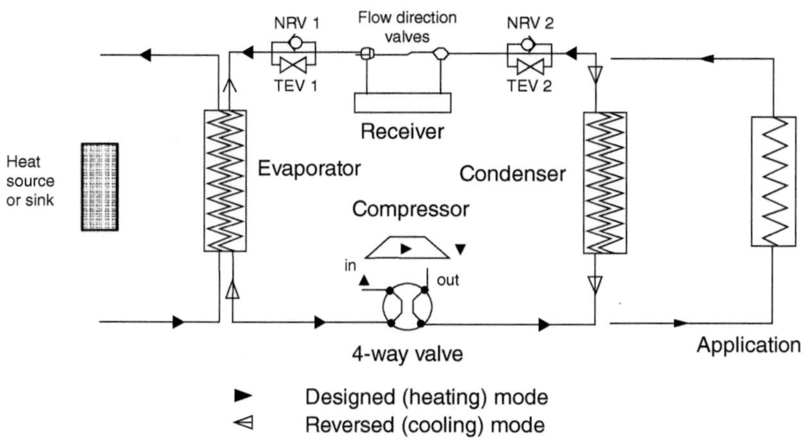

Figure 2.18 Reversible heat pump operation. Source: Leelananda et al. [71].

2.5.1.1 Cooling Mode

Point A: Here the refrigerant is very cold and in liquid form as it enters the evaporator coil inside the building. The hot air inside the building is forced over this coil and loses its heat to the refrigerant.

Point B: The refrigerant leaving the indoor coil has now absorbed heat and so vapourised and in this gaseous state enters the compressor.

Point C: Leaving the compressor, the pressurised refrigerant now moves to the outdoor condenser coil where it loses heat to the outdoor air and condenses back into a liquid.

Point D: The still warm liquid now enters the expansion valve which allows the liquid to expand – so reducing its pressure and lowering its temperature ready to absorb heat again and repeat the cycle at point A.

2.5.1.2 Heating Mode

Note the reversing valve has rotated by 90° which changes the direction of refrigerant flow so that now at:

Point A: Having just left the expansion valve the liquid is at its lowest temperature and is ready to absorb heat from the outdoor coil, so changing state back into its vapour form, i.e. the outdoor coil is now acting as the evaporator.

Point B: Notice that the compressor still operates in the same direction and increases the vapour temperature and pressure.

Point C: The hot vapour now passes to the indoor coil where it loses heat to the building – i.e. the indoor coil is acting as the condenser – and in losing heat the refrigerant condenses back to its liquid state.

Point D: The refrigerant leaves the indoor coil as a warm liquid ready to be expanded and the cycle repeats.

If a heat pump is used in both winter and summer, the total energy exchange with the external environment may approach zero [72], dependent upon the duration of the seasons. Nonetheless, it should be recognised that mechanical energy is provided to the heat pump during operation and so the heat exchanged with the indoor environment needs to be considered as that producing a 'useful effect'.

Accordingly, and as with refrigeration, the performance of a heat pump is expressed as the ratio of useful heat transferred to work input by the *coefficient of performance* (COP) defined from:

$$COP_{hp} = \frac{h_2 - h_3}{h_2 - h_1} \tag{2.11}$$

where h_1, h_2, and h_3 are the enthalpy conditions on a temperature–enthalpy diagram (Figure 2.20), discussed in Section 1.2.4.3 for heating – and Section 1.2.5.2 for cooling.

Irreversibility's invariably reduce the ideal COP and these are due to a number of reasons: friction loses, the slight cooling of vapour just after it leaves the compressor, pressure loss across the expansion valve, outdoor temperature changes during the time of exchanger operation, and the like. However, the COP is very influenced by the temperature of external environment [72] (Figure 2.19).

2.5.2 Variation of Heat Pump System

A complete classification of heat pump types depends upon the parameter that you are comparing – heat source, configuration, application, output, and the like. Nonetheless, two common classifications methods are grouping according to (i) heat sources or (ii) heat source/heat sink configuration (Figure 2.20). (A heat sink is a thermal energy reservoir which can receive huge amounts of energy without a significant change in its temperature. A heat source is a thermal energy reservoir which can supply huge amounts of energy without undergoing a change in its temperature.)

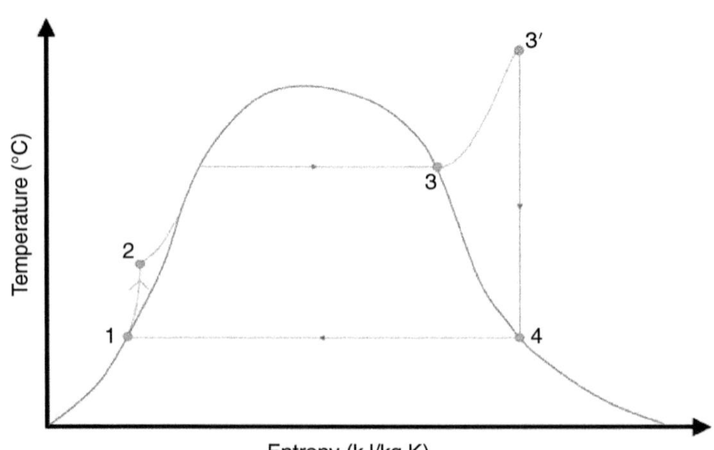

Figure 2.19 *T–s* diagram of the superheated Rankine cycle.

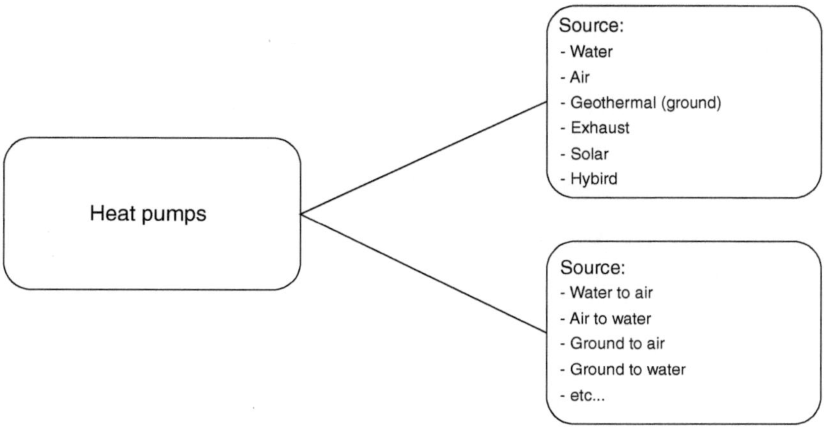

Figure 2.20 Typical heat pump classifications.

In terms of the most common measure of heat pump efficiency – COP – air source heat pumps typically have COPs ranging from 2 to 4 – denoting that they supply 2–4 times more thermal energy than the electrical energy they consume. Water and ground source heat pumps have COP values typically between 3 and 5 but, as noted earlier, the COP of an air source heat pump reduces as the outside temperature drops, so when comparing systems it is important to ensure that their ratings are based on the same outside air temperature. For ground and water source systems, their COPs do not vary so much because ground and water temperature tend to be more constant and stable than air temperatures.

Using the source classification, we'll now consider some common types of heat pump.

2.5.2.1 Air Source Heat Pump System

Air is a common source for heat pumps which are used to meet the HVAC requirements of residential and light commercial applications. Whilst appearing as a free thermal source, we need to remember that in temperate climates frost may accumulate on external heat transfer coils when the temperature falls below 2–5 °C requiring a short-time reversal of the heat pump cycle and hence a loss in overall system efficiency (Figure 2.21).

Regrettably, the performance of an air source heat pump reduces with decreasing external temperature – at time of its greatest need, which then may require an additional supplementary oil-, gas-, or electrically powered device to cope with extremes of weather condition.

2.5.2.2 Ground Source Heat Pump System

A ground source heat pump system uses the ground as a seasonal storage of solar energy for its operation. At a depth of between 1 and 1.5 m there is much less variation in temperature change compared to outdoor air temperature changes, and in addition, the ground temperature changes often lag the corresponding air changes.

Figure 2.21 Air source heat pump. Source: Carroll et al. [73].

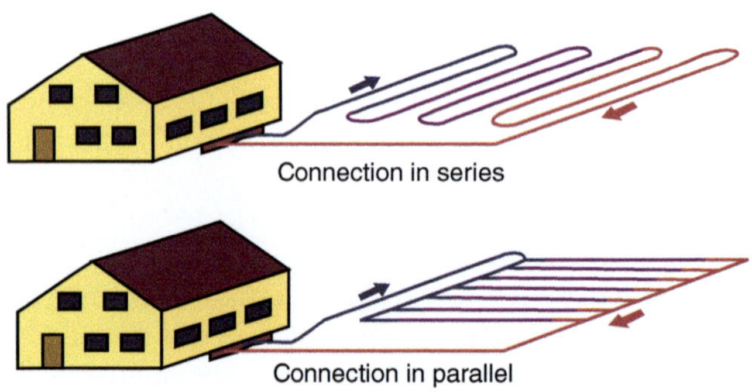

Figure 2.22 Ground source heat pump (indirect) system. Source: Mustafa et al. [74].

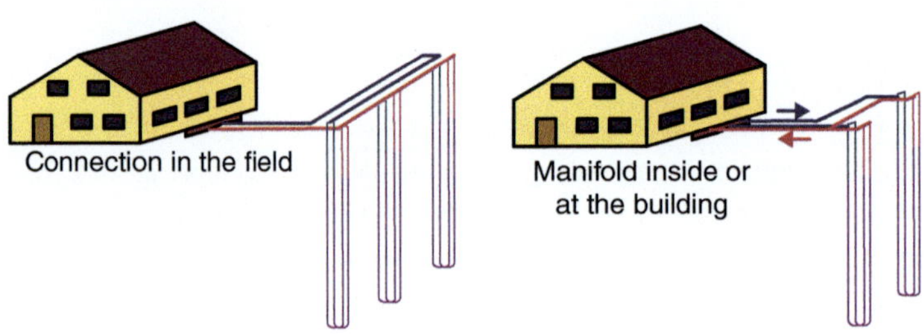

Figure 2.23 Vertical ground source heat pump. Source: Mustafa et al. [74].

Such conditions are therefore very favourable for extracting energy for both heating in winter and cooling in summer (Figure 2.22).

Ground source heat pumps can be designed as a direct system or as an indirect system using a secondary refrigerant as the transfer agent and pipes can be laid horizontally or vertically dependent upon local geological conditions. Both methods tend to be much more expensive that air source systems because of the extent of groundwork and pipework, and coil leakages may lead to maintenance considerations in the longer term (Figure 2.23).

Heat is extracted from the ground typically using glycol as the working fluid, but operation is such that several hundreds of square metres of pipework are required just to heat a single home, so in urban areas where space is usually at a premium, vertical systems may be a more practical solution.

As a variation and in some countries – the United States, Canada, France, for example – geothermal heat sources may be available which considerably raise the COP due to their high and constant temperatures. However, such resources are often localised and sited away from the main areas of high population density. In addition, there may also be a high salt content leading to possible long-term deterioration and fouling problems.

2.5.2.3 Water Source Heat Pump System

Water sources may also be suitable in some areas to act as heat pump sources. If, for example, groundwater is available at a stable temperature between say 5 and 10 °C, then this can be used to drive either an open or a closed heat pump system. With an open system, groundwater (pumped to surface level) is then either discharged to surface water or into a separate rejection well after use. Closed systems typically use a working fluid evaporating in underground HEX coils or brine with a looped pipework systems. The latter have a slightly lower COP but are easier to maintain.

A point to consider is that local water regulations may impose severe constraints due to interaction with the water table and the possibility of ground and water pollution.

Surface water, such as rivers and lakes, also make useful heat sources but can suffer from the problem of freezing in winter.

Sea water is also another alternative (under certain conditions) for medium- and large-scale heat pump installations because at a depth of between 25 and 50 m, the sea temperature is constant at between 5 and 8 °C and so freezing is not generally a problem.

2.5.2.4 Water Loop Heat Pump System

The idea behind the water loop heat pump system is to provide a simple but flexible decentralised approach to HVAC in multi-room buildings, using heat recovery from lighting, occupancy, solar gains, and equipment heating gains, and distributing this energy to where it is needed. Developed in the 1950s in the United States, they became more popular in the 1990s mostly in the United States and Japan.

The principal components of the system are several small, self-contained, reversible heat pump modules which are then connected together by a water loop of uninsulated pipework through which water is continuously circulated and maintained at a temperature typically between 18 and 32 °C.

Operation is successful because many large buildings have internal zones that require cooling during occupied hours whilst, at the same time, the perimeter spaces need heating. The closed water loop pipework provides water to the condenser coils of the interior heat pumps – heat which would be normally rejected to a cooling tower – and this heat is then used as the heat source for the evaporator coils of the perimeter heat pumps. During peak winter conditions if the heat pumps cool the water below 18 °C, an electric heater or a gas or oil-fired boiler is used to maintain the loop temperature. Conversely, during summer peak cooling requirements when nearly all units are rejecting heat to the loop, a supplementary closed-circuit evaporative cooler maintains the loop water at around 32 °C.

2.5.2.5 Exhaust Air System

Exhaust (ventilation) air is another common heat source for heat pumps in both residential and commercial buildings. The heat pump recovers heat from the ventilation air and uses this to provide a base temperature for water and/or space heating. As a disadvantage, the ventilation may be required to operate continuously during the heating season or even throughout the year, and so for some large buildings such systems are often used in combination with air-to-air heat recovery units.

2.5.2.6 Hybrid Heat Pump

As the name suggests, hybrid heat pump systems are a combination of boiler coupled with either an air source or a ground source heat pump and are a logical step in application as the world strives to reduce its reliance on fossil fuels.

Generally, the heat pump is efficient enough to provide year-round heating and hot water but if the external temperature drops too low, then the system intelligently switches to the boiler, so keeping energy usage and costs to a minimum.

2.5.2.7 Solar-Assisted Heat Pumps

A solar heat pump is one which integrates the technologies of thermal solar panels. Typically, the solar panels perform the function of the low-temperature heat source and the heat produced then feeds the heat pump's evaporator (Figure 2.24).

Solar source heat pumps invariably suffer from the same disadvantages as air source heat pumps in terms of extreme variablility, so they usually require (except in locations with high daily irradiance levels) a backup heat source, but in addition, they also have very high capital cost implications.

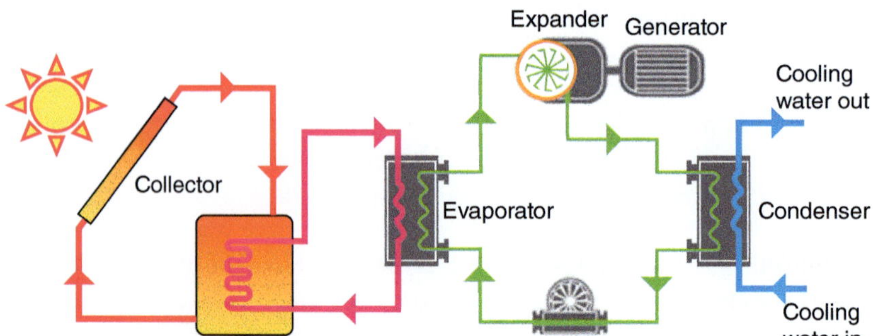

Figure 2.24 Schematic of solar-assisted heat pump. Source: Zhao et al. [75].

2.6 Direct Electrical Conversion Device

Thermoelectric generators (TEGs) have demonstrated their ability to convert thermal energy directly into electrical energy through the Seebeck effect. TEGs have emerged in the last decade as an encouraging alternative among other technologies for green power production because of the unique advantages they present. TEGs are environmentally friendly as they do not contain hazardous materials, they operate with no noise since they do not have mechanical processes or spinning parts and can be made on a large range of substrates such as silicon, polymers, and ceramics. TEGs are also position-independent, have a long working life, and are ideal for use in bulk and compact applications.

When producing energy in a regular power stations, only approximately one-third of the energy released from the fuel actually ends up in the electrical transmission lines leaving the power plant, the rest of the energy is lost in the form of waste heat [76]. Part of the problem is that the steam-powered turbine systems, used to generate the majority of electricity, function by initially burning a fuel to generate heat energy. This is followed by the conversion of this heat energy into mechanical energy within the turbine, and finally turning the mechanical energy into electrical energy with a generator [77].The ability to collect the heat lost in these processes and convert this to usable electrical power would greatly increase the efficiency of the electricity generation process. In addition, since this would also mean that less fuel is used for the same quantity of electricity, we could lower carbon pollution – an advantage to the improvement of the environment. Also, since TEGs generate electrical energy from heat energy without the need for moving parts (such as turbines) this eliminates additional maintenance and replacement costs.

In recent years, TEG systems have formed the focus of much research in the recovery of waste heat due to their incomparable advantages: they produce electrical energy without the need for moving parts thus no maintenance and other extra costs are required. They have no scaling effects and so range in application from micro generation in a restricted position to systems that generate kilowatts. At the same time, one disadvantage for this technology is its low energy conversion efficiency rate; moreover, TEGs require a relatively steady heat source. TEGs operate by exploiting a temperature gradient between the two sides of the generator known as thermoelectric effect or the Seebeck effect. The Seebeck effect is named after the German physicist, Thomas Johann Seebeck, who was believed to first explain it in the 1820s. However, recent work reveals that the Seebeck influence was first observed 27 years before this [78].

TEGs can be used in a wide variety of applications, such as WHR [79] and solar energy applications [80]. Indicative of this flexibility, one study has implemented a multi-generation device coupled with a TEG installed on a biomass gasification facility [81]. Whilst in another, TEGs have been applied to human energy harvesting. In order to harvest body heat, a flexible TEG which uses eutectic gallium indium liquid metal together with a high thermal conductivity elastomer was designed to be used for wearable electronics [82]. In yet another application aimed at portable power generation, a triple micro combustor was designed and advanced, which considerably enhanced the heat transmission from hot gases to thermoelectric modules (TEMs) [83].

2.6.1 TEG – Working Principle

TEG systems consist of three main elements: A HEX which takes and absorbs the heat and transfers it into the TEMs that generate electricity when a temperature difference exists between their respective ends, and a heat sink to dissipate the additional heat from the TEMs.

The TEM itself contains many pairs of thermoelectric couples, each couple combining a pair of p- and n-type elements, and the temperature differential between the two sides of the generator is what drives the function of the TEGs showing the principle behind this. If one side of a piece of metal is heated whilst simultaneously cooling the other side, the electrons within the metal atoms on the hot side will have more energy than the corresponding electrons on the colder side. This implies that the hot electrons will have more kinetic energy than those on the cold side and as a consequence, the hot electrons will tend to travel further towards the cold side than the cool electrons travel towards the hot side. So eventually, the cool end of the thermoelectric device becomes negatively charged and the hot end becomes positively charged [51]. The downside to this method is that the voltage produced is very low and unfortunately this cannot be resolved simply by connecting groups of metal parts together in series. This is because any wires used to connect them are also made of metal, and they would also produce a voltage but in the wrong direction and so oppose the voltages of the metal pieces.

The most successful approach to tackling this problem is to create a substance that can conduct electricity by utilising substitute charged particles, i.e. particles that behave as positively charged particles that then migrate from the hot side to the cold side. In which case, if a chain of these new materials is connected together, the voltages will be applied in sequence, enabling a reasonable amount of power to be produced for the user. Products with this beneficial property are referred to as semiconductors [84]

A further problem regarding thermoelectric devices is that materials that enable electrons to flow efficiently are often also good heat conductors, meaning that the temperature differential controlling the cycle is very quickly lost so causing performance to decrease. Researchers are therefore challenged to identify products of low thermal conductivity but strong electrical conductivity and have found one way to do this is by utilising metal alloys. These can be used to form a grid of atoms of varying sizes which slow down the heat flow but still enable electrons to travel freely. Similarly, certain combinations of nanoparticles can be used to trap and slow down the heat. In general, the more connections or gaps that occur between the objects, the slower the heat can travel.

2.6.2 The Seebeck Effect

The Seebeck effect explains the build-up of a potential gap (ΔV) through a semiconductor (or conductor) because of the diffusion of the charge carriers over a temperature gradient ($\Delta T = T_{hot} - T_{cold}$), encountered by the substance if one side of it is heated or cooled. Charges move from the hot to the cold side, leading to a difference in the amount of charge carriers and which is balanced by a developed resultant internal electric field. The type of carrier for the majority charge specifying the sign of this potential difference which, by nature, is defined by the potential of the cold side in relation to the hot side [84]. The sign of the Seebeck coefficient therefore indicating whether the substance in a conductor is p-type or n-type material – the positive sign noting p-type, a negative denoting n-type material.

The Seebeck coefficient is a measure of the effect and is defined as the relationship between the established theoretical differential and the implemented temperature gradient, as can be seen in Eq. (2.12) [85]

$$S = -\frac{\Delta V}{\Delta T} \tag{2.12}$$

The temperature gradient (ΔT) has to be created between the sample's two sides, and the resulting voltage (ΔV) has to be determined by taking electrical connections from those two sides.

2.6.3 The Peltier Effect

As stated earlier, Thomas Seebeck, the German physicist, conducted various electricity experiments in 1821 and discovered that electricity can flow through a circuit consisting of two different conductors provided the junctions at which these conductors were linked were held at differing temperatures. However, Seebeck was unable to explain the exact scientific theory behind this mechanism and wrongly deduced that flowing heat produces the same outcome as flowing water.

In 1834, whilst studying Seebeck's effect, Gene Peltier, a French physicist, discovered that heat could be absorbed at one junction of dissimilar metals and released at the other junction of the same circuit, and in the 1850s, William Thomson Lord Kelvin was able to reproduce and validate the findings of both Seebeck and Peltier and went on to establish the relationship between them. Nonetheless, this finding was confined to a laboratory-based experiment until 1930, when Russian physicists began re-investigating earlier thermoelectric effect experiments which then led to the creation of practical thermoelectric instruments.

In practice, the Peltier effect is assumed to be the opposite of the Seebeck effect [86]. The electrical current passing across the junction of the two materials produces or absorbs heat in unit time at the junction to offset the gap in the chemical structure capacity of the two materials. All of which influenced the development of the electronic refrigerator, known as the Peltier cooler.

The Peltier effect occurs because the difference in total energy of the electrons involved in the propagation of electrical current in the numerous conductors is significant. Several factors, such as the energy spectrum of electrons, their movement within the conductor, and the atoms scattered at the junction between two separate conductors, all influencing the conductor.

The electrons travelling from one electrode to another (depending on the direction of the electrical charge stream), then transfer their extra energy to the underlying atoms (exude) or absorb energy from the surrounding atoms. Similar to the Seebeck effect, the Peltier effect has the setback of low efficiency, but another drawback is that the flowing current itself tends to generate a large volume of heat that accumulates in the cycle of heat dissipation.

This results in an excessive volume of heat in larger-scale applications, which could mostly be taken care of by external fans but, as a consequence, more electricity would be consumed, making the application very expensive and so eliminating its wide-scale implementation. Furthermore, if the components of the Peltier device are too cold, they may cause a short circuit due to condensation.

The main advantage of the Peltier effect is that it enables the production of cooling and heating systems that do not need moving parts; thus, these systems are less likely to fail compared with conventional coolers and heaters and so may not need maintenance. Also when working, Peltier instruments operate silently and can potentially exceed temperatures as low as −80 °C, allowing the Peltier effect to be productively used at a microscopic level where conventional cooling methods will not work [87].

2.6.3.1 Applications of the Peltier Effect

Commonly used appliances that utilise the Peltier effect include a Peltier heater, a heat pump cooler, and a solid-state refrigerator, all possible because as electrical current flows through a Peltier device from one side of the panel to another it can act to either heat or cool because all Peltier devices work by using electric current to transfer heat from one side of the system to the other towards the temperature gradient. Peltier effect devices have been used in dehumidifiers for the elimination of water from the air (water extraction) and for DNA synthesis, where a thermal cycler utilises the effect for the DNA synthesis process. In fact, astronauts also employ the Peltier effect to counteract the uneven direct sunlight influences on their body, which serves to dissipate heat from intense sunlight on one side of the astronaut to the other which, not being in direct sunlight, is very much colder [88].

2.6.4 Thomson Effect

Thompson suggested that when the current passes through unequally heated conductors, the thermal energy is absorbed or produced in the metallic structure [89] or, in other words the Thomson effect is the generation of 'reversible' heat by an electrical current transmitted through a conductive material subjected to a temperature gradient [90]. The Thomson effect describes the heat escape from a substance with a current flowing through it, and this heat flow is easily detectable. It is different

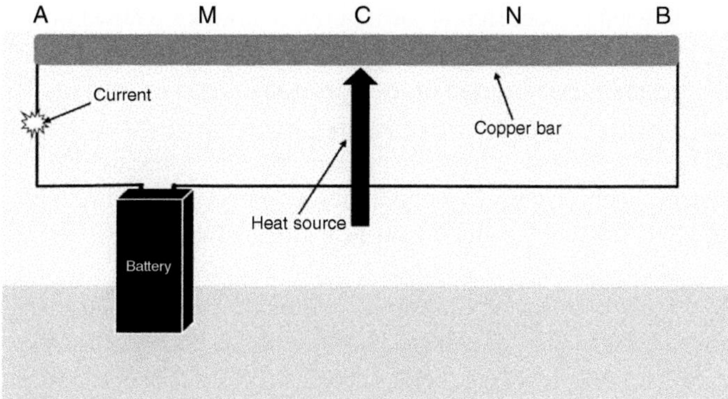

Figure 2.25 Thomson effect demonstration.

from the results of Peltier and Seebeck, by which only the cumulative impact of two separate materials can be observed.

Figure 2.25 displays the copper bar (AB) heated at the middle (point C) with current flowing from point A to point B. With no current flowing, points M and N (both equivalent distant from point C) are both at the same temperature, but with a current flowing from point A to point B, point N reaches a higher temperature than point M and, similarly, point B reaches a higher temperature than point A. We can therefore deduce that heat is generated from point A to point C, and heat is absorbed from point C to point B. This is known as positive Thompson effect. Similar effects are observed in the case of antimony, silver, zinc, and cadmium. However, in the case of iron, with the bar heated at point C and the current still flowing from point A to point B, the temperature at point M becomes higher than point N, which implies that the opposite effect is occurring, i.e. that heat transferring from point A to point C and from point C to point B. And this is termed as negative Thompson influence. Similar effects are found with platinum, bismuth, cobalt, nickel, and mercury [89].

As a measure of this effect, the amount of heat energy absorbed or evolved when 1 A current flows for one second (i.e. 1 C) in a metal between two points which differ in temperature by one degree Celsius is known as the Thomson coefficient [91].

2.6.5 Joule Heating

James Prescott Joule first researched and described heating phenomenon around 1840. Originally, Joule set out to explore how the recently invented electric motor could be made more efficient than the steam engines in operation at the time and this led him to perform a series of studies on the production, transfer, and use of energy and mechanical work that eventually contributed to the first law of thermodynamics.

The study of the interaction between an electrical current that passing through a conductor and its increase in its temperature being one of these studies. The procedure consisted of a wire submerged in water and attached to a battery terminal, and when the circuit was energised the increase in water temperature was measured.

Analysis of the recorded data led to the original form of the relationship now known as Joule's first law, that 'the heat developed (per second) in the wire is proportional to the resistance of the wire and to the square of the current'.

Joule heating (also known as ohmic heating) describes the method of transforming the energy of an electric current into heat as it passes through a resistance. No matter whether the electric current passes through a solid or liquid with limited conductivity, electrical energy is still transformed into heat through resistive material losses.

The heat is produced on the microscale because the conduction electrons transfer energy to the conductor's atoms by way of collisions, and with an electrical current, I, flowing through a conductor, the heat $\dot{Q}_J$ generated within it (by virtue of the Joule effect) is given by:

$$\dot{Q}_J = I^2 R \tag{2.13}$$

where R represents the electrical resistance [92].

Several home appliances use this heating function – electric irons, electric toasters, and the like. Applications of Joule's heating effect also include electric fuses, electric furnaces, and incandescent lamps. This latter application consists of a tungsten filament which is placed inside a glass bulb and heated by the current to the point where the filament glows white hot. Electric discharge lamps, electric welding, and electric arc also utilise the heating effect of current.

2.6.6 Theoretical Principle

TEGs offer a perfectly competitive, solid-state system for energy conversion and by using thermoelectric materials can be an alternative to conventional heat generation engines, especially in lightweight heat recovery systems. The maximum efficiency for the conversion of thermoelectric energy is typically presented in terms of the temperature of each heat reservoir and a figure of merit (zT), and since a rise in zT would contribute to an improvement in the performance of thermoelectric technology, significant attempts have been made to incorporate and construct products with high zT values.

2.6.7 Figure of Merit

Successful TE products should possess the following:

- High electrical conductivity to minimise Joule heating.
- Considerable Seebeck coefficient to maximise the conversion of heat to electrical power or electrical power to cooling performance.
- Low thermal conductivity to avoid thermal conduction across the material.

And these three properties are best represented as a single metric that quantifies the total performance of the thermoelectric unit by a 'thermoelectric figure of merit' or 'Z'. Now as 'Z' has a unit of temperature per degree, it is better to use, as a more practical dimensionless figure of merit, $z{\cdot}T$, where T is the average working temperature. This is a critical parameter which describes the magnitude of any TE devices' maximum power conversion efficiency [93].

In which case, the figure of merit can be calculated from Eq. (2.14) [94]:

$$zT = \frac{S^2 \sigma T}{K_{total}} \tag{2.14}$$

In this equation, S represents the Seebeck coefficient, σ is the electrical conductivity, T is the average temperature, and K_{total} is the total conductivity and is the summation of K_l and K_e which are the thermal conductivities due to the lattice and due to the electrons, respectively.

The figure of merit is dimensionless and in order for TEGs to have a high efficiency, the value of zT needs also to be high and in this respect, the quantity ($S^2\sigma$ – in the numerator) and known as the power factor (PF) is a key element for maximising the figure of merit zT.

2.6.8 Fermi Level

The PF is used as a measure of thermoelectric efficiency, the electrical conductivity being comparative to the density of the carrier and the Seebeck coefficient being inversely proportional to this. Hence optimisation of the Fermi transition rates is very important for optimising thermoelectric efficiency. B. Ohtani [95] describes the Fermi level in the form of the quantity of energy of electric equilibrium for a solid substance. Additionally, the Fermi energy is assumed to be the energy of the most energetic electron at absolute zero temperature ($T = 0$ K). The PF, and thus the merit figure, both highly depend on the Fermi-level location, the PF being balanced when the Fermi level is located near the bottom of the conductive band.

To explain, atoms, in their simplest terms, consist of a central nucleus surrounded by a cloud of fast-moving electrons – the number of which depending upon the element, hydrogen (the lightest element) having 1 electron, helium 2, lithium 3, barium 4, and so forth right up to elements like uranium with 92 electrons and other even large atoms. In free space these electrons each have clear energy levels – grouped into shells – the lowest energy levels or shells (those closest to the nucleus) filling first whilst the outer shells may or may not be completely filled again depending upon the element, and which with uranium for example, has an shell structure (electrons per shell) of 2 : 8 : 18 : 32 : 21 : 9 : 2. The highest or outer shell is termed the valance shell and it is those electrons that make up most chemical bonds and chemical reactions. However, when atoms are arranged in a solid they are in close proximity with each other and these energy levels interact and effect each other so that they appear more like a continuous energy band rather than as discrete separate energy levels. This also means that the valence shell or bands interact to form a new hybrid shell the conduction band representing a band of possible but vacant electron states, and which, in semiconductors is separated from the valance band by a band gap. Now in solids the ability of an electron to act as a charge carrier depends on the availability of vacant states i.e. its ability to move from the valance to the conduction band. The Fermi level therefore represents the energy level of the highest occupied band

and typically sits between the valency shell or band and the slightly more energetic conduction band.

Therefore, if the PF plotted against the Fermi level, the highest values appear when the Fermi level is located close to the bottom of the conduction band, the differences in position (either slightly above or below the bottom) being attributed to the varying dependency of the electrical conductivity and the Seebeck coefficient on the Fermi level.

To get a good-high PF requires doping the semiconductor (adding impurities to the pure semiconductor atoms), so that the Fermi level can be changed to optimise the PF. In addition, since high conductivity at low Fermi energy is also required, this implies that there should be many free channels in the conductive band so that there is little need to move the Fermi level to within the conductive band which would then reduce the Seebeck coefficient [96]. In other words, all energy states lower than E_f are filled and all states beyond E_f are vacant at $T = 0\,\text{K}$.

Now as noted earlier, in insulators and semiconductors, the Fermi level lies in the energy gap between the valence and the conduction bands, the exact position of E_f which is given by the Fermi distribution function as shown in Eq. (2.15) [97].

$$f(E) = \frac{1}{1 + \exp\left(\frac{E-E_f}{kT}\right)} \tag{2.15}$$

where, E_f is energy such that the Fermi distribution function $f(E)$ accurately determines the distribution of electrons between the usable states in the valence and conduction bands, E is the electronic charge, k is Boltzmann's constant, T is the average absolute (Kelvin) temperature of the sample, and E_f is expected to occur in the middle of the energy gap for intrinsic semiconductors (the pure form of the semiconductor) – indicating an equal probability of electrons in the conductive band and holes in the valence band at a given temperature [98].

Nevertheless, in extrinsic semiconductors (a semiconductor in which impurities are purposely introduced to make it conductive), E_f occurs (at low temperatures) either inside the conduction band (n-type) or within the valence band (p-type) depending on the sign and concentration of the majority carriers and the absolute temperature.

2.6.9 Nano-Sizing

Owing to the fundamental strong link between electrical conductivity and thermal conductivity in most materials, strong zT values have been difficult to achieve. So achieving high electrical conductivity to decrease the TE unit's temperature resistance (Joule), but ensuring low thermal conductivity in order to preserve a consistent differential in temperature between the hot and the cold side has been challenging [99].

However, another approach to raising the figure of merit is by reducing the overall thermal conductivity (K_{total}). In order to achieve this, researchers have succeeded in artificially structuring materials on a nano scale and so designing the lattice's thermal conductivity and lowering it from the value it would usually have in bulk materials. This is accomplished by introducing a wide number of grain boundaries and other dispersion centres that essentially disperse phonons but are inefficient in electron scattering. Essentially, it is an attempt to greatly reduce the lattice's thermal conductivity whilst not drastically impacting its electrical conductivity, and this has recently prove to be quite fruitful [100].

Initially, there was a steady progress until around 1970, when figure of merit of 1 was reached, then there was a period where there was practically no change. But once the idea of utilising nanotechnology to advance thermoelectric performance was introduced and enforced, the figure of merit values increased significantly. This is because the thermal conductivity of the lattice could then be designed to lower values [101].

2.6.10 Efficiency of TEG

The efficiency of a TEG can be expressed as the ratio of the produced electrical energy (W_{elec}) to the thermal energy entering the hot face (Q_h), and which for a TEM using a generator can be deduced from Eq. (2.16) for optimal electrical charge [97].

$$\eta_{\text{TEmax}} = \frac{W_{elec}}{Q_H} = \frac{\Delta T}{T_H} * \frac{\sqrt{1+zT}-1}{(\sqrt{1+zT}) + \frac{T_c}{T_H}} \tag{2.16}$$

where zT is usually calculated at the mean temperature of the hot side temperature (T_H) and the cold side temperature (T_c). It should be evident from the equation that an increase in zT will lead to a boost in energy conversion efficiency.

Moreover, the maximum power generated by the TEG can also be predicted via Eq. (2.17) [98]

$$P_{\max}(W) = \frac{\Delta T}{T_H} * \dot{Q} * \frac{\sqrt{1+zT}-1}{(\sqrt{1+zT}) + \frac{T_c}{T_H}} \tag{2.17}$$

where $\dot{Q}$ represents the rate of heat transfer through the TEG (W) and may be defined as the ratio of the temperature difference to the TEG thermal resistance (R_{TEG} $°C/w$) as shown in Eq. (2.18) [102]:

$$\dot{Q}(W) = \frac{\Delta T}{R_{TEG}} \tag{2.18}$$

A TEG can therefore produce its maximum power when it operates at half the open circuit voltage and at half the short circuit current, in which case an alternative means of measuring a TEG's potential strength may be written as Eq. (2.19) [98]:

$$P = VI \rightarrow P_{max} = \frac{1}{2}V_{Open\ Circut} * \frac{1}{2}I_{Short\ Circut} \rightarrow P_{max} = \frac{1}{4}V_{\text{OC}} * I_{\text{SC}} \tag{2.19}$$

2.7 Thermal Storage

The policy position of the European Union (EU) is to shift forward into a low-carbon economy with a cut of at least 40% in domestic-related greenhouse gas (GHG) emissions before 2030 and investigations have shown that a vast majority of GHG emissions are related to the production of energy. As an example, and as explained by European Association for Storage of Energy (EASE) [103], more than half of the energy consumption in the EU comes from the cooling and heating of the buildings. Additionally, a large amount of energy is wasted by devices which, although implemented for the production and utilisation of energy applications, are not used efficiently [104]. Accordingly, the use of energy storage systems has been identified as one of the most important areas of research and development to improve the energy efficiency of industrial and domestic appliances and for helping to achieve the goals which are defined in the EU 2020 targets and the European Commission (EC) Energy Roadmap 2050 [105].

As the name implies, thermal energy storage works by adjusting the state of a storage system in terms of heating or cooling its TES medium. The use of TES for domestic and industrial processes allows for the accumulation of energy from processes such as WHR and power generation, to then be used, at a later stage, for heating or cooling, and the like. Moreover, utilising TES within a system allows for better balancing between the supply and demand of energy [106], so reducing running costs and the time for any ROI [107]. In addition, combing and integrating TES with renewable energy systems such as solar and power to heat can provide a suitable approach for low-carbon energy harvesting and usage [108].

As explained by Koohi and Rosen [109], energy storage systems can be split into different categories, namely, thermal, thermochemical, magnetic, hydro, flywheel, hydrogen, compressed air, and liquefied air energy storage systems. As can be seen from Figure 2.26, each system is also divided into further classifications, each of which use a different medium to store energy.

Figure 2.27 illustrates an overview of different thermal energy storage methods which, as can be seen, classifies thermal energy storage into chemical and physical processes, the later by sensible and latent heat [109–114]

When selecting and designing a TES system, several important parameters such as energy storage density, mechanical strength, chemical stability, and cyclability properties need to be considered, but having indicated that, thermal conductivity of the TES system has also been shown to be an important factor since this indicates the dynamics of change between charging and discharging. As indicated by Badenhorst [115], a low thermal conductivity means a rapid reduction in energy storage and generation. Moreover, other criterions such as duty cycle, peak load, operating temperature, and ease of system integration have also been indicated as key factors when designing a TES system.

Table 2.10 indicates the selection of different parameters in terms of different requirements and reasons for TES systems [110, 116].

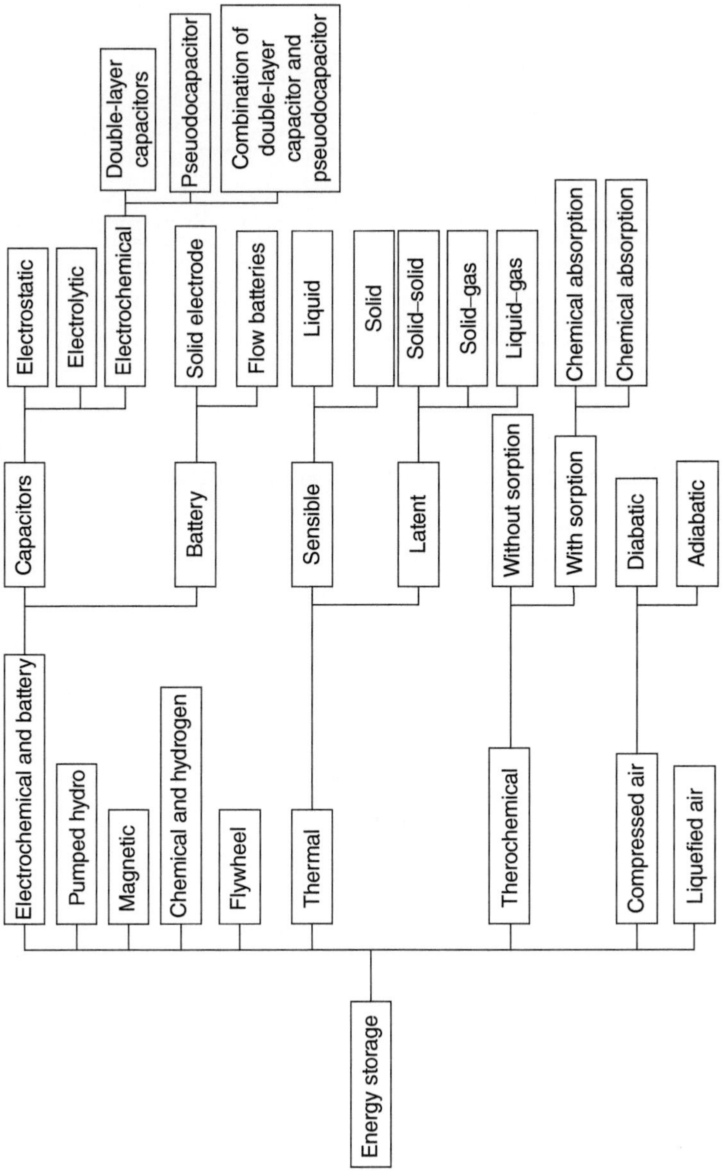

Figure 2.26 Classification of energy storage types. Source: Koohi-Fayegh and Rosen [109].

2.7.1 Sensible Heat Storage

Sensible heat is described as the most common and feasible approach to TES [110, 114]. Sensible heat is described as the heat that is transformed to change the temperature of a storage medium without changing its phase. Figure 2.28 shows the linear relationship between the temperature and stored heat of a storage medium based on sensible heat. Sensible heat storage therefore only offers low energy density

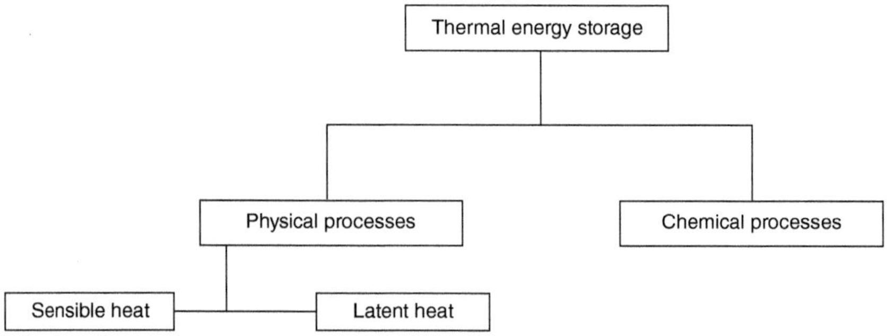

Figure 2.27 Methods for thermal energy storage.

Table 2.10 Different requirements and reasons for TES system properties selection.

		Requirements	**Reason**
Material	Chemical	Long-term chemical stability	Keeping the initial thermochemical properties aligned with the cycling periods during their lifetime
		Compatibility with container materials and low reactivity to heat transfer fluids (HTFs)	Ensuring long lifetime of the container and the surrounding materials in case of leakage
		No fire and explosion hazard	Ensuring workplace safety
		No toxicity	Ensuring handling safety
		No phase separation/incongruent melting	Avoiding changes on the stoichiometric composition of melt
	Kinetic	Small or no sub-cooling	Leaving the same melting/solidification temperature and avoiding heat release problems
		Sufficient crystallisation rate	Meeting the recovery system heat transfer demands
	Physical	High density	Minimising the volume occupied by the TES material
		Low vapour pressure	Diminishing the mechanical and chemical stability requirements of the container or vessel
		Small volume changes (low density variation)	Possibility of using eutectic mixtures
	Thermal	Favourable phase equilibrium	Providing significant sensible heat storage

Table 2.10 (Continued)

		Requirements	Reason
		High specific heat	Enhancing the heat transfer within the TES material by providing the minimum temperature gradients
		High thermal conductivity in both solid and liquid states	Ensuring the success of the charging and discharging processes within the operation conditions
Material/ system		Melting/solidification temperature in the desired operating temperature range	Providing significant latent heat storage in small volumes
		High latent heat of transition per unit volume near temperature of use	Ensuring the complete melt of the TES material and their homogeneity
		Congruent melting	Being cheaper than other options
	Economic	Abundant and available	Avoiding replacements and maintenance during the lifetime of the TES system
		Long lifetime Cost-effective	Being competitive in front of other options
	Environmental	Low manufacturing energy Easy recycling and treatment Low CO_2, footprint and use of non-polluting by-products	Reducing the environmental impact of the systems and accomplishing sustainable regulations and trends
	Technological	Operational strategy Integration into the facility Suitable heat transfer between the HTF and the storage medium (efficiency)	Optimising the processes by adapting them to limiting factors such as maximum loads, nominal temperatures, and specific enthalpy drops in load Enhancing the heat transfer from the TES material to the HTF and vice versa

Source: Adapted from Mehling and Cabeza [110]; Gasia et al. [116].

and thermal energy runaway, which can be a major disadvantage compared to other technologies [111].

Absorption and release of heat energy in the storage medium can take place through radiation, convection, and conduction, and is described by the following formula [114–117]:

$$Q = \dot{m}C_p(T_f - T_i) \tag{2.20}$$

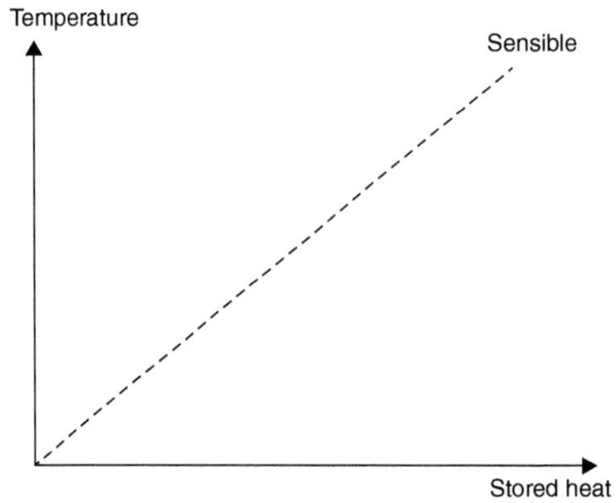

Figure 2.28 Temperature increase of a storage medium during sensible heat storage. Source: Based on Elias and Stathopoulos [111].

where Q is the quantity of the sensible heat storage of a material (kW); $\dot{m}$ is mass of storage material (kg/s); C_p is specific heat (kJ/kg/°C), and T_f is the difference between final temperature (T_f) and initial temperature (T_i) (°C).

Water, oil, and ceramics are the most widely used sensible heat storage mediums and are employed in both domestic and industrial settings, for example, as hot water storage for district heating or energy storage within a building envelope [111]. Other sensible thermal energy storage methods include the use of geothermal energy through a borehole, an aquifer, a cavern, ducts in the soil, pit hot water storage (hydro-accumulation), and rock-filled storage (rock, pebble, gravel) [109].

2.7.2 Latent Heat Storage

Latent heat storage systems on the other hand, take advantage of a phase change phenomenon in a specific temperature range for the storage medium to absorb and later transfer heat. The most commonly used materials for latent heat storage being water, ice, paraffin wax, and molten salt [111]. These TES systems can be split into two main types – direct and indirect heat transfer methods [108]. The direct technique includes the transfer of heat between a heat transfer fluid and the latent heat storage material, whereas the indirect method comprises a solid heat transfer border between the two.

Phase change of a medium occurs through evaporation, condensation, crystallisation, or meting, all processes without affecting the aggregation state of the solid or liquid phase. This process therefore enables the storage of a large quantity of thermal energy based on the properties of the employed material. As indicated by Mehling and Cabeza [110], a very slight change of volume of less than 10% occurs with no change in pressure when the storage medium experiences a temperature change. This means that the phase change between liquid and solid state of the storage material occurs within a constant temperature. Hence, once the material has undergone a phase change, its temperature is kept within a constant range, which is also known as the phase change temperature, but once this phase change is completed and with a further transfer of heat, the heat storage turns back into sensible heat phase as shown in Figure 2.29.

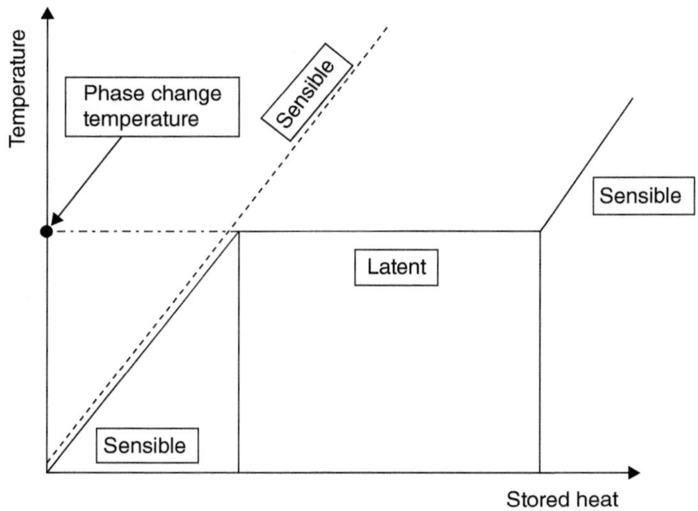

Figure 2.29 Temperature increase in respect to latent heat storage.

The latent stored heat has been shown to be equal to the enthalpy change that occurs between the two states, for example between solid and liquid phase as shown by Eq. (2.21).

$$\Delta Q = \Delta H = m\Delta \mathrm{h} \tag{2.21}$$

When the phase transition temperature is reached, the molecular structure within the phase change material (PCM) changes from solid to liquid form. The phase transition temperature is dictated by several factors such as the rate of cooling and the amount of particle impurities within the PCM, but in general, PCMs that go through solid to solid phase change are known to offer higher storage capacity whilst having smaller volume change. This also means that there will be no leakages from the PCM and hence no material encapsulation will be required. These beneficial properties therefore make use of this type of PCM more likely than say solid to liquid PCMs [106]. However, that being said, the energy density in liquid storage systems can be higher than that of a solid storage system, since a phenomena in relation to chemical bond transformation within the material structure during phase change can occur [111]. In which case the latent thermal capacity of a material can be demonstrated to be [114]:

$$Q = m\, C_p dT(s) + mL + mC_p DT \tag{2.22}$$

where m represents the mass of PCM in kg, dT is the temperature change, and L is the fusion enthalpy. The equation therefore demonstrates the correlation between the available heat within the PCM at the solid phase, the liquid phase with additional sensible heat and the fusion latent heat.

The most popular PCMs used for WHR, buildings, and solar systems are mainly based on fatty acids, hydrochloric hydrates, and paraffin waxes [117]. When designing a latent heat TES system, several factors such as geometric and operational parameters as well as exergy and energy performance need to be carefully considered. These parameters can include the working fluid mass flow rate, PCM number, and specific properties such as melting temperature, types, and amounts of additives to the PCM, heat exchange surface dimensions, and reference temperatures related to heating and cooling of the PCM.

As Prasad et al. [118] explain, the amount of thermal energy through charging and discharging processes can be determined from the following correlations.

Charging process:

$$\eta_{charging} = \frac{\int \dot{Q}_{stored} dt}{\int \dot{Q}_{supplied} dt} = \frac{Q_{stored}}{Q_{supplied}} \tag{2.23}$$

Discharging process:

$$\eta_{charging} = \frac{\int \dot{Q}_{used} dt}{\int \dot{Q}_{stored} dt} = \frac{Q_{used}}{Q_{stored}} \tag{2.24}$$

Overall process:

$$\eta_{charging} = \frac{\int \dot{Q}_{used} dt}{\int \dot{Q}_{supplied} dt} = \frac{Q_{used}}{Q_{supplied}} \tag{2.25}$$

The energy performance of the heat storage can be evaluated through the ratio of the total energy over the maximum energy stored in the system and by:

$$\text{Ratio}_{heatstorage} = \frac{\int \dot{Q}_{stored} dt}{Q_{maximum}} = \frac{Q_{stored}}{Q_{maximum}} \tag{2.26}$$

On the other hand, the performance of energy release by the system can be derived through the ratio of the total amount of released heat over the maximum energy in the system:

$$Ratio_{heatstorage} = \frac{\int \dot{Q}_{released} dt}{Q_{maximum}} = \frac{Q_{released}}{Q_{maximum}} \tag{2.27}$$

We should however note that energy performance analysis only indicates the quantity of energy transmitted during the charging and discharging processes, rather than the quality of the transmitted energy, but energy quality during charging and discharging can also be indicated through exergy analyses and this way the design of a TES system can be optimised.

Charging process:

$$\psi_{charging} = \frac{\int \dot{E}x_{stored} dt}{\int \dot{E}x_{supplied} dt} = \frac{Ex_{stored}}{Ex_{supplied}} \tag{2.28}$$

Discharging process:

$$\psi_{discharging} = \frac{\int \dot{E}x_{used} dt}{\int \dot{E}x_{stored} dt} = \frac{Ex_{used}}{Ex_{stored}} \tag{2.29}$$

Overall process:

$$\psi_{overall} = \frac{\int \dot{E}x_{used} dt}{\int \dot{E}x_{supplied} dt} = \frac{Ex_{used}}{Ex_{supplied}} \tag{2.30}$$

The exergy efficiency can then be determined through the number of entropy generation following Eq. (2.30).

$$\psi = 1 - N_s \tag{2.31}$$

2.7.3 Thermochemical Storage

Thermochemical energy storage (TCES) involves a fully reversible chemical reaction and allows for a higher energy density, operating temperature, and greater exergy efficiency to be achieved compared to other types of TES systems. The thermochemical process includes restructuring of the molecular chemical bonds during the charging and discharging processes which, in turn, increases the energy capacity [119, 120]. As you might expect, selection of the thermochemical material is a key factor when developing a TCES system, and TCES materials needs to be economical, abundant, stable, cyclical, fast and highly kinetically reactive, highly conductive, and non-toxic with no side reaction [118]. The thermochemical reaction itself needs to be a reversible process in which two substances are involved in a chain chemical reaction once the necessary quantity of thermal energy is applied. One problem however, is that over time, the efficiency of the chemical reaction becomes slower which can result in lower energy capacity.

$$\mathrm{AB} + \mathrm{Q} \leftrightarrow \mathrm{A} + \mathrm{B} \tag{2.32}$$

Research into the effectiveness of different TCES storage systems has been carried out through experimental analyses on gas-to-gas, solid-to-gas, and sulfuric-based thermoelectric energy storage systems for high-temperature applications at or above 300 °C [118]. The effect of air pollution on the use of TCES for buildings were investigated and ageing tests were performed to investigate the usability of the storage material for long-term application and usage. Furthermore, absorption tests were performed under several dynamic and static conditions and for different pollutants. Finally, analyses was carried out to indicate how different environmental conditions affect any loss in heat storage capacity.

Figure 2.30 represents different applications of solid, liquid, and thermochemical storage systems, and shows the various technologies being used to provide high-energy efficiencies grouped by the method of energy storage.

2.7.4 Phase Change Materials

PCMs are considered to be an ideal choice for TES systems because of their ability to absorb, accumulate, and then release large amount of thermal energy from a small unit of mass [120]. They are used in numerous applications [112] and have proved to be a viable solution for WHR, solar system applications, and energy savings in building applications [120].

PCMs are mainly used in latent heat TES systems because they are able to help eliminate the gaps between system demand and energy supply – their specific characteristics offering enhanced energy performance together with a means to achieve

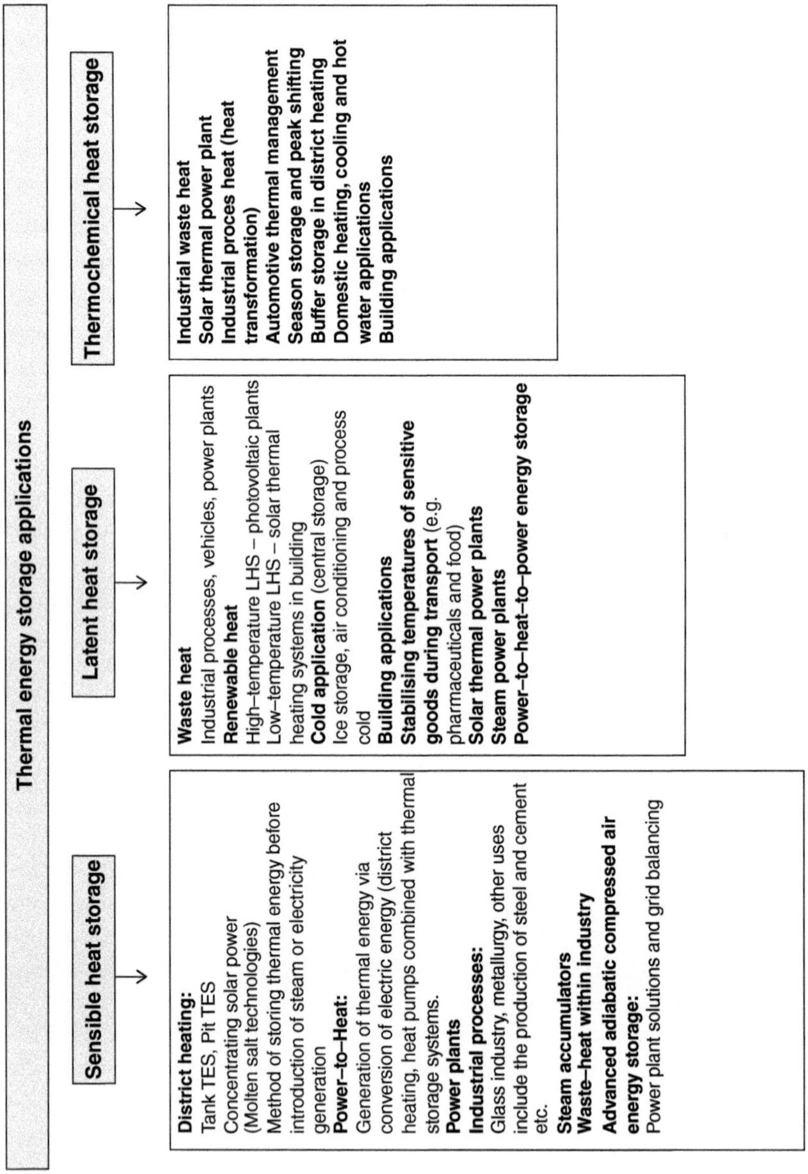

Figure 2.30 Thermal storage applications.

efficient and effective improvements in the conservation of system energy [119]. The advantages of using PCMs, and their attraction for use within latent heat TES systems include their ability to supply thermal energy at an almost constant uniform temperature, their high-energy density for comparatively low material volumes and weights [119], and their lower cost.

Research has shown that the melting enthalpy and the melting temperature or melting point of the material are the main parameters that determine the performance of a PCM-based TES system [110, 121]. So, in simple terms, when selecting a PCM-based system, the melting point of the material needs to be higher than the maximum ambient temperature and lower than the temperature supplied by the heat source.

However, in view of the previously noted key parameters, selecting an appropriate TES PCM can be a challenging task and so requires special consideration, and an ideal PCM should meet certain parameters and contain the characteristics listed below [110–122]

- The phase transition temperature must satisfy the TES application
- Contain a high storage capacity and offer a high melting enthalpy
- Possess good thermal conductivity to ensure a high rate of charging and discharging
- Present adequate chemical and thermal properties to provide robust storage capacity
- Be safe, non-toxic, non-flammable or non-explosive and environmentally friendly
- Be easily integrated with other construction materials or components
- Offer minimal volume change between phase transitions and contain a well-balanced rate of crystallisation
- Produce low vapour pressure and be completely reversible during each cycle of storage
- Offer good recyclability, low cost, and be available and abundant.

It should therefore not be surprising, considering this extensive and sometimes conflicting list, that until now there are no PCMs that can fully meet all these parameters [111, 121]. Each available PCM offers certain pros and cons so that each material must be individually considered for a specific TES application.

To assist with this choice, PCMs are classified into different categories based on their characteristics of phase transition, chemical nature, and temperature range. Widely used PCMs are mainly paraffin or organic non-paraffin based and include, fatty acids, alcohols, polymers, and glycols but they can also comprise a combination of organic–organic, inorganic–inorganic, or organic–inorganic materials. Again these can also be classified by their properties of phase transition change and then in terms of their solid–liquid, solid–solid, and solid–gas criteria [111]. Moreover, since each PCM can be utilised for different applications, they can also be classified by particular shape stabilisation properties. Figure 2.31 shows different classifications for latent TES mediums [120, 121, 123] (Table 2.11).

2.7.5 Organic Material

In general, organic PCMs can be split into three different groups: fatty acids, paraffins, and organic mixtures, but organic-based PCMs offer several advantages that makes them more favourable than inorganic substances. For example, they have greater chemical/thermal stability, are easily recycled, are non-corrosive and offer larger latent heat of fusion values. That being said, they do have some disadvantages (compared to inorganic materials) and these include: being prone to flammability,

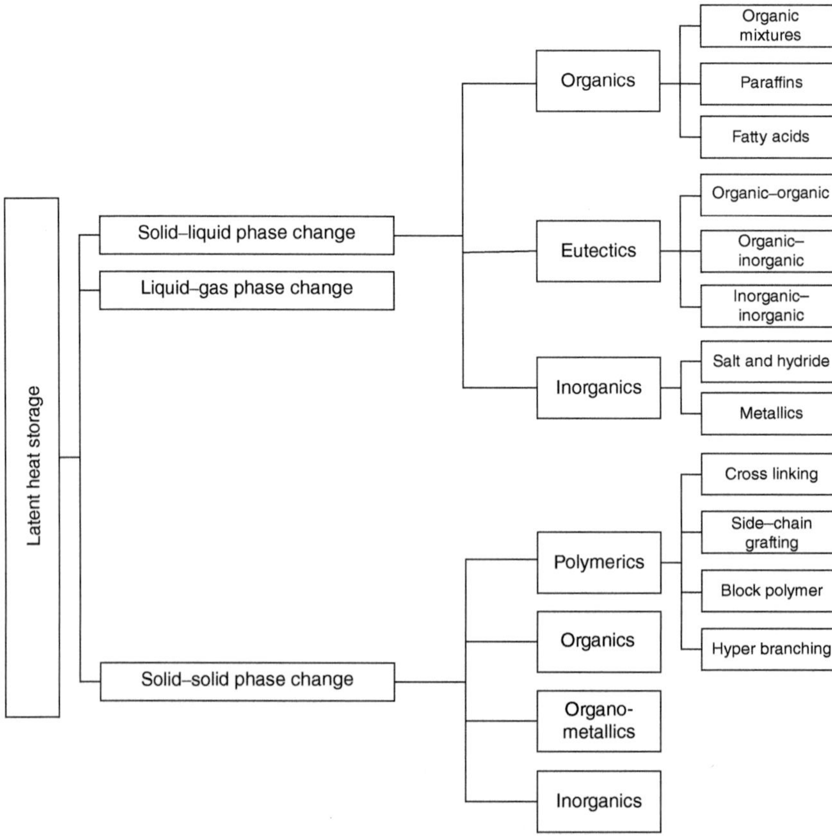

Figure 2.31 Outline of phase change transition of PCMs used for latent heat storage systems.

Table 2.11 Chemical properties of different types of PCMs.

Phase change materials		
Organic	**Inorganic**	**Eutectics**
1. These are the polymers of long chains of carbon and hydrogen molecules, e.g. Paraffin (C_nH_{2n}). 2. No supercooling 3. Available in wide ranges of temperatures 4. Physically and chemically stable 5. High heat of fusion 6. Ability to melt congruently 7. Flammable 8. Exhibits low thermal conductivity 9. Low volumetric energy intensity	1. High heat of fusion 2. High thermal conductivity 3. Low volume change 4. Easily available 5. Non-flammable 6. Supercooling 7. Lead to corrosion 8. Not melting congruently 9. High volumetric energy intensity	1. Sharp melting point 2. High volumetric thermal storage 3. High cost 4. Limited data availability about properties

Source: Modified from Rathore and Shukla [112].

offering lower enthalpy, having lower thermal conductivity, and presenting less compatibility with encapsulation [111].

Paraffin for instance has been used as the ideal choice of organic PCM and has been applied to many thermal energy storage systems [121, 124], the main reason being that the material offers high fusion heat, has a low vapour pressure, allows constant conductivity cyclic stability, and is chemically inert without being toxic or corrosive. In fact, paraffin has been found as an optimal PCM for energy saving within buildings by improving the energy performance of double-glazed windows [125]. Conversely, the material still has certain limitations such as a low thermal conductivity and high volume increase during phase change [123, 124].

Fatty acids or as called non-paraffin PCMs are another popular organic material which are constructed from animal or plant fat. This type of PCM is entirely organically based and offers enhanced physical and thermal properties compared to paraffin. To offset this, the material does require specific encapsulation techniques for its application since it is prone to leakage. Figure 2.32 illustrated the advantages and disadvantages of different organic PCMs.

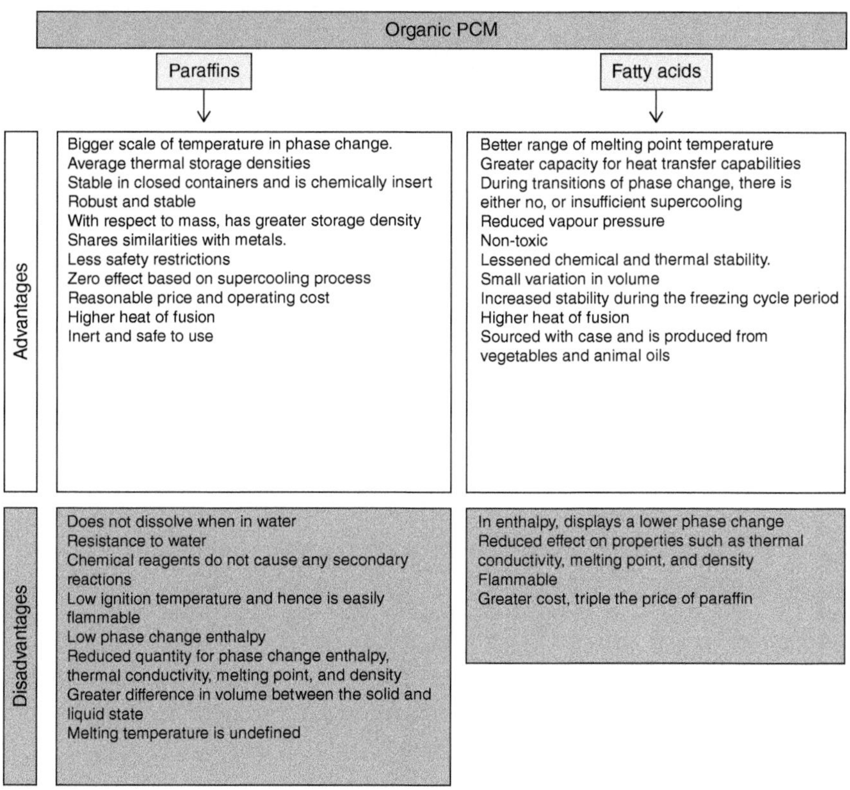

Figure 2.32 Advantages and disadvantages of different organic PCMs.

2.7.6 Inorganic PCMs

Inorganic PCMs are another ideal option for storing and releasing large amounts of latent heat, their success being achieved by a combination of specific techniques employed to allow storage in different molecular transformations. The techniques include transformation of the material through crystallographic, amorphous, and crystal structures, and order–disorder and magnetic field changes. Compared to organic PCMs, employing inorganic materials is often more cost-effective, offers higher phase change enthalpy, and better conduction of thermal energy. Despite these, certain disadvantages do exist such as being corrosive, having potential phase boundary separation, and offering instable thermal output limits, the consequences of which mean that the use of these materials has been mainly restricted to high-temperature WHR from industrial plants [106, 111]. Another source of popular inorganic PCMs has been metallic foams. These offer high thermal conductivity whilst maintaining good stability, high coefficient ratios, and low density [126].

2.7.7 Eutectic PCMs

Inorganic eutectic PCMs are a mixture of two (or more) different types of PCMs and are available in organic–organic, organic–inorganic, or inorganic–inorganic combinations. For example, the following are illustrations of the different material compositions that can be used as an eutectic mixture:

- Organic–organic: Myristic-stearic, Lauric palmitic, Capric-laurie
- Organic–inorganic: Salt-based $BACl_2$-KCl-NaCl, $MgKCl_2$-KCl, LiOH-KOH
- Inorganic–inorganic: Urea NH_4Br, naphthalene benzoic acid

One advantage of a eutectic material is its rapid transformation ability, both in its changes to the liquid phase during heat absorption and also changing back into a crystal mixture during the freezing stage. This is why mixtures such as water–salt are such a viable solution for dealing with energy storage at sub-zero temperatures [110]. Alternatively, organometallic, solid–solid PCMs are another type of eutectic PCMs which are structured from different layers of organic and inorganic materials into a sandwich-like crystalline format [106]. These materials then offer a unique method of energy storage since heat absorption is achieved only through the organic or soft segment of the material which are made of alkyl chains. The inorganic section of the PCM remain thermally inert and mainly stays structurally intact. During energy absorption, the alkyl chains experience a transition from solid phase or crystalline form into liquid or amorphous form in a disordered state, and this particular property enables the solid–solid organometallic PCMs to be stable and fully reversible with minimal volume change during the transition even at relatively at high temperatures.

2.7.8 PCM Methodologies

2.7.8.1 Encapsulation of PCMs

Encapsulation is a technique that enhances the functionality of PCM-based TES systems by helping to eliminate risks associated with leakage or the interaction of the active element of a material with the outside environment. It also helps eliminate issues associated with phase separation and increases lifespan [111, 127]. Encapsulation is therefore an effective way of increasing both the efficiency and the performance of PCM-based TES systems. For example, by encapsulating PCM and modifying the configuration of the system and by adding finned tubes, both the heat transfer and the heat absorption can be increased. Regrettably though, encapsulation can also allow the introduction of relative impurities, such as nano particles to the active substance of the PCM – which in turn, may subsequently increase the thermal conductivity of the material. Notwithstanding all of this, we should note that in most applications, except for a few materials such as ice PCMs need to be encapsulated to avoid leakage – particularly in their liquid phase, and this is an important factor towards protecting the environment [110].

There are several encapsulation techniques, and each can be categorised by their relative applications, advantages, and disadvantages [128]. Further, the encapsulation methods can also be grouped into nano, macro, and micro categories and each of these can include different methods of production [110, 123, 124]. Figure 2.33 shows advantages and disadvantages of each encapsulation technique applications [120].

2.7.8.2 Microencapsulated PCMs

Microencapsulation is a process by which individual particles of an active agent are stored within a shell, surrounded or coated with a continuous film of polymeric material to produce particles in the micrometre to millimetre range, and its use can increase the heat transfer rate of PCMs by enhancing their surface area to volume ratio. The method allows continuous matrices of the materials to be produced so making them suitable for use in building applications of passive cooling, heating, and energy-saving purposes. Microencapsulation can be developed through either physical or chemical means and can be used for several applications such as heat transfer fluids, heat insulation materials, phase change slurry, and textiles [129, 130]. Equation (2.33) indicates the potential energy saving within a building when PCM TES for cooling and heating is used. In addition, Eq. (2.34) also shows the percentage improvement in peak heat flux reduction when the same technique is employed [114].

$$\begin{aligned}\%\ \text{Energy savings} &= \frac{\text{Energy consumption without PCM} - \text{Energy consumption with PCM}}{\text{Energy consumption without PCM}} \\ &\quad \times 100 \end{aligned} \tag{2.33}$$

$$\begin{aligned}\%\ \text{Peak heat flux reduction} &= \frac{\text{Maximum heat flux without PCM} - \text{Maximum heat flux with PCM}}{\text{Maximum heat flux without PCM}} \\ &\quad \times 100 \end{aligned} \tag{2.34}$$

Encapsulation techniques	Macroencapsulation (1 mm to >1 cm)	Microencapsulation (1 μm to >1 mm)	Nanoencapsulation (< 1 μm)
Applications	Building envelope: Reduces the building energy consumption and improves the thermal comfort	Food industry: Improves product life of foods Beverage production Transportation of drugs to the target site in the body Textile industry: Improving product finishes. Commonly used in agricultural and environmental applications	Delivers drugs to the desired location as required Extends products stability and lifespan Incorporated fragrances for aromatic clothing
Disadvantages	Low thermal conductivity or heat transfer elements Solidification can occur at the boundaries	Material strength can be changed Direct to super cooling properties Complicated encapsulation techinque Solidification occurrence can be interrupted during the process	Production costs can be higher Treatment may be needed for purification
Advantages	Material adaptability is enhanced with the environment. Volume change properties are less intensive. Can be employed as an effective heat exchanger through preventing the exposure of the material with the environment. Manufacturing and production costs can be lower. The mechanical strength of the material is improved, thanks to the usage of the shell.	Volume of the material remains unchanged during phase change Can be combined with building materials without any complication Chemical strength of the material is increased Thermal stability and reliability are largely improved Avoids the leakage from PCM by developing a wall around the material	The production and manufacturing costs can be higher Further treatment of the material may be needed for higher purification

Figure 2.33 Encapsulation techniques: applications, advantages, and disadvantages. Source: Based on Magendran et al. [120].

Microencapsulation as a technique involve the closure of unstable, sensitive solid or liquid particles with diameters between 1 and 1000 μm with a solid layer [110, 131] to form solid particles. This protects the energy storage material so preventing long-term degradation and evaporation processes and ensuring better solubility [120]. Research has shown that the newly formed microstructure of the material delivers better heat transfer and solubility.

The following techniques can be employed to develop microencapsulation:

- Chemical: mini-emulsion polymerisation, suspension polymerisation;
- Physicochemical: ionic gelation, coacervation;
- Physico-mechanical: vibrational nozzle, centrifugal extrusion, solvent evaporation, and spray drying.

Figure 2.34 indicates different shapes of microstructures that can be coated using microencapsulation technique illustrating how it can be applied to any form – shapes in matrix form, simple and irregular cores, or multi-wall and multi-core configurations [132].

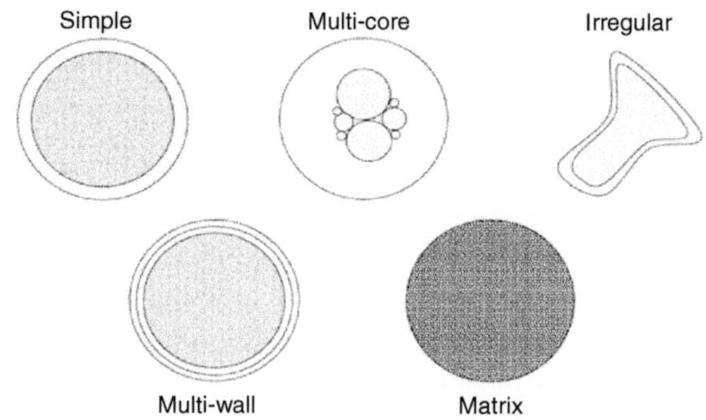

Figure 2.34 Different types of microcapsules. Source: Huang et al. [132].

Shell Material

- ❖ Leak proof
- ❖ High thermal conductivity
  - To improve heat transfer
- ❖ Anti osmosis properties
  - To ensure a longer life span
- ❖ Retain all thermo-physical characteristics at micro and nano level
- ❖ High mechanical and thermal strength
  - To withstand high transition temperatures
- ❖ Inert
  - To not react with the PCM

Figure 2.35 Benefits of microencapsulation of PCMs. Source: Based on Rathore et al. [114].

Figure 2.35 illustrates the advantages of microencapsulation and of using a shell material to encapsulate PCMs [114].

However, in addition to these benefits, microencapsulation can also be used to enhance the TES capacity of an organic PCM by using an inorganic substance as the shell material. Generally, the thermal conductivities of organic PCMs are lower than those of inorganic materials but by combining these through encapsulation of the overall heat transfer of the substance can be improved. Further increases in the thermal conductivity of the shell is also possible by the addition of carbon-based materials. Conversely and notwithstanding these benefits, applying microencapsulation can however also reduce a material's capacity for latent heat absorption. This is mainly because by introducing a shell to the PCM, its microstructure proportion, and subsequently its energy storage operating point can become uniform and so limited [132].

Huang et al. [132] concluded that suspension polymerisation and coacervation are suitable for the preparation of large particle MPCMs with high core content and concluded that:

- The spray drying method has the advantages of being both economical and easy to scale-up to produce microencapsulated phase change materials (MPCMs), however, it can also produce uneven results.

- The sol–gel method is suitable for the synthesis of inorganic shells.
- The process of interfacial polymerisation is simple, the reaction rate is fast, and the encapsulation efficiency is high, but the activity requirement for the reactive monomer is also relatively high (Table 2.12).

2.7.8.3 Macroencapsulation of the PCMs

Macroencapsulation is a process in which the PCM is encapsulated within a shell substance to form larger containers than microencapsulation (5 mm [133], 1 mm [134]). The shape and geometry of the encapsulating shell can be in the form of cubes, pouches, cylinders, tubes, or any other configuration, and so is ideal for incorporation into building envelope materials in different forms, sizes, and shapes. The actual structure and physical configuration of microencapsulation being different for different TES devices as shown in Figure 2.36 [119].

2.7.8.4 Nanomaterial PCMs

Mahdi et al. [127] demonstrated that PCM thermal transfer improvements can be made either by specifically inserting solid high thermal conductivity components such as metallic fins, foam, and heat pipes in the material; or indirectly by enhancing the heat transfer-physical properties, such as thermal conductivity, real heat, and latent heat of the PCM itself using micro- and nano-additives. These additives can be in the form of carbon nanotubes, metal nanoparticles, or carbon nanofibers where particle dimensions (for nanomaterials) range from 1 to 1000 nm [134] and, as noted subsequently, their production methodology uses either a top-down or a bottom-up technique [120]:

Top-Down Technique: This approach involves the use of specific method to deconstruct and so reduce the size and shape of larger materials to that required by the particular nanomaterial application. The methods used for this technique are namely emulsification and emulsification solvent evaporation.

Bottom-Up Technique: In this method, materials are formed from basic building blocks by the self-assembly of molecules determined by elements such as concentration, ion properties, temperature, and pH. Techniques used are namely such as coacervation and nanoprecipitation but integration complexity are also part of this strategy.

Nanoparticle Diffusion: (also known as nano-PCM) Nanoparticle diffusion is another process that can increase the thermal conductivity of PCMs [128]. For example [124], paraffin mixed with a type of nano-SiC (silicon-carbide) led to the increase of thermal conductivity for a photovoltaic/thermal (PV/T) panel, but other articles [126] also discuss the practicality of using other highly thermally conductive nanoparticles and draw upon studies around materials such as carbon-based fibres and nanotubes, aluminium and silver nanoparticles, and porous metallic foams.

For example with carbon materials, there are several techniques to obtaining the required characteristics of particle size, pore size distribution, and surface

Table 2.12 The preparation of smaller particle microcapsules is suitable for emulsion polymerisation.

Technique	Forming process	Advantages	Disadvantages	Shelf life	Quality (mechanical and thermal performances)	Reliability
Physical	Spray drying	• Low cost • Easy to scale up	• Agglomeration of particles • Remaining uncoated particles	Short	Poor	Poor
Physical–chemical	Solvent evaporation	Low cost	Small production	Medium	Poor	Poor
	Simple coacervation	• Versatile • Flexible operation	Agglomeration	Short	Poor	Poor
	Complex coacervation	Efficient control of the particle size	Agglomeration Difficult to scale-up	Short	Poor	Poor
	Sol–gel	Inorganic shell with high thermal conductivity	Still under research	Medium	Good	Good
Chemical	Interfacial polymerisation	• Versatile • Degradability • Mechanical resistance	Poor mechanical properties	Short	Poor	Poor
	Suspension polymerisation	Good heat control of the reaction	Few monomers are water soluble	Long	Good	Good
	Emulsion polymerisation	High molecular weight polymer	Purify polymer from the surfactant	Long	Good	Good

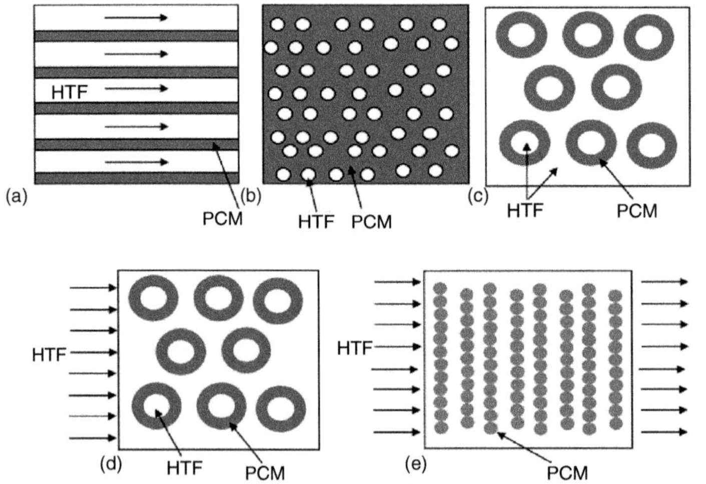

Figure 2.36 Different types of TES devices: (a) flat plate; (b) shell and tube – internal flow; (c) shell and tube – parallel flow; (d) shell and tube – cross flow; (e) packed bed. Source: Li et al. [119].

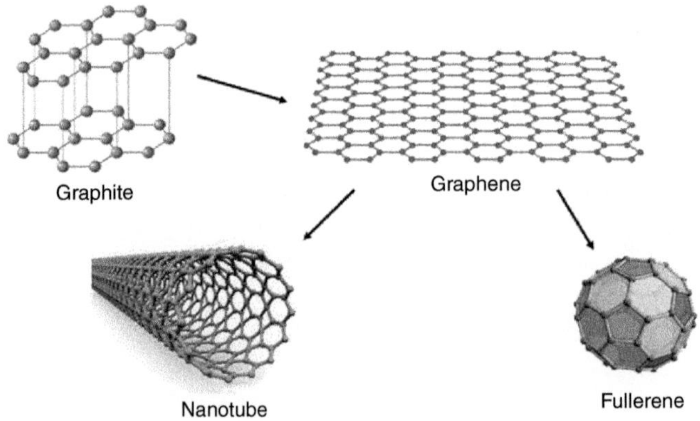

Figure 2.37 Graphite and three comprising nanomaterials. Source: Badenhorst [115].

area. Additionally, in an expanded form, graphite can be separated into very fine particles with a thickness of 10–100 nm, known as graphite nano-platelets (GNPs) or exfoliated graphite. As shown in Figure 2.37, the graphene (a single layer of atoms arranged in a two-dimensional honeycomb lattice) is classified as a pure graphite layer, and is derived from breaking a graphite crystal until a single layer of this element form is created [115]. A nanotube, in simple terms, is made of graphene sheet rolled into a cylindrical shape with edges knitted together, whereas a fullerene is produced by mating graphene segments to each other into a spherical shape – these forms each displaying unique characteristics differ from graphite, even though they have the same atomic configuration.

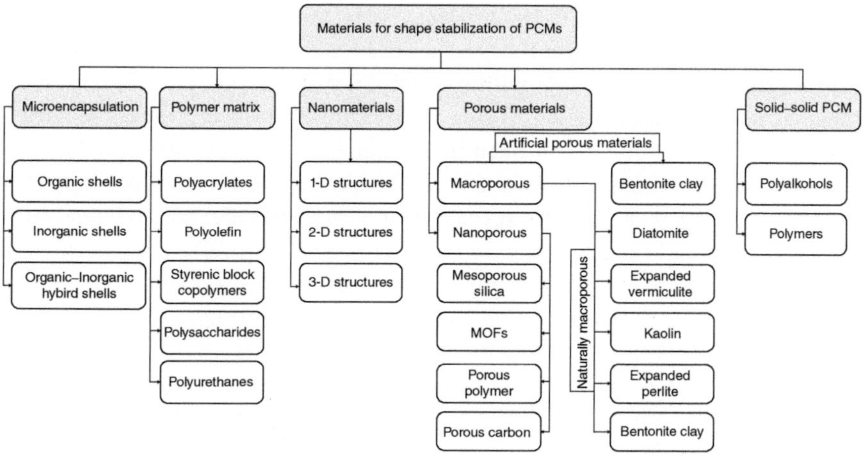

Figure 2.38 Supporting materials used for shape stabilisation of PCMs. Source: Alva et al. [121]; Umair et al. [122]; Huang et al. [132].

A further example of using nano-PCMs was for the cooling of a PV/T system constructed to increase its power output and enhance heat transfer from the panels [131].

2.7.8.5 Shape Stabilisation

Shape stabilisation is also important to avoid leakages and to improve the energy storage ability of PCMs [121, 122, 124, 132] and form-stabilised or shape-stabilised PCM (SSPCM) consists of a working material with a supporting component. The working material absorbs or removes latent heat during the melting or solidifying cycle, whilst the supporting substance stops the melting process from leaking material so that the entire system stays in a constrained state [122]. A variety of materials are used to stabilise the various forms of PCMs which include porous materials, nanomaterials, and polymers, and an outline of the supporting materials is shown in Figure 2.38. It should however be noted that the addition of any supporting material eventually results in a reduction of energy storage density [135].

2.8 Design Development Methods

2.8.1 Introduction

The design of a WHR system relies on heat transfer fundamentals i.e. heat recovery technologies – such as HEXs, boilers, condensers, burners, heat pumps, and others – all primarily designed and sized using heat transfer theory. And even though multiple tools and software packages have been developed to make the design process easier and more user-friendly, it, nonetheless, remains crucial to understand the thermodynamic processes involved. Now when analysing any WHR system, two steps are usually considered. Firstly the variables that may affect the heat transfer need to be identified and reasonable assumptions need to be made about the importance of these variables to the process. Then, the appropriate physical laws are considered and those thought relevant are used to describe and predict the heat transfer mechanisms taking place within the WHR system. The most difficult part therefore in designing and predicting the performance of a system in understanding the physics and identifying the implied phenomena. Experience has shown that a significant number of errors made in the predictions or the design are due to the omitted assumptions regarding unexpected heat or mass transfer mechanisms during any process. Therefore, great attention must be paid during the selection of all the relevant heat transfer laws and their implications during the thermal design stages.

Only then, once the heat transfer phenomena involved during the WHR process have been carefully identified can the fundamental laws of heat transfer can be used to size the installation. Accordingly, the basics of heat transfer fundamentals must be known and several main principles are covered in the following material. To begin with, the first law of thermodynamics (also known as the conservation of energy) stipulates that the change of internal energy (ΔE_{system}) for a system equals the total energy entering the system (E_{in}) minus the total energy leaving the system (E_{out}), expressed in the form of the equation:

$$\Delta E_{system} = E_{in} - E_{out} \tag{2.35}$$

where E is the energy in Joules (J).

However, we also need to note that during any transient behaviour periods for a system (such as the start-up), the variation of internal energy of the system is not zero. Whilst contrastingly, during steady-state operating conditions the energy leaving the system is equal to the energy entering the system and so there is no variation in internal energy.

Now energy provided by, or given to the system can be transmitted through two media: heat transfer and work. In the case where no work is provided and the energy variation is only due to heat transfer, the first law of dynamics is adapted to a heat balance equation given as:

$$\Delta E_{system} = Q_{in} - Q_{out} + E_{generated} \tag{2.36}$$

where Q is the net amount of heat transfer (J) and $E_{generated}$ is the internal heat generation for the system (J). For steady heat flow cases, such as water being heated in a pan, the energy given to the system during the heating of a constant mass without phase change is:

$$Q = m_{system} c_p \Delta T \tag{2.37}$$

where Q is the thermal energy provided to the constant mass system during the heating (J), m_{system} is the mass of the system (kg), c_p its specific heat (J/kg K), and ΔT the difference in temperature before and after the heating occurs (K). To illustrate heat transfer, the concept of a thermal resistance is used, and by electrical analogy, the thermal resistance is the difference of a potential (or driving force) divided by the energy flux going through this resistance. The electrical–thermal analogy is described in Table 2.13.

Table 2.13 Electrical–thermal analogy.

Variable	Electricity	Thermodynamic
Resistance	$R_{elec:}$ Electrical resistance (Ω)	$R_{thermal}$: Thermal resistance (K/W)
Energy flux	i: Current (A)	$\dot{Q}$: Heat transfer rate (W)
Difference of potential	$U = e_1 - e_2$: Voltage (V)	ΔT: Difference of temperature (K)
Relation	$R_{elec} = \frac{U}{i} = \frac{e_1 - e_2}{i}$	$R_{thermal} = \frac{\Delta T}{\dot{Q}}$
Diagram	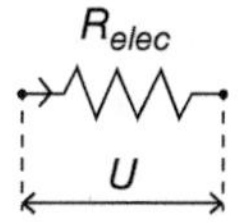	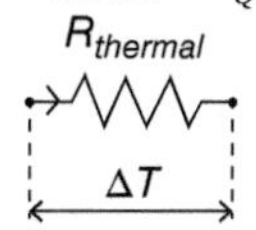

This relationship between the thermal resistance, the heat transfer rate, and the system difference in temperature is called Newton's law of cooling, and to estimate the thermal resistance this fundamental law is often arranged in terms of a heat transfer coefficient (h) as:

$$\dot{Q} = \frac{\Delta T}{R} = hA\Delta T \tag{2.38}$$

where $\dot{Q}$ is the heat transfer rate (W), h the heat transfer coefficient (W/m^2 K), and A the heat transfer area (m^2). Newton's law of cooling is very commonly used in thermal design as it gives a general relationship that can then be adapted according to the heat transfer mode taking place. The different heat transfer modes are usually classified by three mechanisms: conduction, convection, and radiation. Conduction is the transfer of energy without material displacement and is typically taking place in solids where the molecules are fixed. The conductive transfer of energy is expressed by the Fourier's law in three dimensions as:

$$\vec{\dot{Q}}_{conduction} = -kA\vec{\nabla}T = -kA\left(\frac{\partial T}{\partial x}\vec{i} + \frac{\partial T}{\partial y}\vec{j} + \frac{\partial T}{\partial z}\vec{k}\right) \tag{2.39}$$

where $\dot{Q}_{conduction}$ is the conduction heat transfer rate (W), k is the thermal conductivity of the material (W/m K), A is the heat transfer area (m^2), and T is the temperature (K). If the conduction heat transfer is taking place in one direction only, then Fourier's law can be reduced to:

$$\dot{Q}_{conduction} = -kA\frac{dT}{dx} \tag{2.40}$$

which means that the temperature variation in the material is linear on the x-axis. In contrast to conduction, convection describes heat transfer with molecular movement or displacement, e.g. where heat transfer occurs between a solid material and an adjacent liquid or gas that is in motion.

Convective heat transfer uses Newton's law of cooling (given in Eq. (2.38)) to describe the temperature profile for the heat transfer rate in the system, the heat transfer area, and the convective heat transfer coefficients, and to estimate the convective heat transfer coefficient, numerous correlations are available and depend on the type of convection taking place.

Firstly, convection is divided into two actions – (i) natural convection and (ii) forced convection. In natural convection, fluid motion occurs naturally due to some physical phenomena such as density gradients (differences in density) or buoyancy; and the heat transfer coefficients for natural convection are generally lower than those for forced convection. In forced convection the fluid motion is forced by an external mean, such as a pump. Forced convection itself is also further divided in two domains. For flows where the fluid surrounds a solid material, such as an air stream flowing around a vertical heated pipe, the forced convection is called 'external forced convection'. Alternatively, when the fluid in motion is itself surrounded by a solid material, the forced convection is designated as 'internal forced convection'. For instance, water flowing inside a heated pipe is a case of internal convection. For these three identified types of convection, namely natural convection, internal forced convection, and external forced convection, correlations to estimate the heat transfer coefficient are strongly dependent on the fluid velocity and the fluid regime. For low velocities, the fluid tends to flow without lateral mixing, i.e. there are no cross-currents perpendicular to the direction of flow, nor eddies nor swirls of fluid. Hence the motion of the fluid molecules is very orderly, moving in straight lines or layer which do not mix and remain in the same relative position. This is the most stable state for a moving fluid without turbulence and is described as 'laminar flow'. However, at high velocities, the fluid experiences the onset of a more chaotic flow patterns, creating turbulences which forces the different layers to mix, the flow becomes less organised and is designated as 'turbulent flow'. At medium velocities, the flow transits from laminar to turbulent and the fluid motion becomes more difficult to describe because changing turbulent disturbances begin to appear and so the flow switches from laminar to turbulent in an unpredictable way. This regime is called 'transitional flow'. To better understand these flow patterns, in 1883 Professor Osborne Reynold studied these effects and, in honour of his work, we now describe the governing parameter for the onset of this effect with was a dimensionless constant called the Reynolds number (*Re*) and defined by:

$$Re = \frac{\rho v D}{\mu} \tag{2.41}$$

where ρ is the fluid density (kg/m^3), v the mean velocity of water inside the tube (m/s), D the diameter of the section (m), and μ the fluid dynamic viscosity (Pa s). It should be noted that the Reynolds number, and thus the turbulences of a flow, depends on the system geometry and for fluids circulating in a pipe, the smaller the pipe diameter, the higher the turbulence.

In order to identify which correlations to use with which category, limits have been set to describe the three turbulence regimes, and the commonly agreed regime limits are given by [136–138]:

$Re < 2300$ Laminar flow

$2300 \leq Re \leq 10\,000$ Transitional flow

$10\,000 \leq Re$ Turbulent flow

Convective heat transfer coefficients therefore usually depend on a fluid's Reynolds number, but in addition, in forced convection, a further significant parameter is the friction factor which describes the solid–fluid interaction. This is because the roughness of a surface strongly affects the fluid dynamics and can generate fluid turbulences close to the wall. In fact, turbulences actually enhance the heat transfer by increasing the mixing of the different fluid layers, hence, heat transfer coefficients are higher for turbulent flows compared to laminar and more organised flows. However, even if from a heat transfer viewpoint it is tempting to increase the turbulences of a flow by adapting the geometry, increasing the fluid velocity, and/or decreasing the flow cross section, a balance must be found with the increased pressure drop this will also induce. The notion of pressure drop describes the resistance to the flow created by a fluid's viscosity and the frictional forces between that fluid and its solid constraining wall and is also fundamental to the design of equipment. Indeed, to force a fluid such as air to circulate across HEXs, or water to flow through pipes, a necessary magnitude of pressure is required to enable the equipment to succeed in pushing the fluid. So in addition to the size of the plant, higher pressure drops will require more powerful equipment and so further increase in energy consumption, all of which will need to be balanced against the profit made from the heat transfer and the overall performance of the system.

Heat can also be transferred by radiation where energy is conveyed by photons (explained as an elementary particle by the ideas behind wave-particle duality). So in WHR, radiation heat transfer can take place from burners due to the presence of their flames. But in addition, radiation can also take place from any hot surfaces to other solids and or fluids that are not in direct contact with that surface. For a surface at a temperature T_s, the maximum heat transfer rate achievable $\dot{Q}_{radiation,max}$ is given by the Stefan–Boltzmann law:

$$\dot{Q}_{radiation,max} = \sigma_{sb} \times A_s T_s \tag{2.42}$$

where $\sigma_{sb} = 5.67 \times 10^{-8}$ W/m^2 K^4 is the Stefan–Boltzmann constant, A_s the hot surface area (m^2), and T_s the surface temperature (K). Then the amount of actual heat that can be transferred from a hot surface with emissivity ε_s at a temperature T_h to a cold surface at temperature T_c and the heat transfer rate is calculated using:

$$\dot{Q}_{radiation} = \varepsilon_s \sigma_{sb} A_s ({T_h}^4 - {T_c}^4) \tag{2.43}$$

where ε_s is the surface emissivity (dimensionless), and T_h and T_c are the hot and cold surface temperatures (K). It should also be noted that the above equation is only valid if all the radiation from the hot surface is received by the cold surface and, regrettably, applied radiation heat transfer can become very complex due to the spreading of radiations between all the surfaces of any system. Hence to pragmatically consider the rate of radiation given from one surface to the other, depending on their geometry, 'view' factors are used.

Sometimes considered as a form of convection, very useful heat transfers can be achieved by a phase change within a fluid. Indeed, during the phase change process for an element, an important amount of energy, called the latent heat, is either absorbed or released. By reorganising the molecules, elements transiting between solid, liquid, and gaseous states allow the transport or storage of thermal energy and for WHR, the main material states used are those occurring between the liquid and gaseous phases as this transition usually transports a greater amount of energy compared to solid–liquid changes.

The subsequent subsections will show how these heat transfer fundamentals provide a foundation for the development of multiple WHR technologies.

2.8.2 Heat Exchangers

To fully appreciate the processes of sizing HEXs, some knowledge regarding the fundamental laws of conduction and convection are required so as to understand the necessary heat transfer phenomena. Depending on the type of HEX and the geometry being considered, different heat transfer mechanisms are employed for the transfer of thermal energy from the hot fluid to the cold fluid, and in case of the indirect transfer types of HEX, where the hot and cold fluids are separated by a solid wall, two methods are commonly used in their design: the log mean temperature difference (LMTD) method and the effectiveness-number of transfer units (ε-NTU) method. The LMTD technique is generally used to size a HEX based on the required fluid temperatures, whereas the ε-NTU technique is better placed to help estimate the thermal performances of a given HEX. However, before discussing these methods further, we will first introduce the concept of an overall heat transfer coefficient, necessary to appreciate their application.

2.8.2.1 Local Heat Transfer

In HEXs, thermal energy is typically transferred from a hot fluid to a cold fluid separated by a solid material, and for this type of HEX, the heat transfer between the fluid and the solid wall can occur by either internal forced convection or external forced convection. With internal forced convection, the fluid flows inside a closed cavity such as a tube, whilst with external forced convection the contact between the fluid and the solid material is not confined; for example, hot water flowing over a bank of solid cylinders is described by external forced convection correlations. The temperature of the hot fluid T_h and the temperature of the cold fluid T_c can be linked by Newton's law of cooling:

$$\dot{Q} = \frac{1}{R_{tot}}(T_h - T_c) = UA(T_h - T_c) \tag{2.44}$$

where $\dot{Q}$ is the heat transfer rate (W), R_{tot} is the total thermal resistance between the hot and cold fluids (K/W), T_h and T_c are the hot and cold fluid temperatures respectively (K), U the overall heat transfer coefficient (W/m^2 K), and A the heat transfer area (m^2). To estimate the overall heat transfer coefficient and thus the total thermal resistance between the hot and cold fluids, a resistance network analogy is commonly used – the objective of which is to trace the heat flux and identify the different resistances to the heat flowing between the heat source and the heat sink. In numerous HEXs, thermal energy is first transferred by the hot fluid to the wall by forced convection, then transferred by conduction through the solid material,

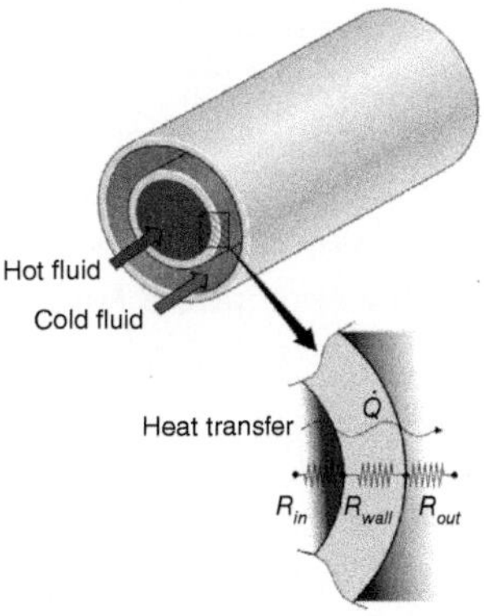

Figure 2.39 Thermal resistance network of a double-pipe parallel flow heat exchanger.

and finally recovered by the cold fluid through forced convection. But other types of HEX employ other forms of heat transfer such as radiation, boiling, or condensation, amongst others. In the simple case of a double-pipe HEX with parallel fluid flows, the thermal resistance network describing the heat transfer modes are presented in Figure 2.39.

The total thermal resistance between the hot and cold fluid being given by:

$$R_{tot} = \Sigma R = R_{in} + R_{wall} + R_{out} \tag{2.45}$$

Between the inside and the outside of the solid material, thermal energy is transmitted by forced convection where the forced convection thermal resistance R_{fc} is given by:

$$R_{fc} = \frac{1}{h_{fc}A_{fc}} \tag{2.46}$$

where h_{fc} is the forced convection heat transfer coefficient (W/m^2 K) and A_{fc} is the forced convection heat transfer area (m^2). However, to estimate the forced convection heat transfer coefficient, some form of correlation must be used, and such correlations typically consider the fluid properties, the fluid regime, and the solid/fluid interactions by including the flow Reynolds number (*Re*) and a friction factor. Similarly, the wall thermal resistance must be evaluated using a suitable equation based on the geometry. For example, for conduction through a thick-walled tube with an internal diameter D_i and an external diameter D_o, the conduction resistance R_{cond} is given by:

$$R_{cond} = \frac{\ln(D_o/D_i)}{2\pi k_l L} \tag{2.47}$$

where k_l is the liquid thermal conductivity (W/m K) and L the tube length (m). Hence in the case of a double-pipe HEX, and taking into account the forced convection resistances for both the inside and outside of the tube, together with the conduction resistance for the tube wall itself, then the overall heat transfer coefficient U and the total thermal resistance R_{tot} can be obtained from:

$$R_{tot} = \frac{1}{UA} = \frac{1}{h_{fc,in}A_{fc,in}} + \frac{ln(D_o/D_i)}{2\pi k_l L} + \frac{1}{h_{fc,out}A_{fc,out}} \tag{2.48}$$

where R_{tot} is the total thermal resistance (K/W), U is the overall heat transfer coefficient (W/m^2 K), A is the heat transfer area (m^2), $h_{fc,in}$ and $h_{fc,out}$ are the forced convection coefficients for the inside and outside of the inner tube (W/m^2 K), and $A_{fc,in}$ and $A_{fc,out}$ are the respective forced convection heat transfer areas (m^2). In other words, to evaluate the local heat transfer, correlations suitable for forced convection and the appropriate HEX geometry must be used. But in addition, other geometrical characteristics such as the presence of fins or other phenomena such as the appearance of fouling can also be added to the thermal resistance network. In particular, fouling becomes more significant over time due to the accumulation of scale deposits or occurrence of corrosion on the heat transfer surfaces thereby creating additional resistances to the flow of heat. In the case of fouling (or scaling), additional resistance is introduced by adding a fouling factor R_f to represent the thermal resistance added because of this effect. In which case the total thermal resistance is now defined from:

$$R_{tot} = \frac{1}{UA} = \frac{1}{h_{fc,in}A_{fc,in}} + \frac{R_{f,in}}{A_{fc,in}} + \frac{ln(D_o/D_i)}{2\pi k_l L} + \frac{R_{f,out}}{A_{fc,out}} + \frac{1}{h_{fc,out}A_{fc,out}} \qquad (2.49)$$

where R_f is a fouling factor (m^2 K/W). An estimate of the fouling factor depends on many parameters such as the fluids used, their composition, the extent timeliness, and practicality of HEX maintenance. But some commonly used values for fouling factor R_f are noted in Table 2.14.

In the above example analysis, we considered a simple double-pipe HEX to illustrate the relationships needed to calculate the overall heat transfer coefficient for the HEX. However, we also need to note that the overall heat transfer coefficient is unique for each HEX i.e. it will be based on many factors such as the type of HEX; the heat transfer process and heat transfer mechanisms; the materials and fluids used; and the specifics of the studied HEX such as the presence of fins, baffles, and the like. All of which will influence the calculation of the overall heat transfer coefficient and to emphasise this variability, Table 2.15 suggests some recommendations to consider when estimating the overall heat transfer coefficient for a given HEX.

Table 2.14 Typical fouling factors R_f for commonly used heat transfer fluids.

Fluid	**Fouling factor R_f (m^2 K/W)**
Water below 50 °C	0.0001
Water above 50 °C	0.0002
Steam	0.0001
Air	0.0004
Fuel oil	0.0009
Liquid refrigerants	0.0002
Gaseous refrigerants	0.0004
Alcohol vapours	0.0001

Table 2.15 Specificities for the estimation of the overall heat transfer coefficient U for different types of heat exchangers.

Type of heat Exchanger	Specifics for the estimation of the overall heat transfer coefficient U
	Classification according to construction
Tubular heat exchanger (double-pipe, shell-and-tube, spiral tube, pipecoils)	We have already considered above simplest and most common form of heat exchanger – the simple tubular heat exchanger, and have noted that the overall heat transfer coefficient will very much depend upon the geometry, the arrangement of tubes – in-line or staggered, the flow rate, the velocity, the density and the viscosity of the fluids, the friction coefficient between the fluid and the solid wall, etc. In the case of spiral tubes and helicoidal pipes, the internal forced convective heat transfer coefficient correlation needs to be adapted because centrifugal forces induce a second flow pattern within the tube that enhances the mixing and so improves the heat transfer. In addition, the conduction resistance is usually through a pipe thickness and therefore uses cylindrical geometry. In the earlier analysis, our estimation of the overall heat transfer coefficient did not take into account the direction of the flow. This will also indirectly affect the relationships being used, but we will considered this later within the sections that cover the log mean temperature difference (LMTD) method and the effectiveness-number of transfer units (ε-NTU) method.
Plate heat exchanger (PHE, spiral, platecoil, printed circuit)	In a plate heat exchanger, the fluid circulates between two plates within a space that is defined by the design of the plate. The overall resistance model is similar to that for the tubular heat exchanger as it usually consists of two forced convection resistances and a conduction resistance through the thickness of the plate. However, unlike a tubular heat exchanger, relationships for forced convection and conduction vary depending on the plate geometry, i.e. the forced convection correlations can vary according to the surface aspect of the plates used. For example, it is common to use corrugated plates to increase the heat transfer and performances of plate heat exchangers. The use of which now means that two main aspects that must be considered in the thermal design process. Firstly any corrugation of the plate increases its heat transfer area. So an enlargement factor ($\boldsymbol{\Phi}$) (which is unique for each type of plate) needs to be included [139] for estimating the heat transfer area when using the log mean temperature difference (LMTD) method or the effectiveness-number of transfer units (ε-NTU) method. Secondly, each type of corrugation aims at improving the forced convective heat transfer coefficient, so again a unique and suitable forced convection heat transfer coefficient correlation needs to be used for the analysis.

Table 2.15 (Continued)

Type of heat Exchanger	**Specifics for the estimation of the overall heat transfer coefficient *U***
Extended surface heat exchangers (plate–fin, tube-fin)	For extended surface heat exchangers, the heat transfer surface is typically increased by incorporating fins and in all such cases, the total heat transfer area estimated using either the LMTD or the ε-NTU method will include the finned area. Hence by using finned surfaces, a more compact heat exchanger can be made, of if the size has already been fixed, it can increase efficiency, but this then implies yet another adaption to the heat transfer relationships. Now for annular fins, the forced convection heat transfer coefficient is not significantly impacted because the fins align with the direction of the flow. However, in the cases of helical and spiral finning tubes, fluid motion is changed, fluid mixing increases which, in turn, increases the forced convection heat transfer coefficient. So for these cases, the forced convection correlation needs adapting based on the fins used and on the geometry of the heat exchanger, and additionally to complicate things even further, with plate–fin heat exchangers, the fins are also likely to use corrugations to further improve their heat transfer rates! Plate–fin heat exchangers are usually found in compact heat exchangers in which the fluid flows within channels between plates. So in such cases, the estimation of the overall heat transfer coefficient for the heat exchanger also requires the consideration of the hydraulic diameter for these small passages. Finally, with extended surface heat exchangers, the use of fins increases the fouling and again this need to be taken into account by choosing an 'adapted' fouling factor whilst estimating the overall heat transfer coefficient for the exchanger.
Regenerative heat exchangers (rotary, fixed-matrix, rotating hoods)	In regenerative type heat exchangers, a heat transfer media is used to store thermal energy and subsequently release it under different conditions. Usually, heat is recovered by forced convection from a hot stream to the heat transfer media and then released according to the same principle. Yet, for regenerative heat exchangers no steady-state condition is reached. Instead, transient behaviour must be studied. Hence to characterise the transient behaviour of the heat transfer media, transient conduction laws need to be applied and so once again, depending on the type of heat transfer media and the geometry of the matrix used, the forced convection heat transfer coefficient will vary.
Classification according to heat transfer process	
Indirect contact heat exchangers	In indirect contact heat exchangers, the heat source and heat sink fluids are separated by a solid material and do not mix, and this is the typical case for most heat exchangers. So our estimate for the thermal resistance network for the exchanger must include all the possible heat paths from the heat source to the heat sink.

Table 2.15 (Continued)

Type of heat Exchanger	Specifics for the estimation of the overall heat transfer coefficient *U*
Direct contact heat exchangers	Whereas with direct contact heat exchangers, the heat source is in direct contact with the heat sink. In which case, the overall heat transfer coefficient is only that for heat transfers that are taking place at the interface between the two materials in contact, say fluid to solid, gas to fluid, or gas to solid. However, in case of a contact between two solids, a thermal resistance at the interface is relevant and should be considered.
Classification according to flow arrangement	
Parallel flow heat exchanger	The overall heat transfer coefficient *U* of the heat exchanger is estimated normally without considering the direction of the flow.
Counter-flow heat exchanger	
Cross-flow heat exchanger	
Classification according to pass arrangement	
Single-pass heat exchanger	The overall heat transfer coefficient *U* of the heat exchanger is estimated using traditional methods.
Multi-pass heat exchanger	In multi-pass heat exchangers, one or both of the fluids go through a number of passes which increases the heat transfer area and improves the performance, yet, the impact of these passes is not considered in the estimation of the overall heat transfer coefficient for the exchanger. Instead, their impact is considered during the estimation of the heat transfer area for both the LMTD method and the ε-NTU method. In particular, when estimating the effectiveness of the heat exchanger (using the ε-NTU method) separate correlations for multi-pass heat exchangers exist and must be used.
Classification according to the Heat transfer mechanisms	
Single-phase heat exchanger	The overall heat transfer coefficient *U* of the heat exchanger is estimated normally.
Two-phase heat exchangers	In two-phase heat exchangers, at least one of the fluids experiences a phase change during the heat transfer process showing that heat can also be exchanged by latent heat transfer. In the case of two-phase heat exchangers, the correlations to use depend on the fluid motion and proportion of each phase. Plus, any interactions between each phase must also be included in the model as it may impact the heat transfer rate. For instance, in cases where the liquid vapourises, the location of the vapour and liquid volume is predominant in the model. At first, vapour bubbles departing inside the fluid increase the mixing of the

Table 2.15 (Continued)

Type of heat Exchanger	**Specifics for the estimation of the overall heat transfer coefficient *U***
	fluid phase which improves the heat transfer locally and if a significant portion of the liquid has already vapourised, then the remaining liquid phase can be forced to flow on the wall. In which case, a thin film of liquid evaporates, and the heat transfer is significantly higher compared to the heating of a complete volume of fluid. Overall, pool boiling, flow boiling, condensation, and forced convective condensation correlations must be used to estimate the overall heat transfer coefficient for two-phase heat exchangers.
Condensers	In condensers, as the name suggests, condensation takes place, but the correlation to use depends mainly on the presence of a fluid motion or not. For example, if condensation of a gas takes place within a tube then forced convective condensation correlations should be used because those correlations depend on a two-phase flow regime and proportion of each phase. However, if condensation of a gas takes place without motion, then condensation correlations are used. Notwithstanding, depending on the difference between vapour and wall of temperatures, two types of condensation can appear: namely filmwise and dropwise condensation, but in practice dropwise condensation cannot be maintained and filmwise condensation then dominates. The reference correlation for filmwise condensation was determined by Nusselt [140] and in the estimation of the overall heat transfer coefficient for condensers, gravity must also be considered. Finally, since the condensation heat transfer coefficient depends so much on the vapour and wall temperatures, assumptions and/or iterations may be necessary to establish a final estimate for the overall heat transfer coefficient.
Vapourisers	In vapourisers, a fluid transits from a liquid phase to a gaseous phase and like condensation, the correlations differ depending on the motion of the fluid. If the fluid vapourises without motion, *pool* boiling takes place. Conversely, if boiling occurs whilst the fluid is in motion, *flow* boiling takes place. In other words, for both pool boiling and flow boiling, the bubble motion influences the heat transfer coefficient. For pool boiling, the most commonly applied correlation was developed by *Rohsenow* [141]. Whereas with flow boiling, different fluid regimes exist, and each regime is described by a different correlation. Again, to estimate the boiling heat transfer coefficient, the liquid and wall temperatures must be known, and therefore, assumptions or iterations will be required to estimate the overall heat transfer coefficient for a vapouriser.

Table 2.15 (Continued)

Type of heat Exchanger	Specifics for the estimation of the overall heat transfer coefficient *U*
Heat pipe-based heat exchangers	In heat pipe-based heat exchangers, the heat sources and heat sinks are separated by heat pipes. On the heat source side, heat can be transferred from the heat source to the heat pipe by conduction, convection, radiation, boiling, and condensation and to estimate this heat transfer coefficient, the appropriate laws must be used. The conveyed thermal energy is then transported by the heat pipe from the *evaporator* section to the *condenser* section – these denote that the heat pipe sections are in contact with the heat source and heat sink, respectively. The thermal energy is then released by the heat pipe and recovered by the heat sink. In addition to the heat source and heat sink heat transfer thermal resistances, the heat pipe(s) equivalent thermal resistance must also be estimated, and to do so, conduction, boiling, and condensation resistances inside the heat pipe must be determined [138, 142, 143]. Finally, the equivalent thermal model – considering both heat source and heat sink thermal resistances, and the heat pipes equivalent resistance – leads to the estimation of the overall heat transfer coefficient.

The analysis presented earlier describes the overall heat transfer coefficient for a HEX, but this approach is not sufficient to fully connect the different inlet and outlet fluid temperatures and the heat exchange area. So in practice, to design and size of HEXs and additional methods for describing the HEX processes are employed, all of which rely on an estimation of the overall heat transfer coefficient for the HEX being studied. Of these, the two main methods (for sizing a HEX) are the LMTD method and the ε-NTU method and these are detailed subsequently.

2.8.2.2 LMTD Method

The LMTD method is the first of these commonly used techniques and links the inlet and outlet temperatures of the fluids to the heat transfer area. Hence, for cases where the fluid inlet and outlet temperatures are known and/or defined by the user, the LMTD method can estimate the HEX area needed to achieve the desired performance.

According to the first law of dynamics for a system thermally isolated from its environment, when heat is transferred from a hot fluid to a cold fluid all the heat lost by the hot fluid is recovered in full by the cold fluid. So for both the hot and cold fluids, the heat transfer rate taking place can be related to the inlet and outlet temperatures of each fluid as:

$$\dot{Q} = \dot{m}_h c_{p,h}(T_{h,in} - T_{h,out}) \tag{2.50}$$

$$\dot{Q} = \dot{m}_c c_{p,c}(T_{c,out} - T_{c,in}) \tag{2.51}$$

where $\dot{Q}$ is the heat transfer rate (W), T_h and T_c are the hot and cold fluid temperatures, respectively, (K), $\dot{m}_h$ and $\dot{m}_c$ are the hot and cold fluid mass flow rates (kg/s), $c_{p,h}$ and $c_{p,c}$ are the hot and cold fluid specific heats (J/kg K), $T_{h,in}$ and $T_{h,out}$ are the inlet and outlet temperatures for the hot fluid (K), and $T_{c,in}$ and $T_{c,out}$ the inlet and outlet temperatures for the cold fluid (K). In addition, for convenience, the mass flow rate and specific heat for a fluid are often combined into its heat capacity rate, given by:

$$C_h = \dot{m}_h c_{p,h} \tag{2.52}$$

$$C_c = \dot{m}_c c_{p,c} \tag{2.53}$$

Usually, the fluid heat capacities on the hot and cold sides are not the same in which case the temperature differences between the hot and cold fluids on each side of the HEX are not proportional. However, when the same fluid is used for both hot and cold streams, the heat capacities of both hot and cold fluids are equal, which then means that the respective inlet and outlet temperature differences for hot and cold fluids will be proportional. Nonetheless, along the length of a HEX, the temperature difference between the hot and cold fluids is not linear. In fact, in the case of a parallel flow HEX, we can appreciate that a much higher rate of local heat transfer is taking place at the inlet of the exchanger because this is where the hot fluid is at its maximum temperature and the cold fluid is at its minimum. So, when looking at the analysis, as tempting as it might be to use arithmetic mean of the inlet and outlet temperatures to estimate the respective average fluid temperatures, a quick look at fluid temperature profiles shows why such a simple estimate would be in error. To illustrate the importance of this, the typical temperature profile trend lines for both hot and cold fluids in the case of a double-pipe HEX with (a) a parallel flow and (b) a counter-flow arrangement is illustrated in Figure 2.40.

Also, for design purposes, it is helpful to relate the heat transfer taking place within a HEX to both the heat transfer area and the fluid temperatures. Therefore, an expression in the form of Eq. (2.54) would be desirable:

$$\dot{Q} = UA_s \Delta X \tag{2.54}$$

where ΔX is a variable representing the hot and cold fluid temperatures – the objective here being to adapt Newton's law of cooling for HEXs to find such an expression – for the differential analysis as detailed in Figure 2.41.

For the volume of fluid passing through the area dA_S being studied, a quantity of heat is transferred from the hot fluid to the cold fluid, the rate of which ($\delta\dot{Q}$) can be related to the mass transfer rates and specific heats of both fluids from:

$$\begin{cases} \delta\dot{Q} = -\dot{m}_h c_{p,h} dT_h \\ \delta\dot{Q} = \dot{m}_c c_{p,c} dT_c \end{cases} \tag{2.55}$$

Now by isolating the temperatures, these become:

$$\begin{cases} dT_h = \dfrac{\delta\dot{Q}}{-\dot{m}_h c_{p,h}} \\ dT_c = \dfrac{\delta\dot{Q}}{\dot{m}_c c_{p,c}} \end{cases} \tag{2.56}$$

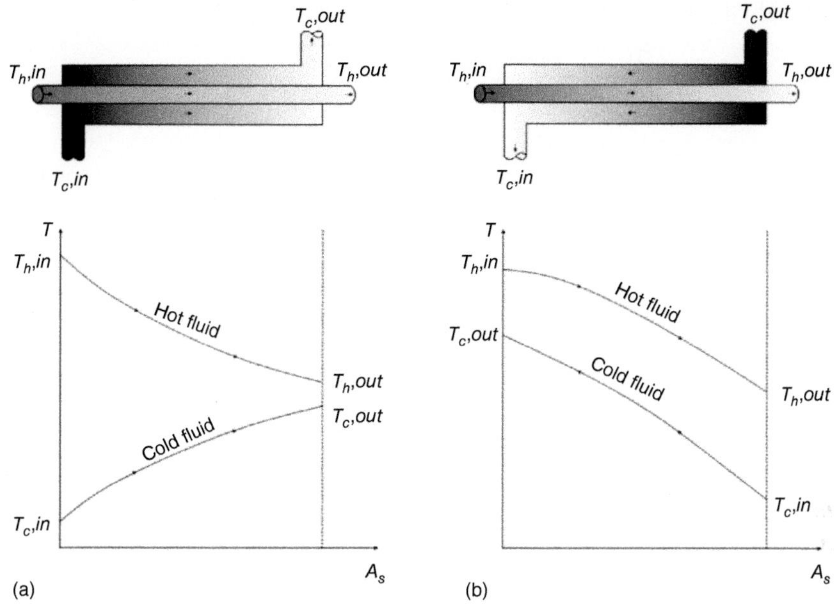

Figure 2.40 Typical fluids temperature profile in a double-pipe heat exchanger with (a) parallel flow and (b) counter-flow disposition.

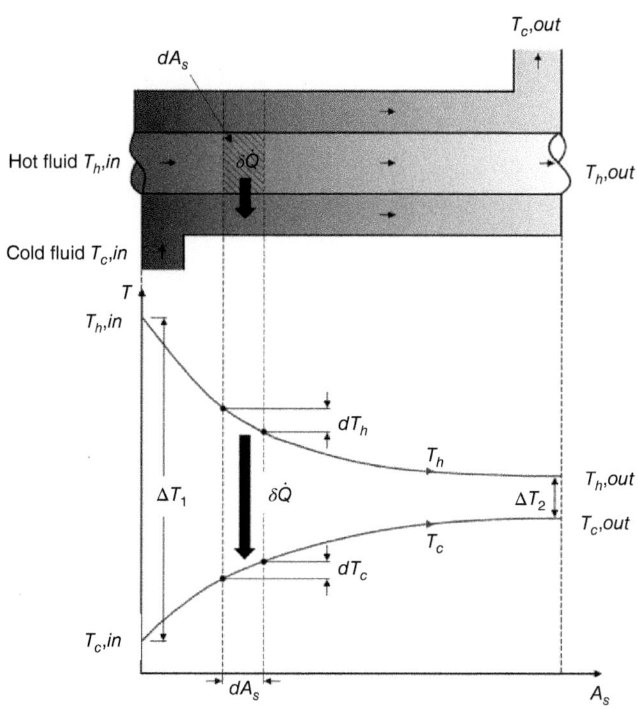

Figure 2.41 Differential analysis on a control volume of a heat exchanger.

Hence the difference between the above two equations leading us to:

$$dT_h - dT_c = \frac{\delta\dot{Q}}{-\dot{m}_h c_{p,h}} - \frac{\delta\dot{Q}}{\dot{m}_c c_{p,c}} \tag{2.57}$$

Which can be rearranged as follows:

$$d(T_h - T_c) = -\delta\dot{Q}\left(\frac{1}{\dot{m}_h c_{p,h}} + \frac{1}{\dot{m}_c c_{p,c}}\right) \tag{2.58}$$

But we know (on the studied control volume), that the heat transfer rate can also be described by the Newton's law of cooling from:

$$\delta\dot{Q} = UdA_s(T_h - T_c) \tag{2.59}$$

So by replacing the previous heat transfer rate by this expression, we get:

$$\frac{d(T_h - T_c)}{(T_h - T_c)} = -UdA_s\left(\frac{1}{\dot{m}_h c_{p,h}} + \frac{1}{\dot{m}_c c_{p,c}}\right) \tag{2.60}$$

Hence, using this form of local analysis, it is possible to relate the fluid temperatures to the heat transfer area, and then to generalise this expression to one for the overall heat exchange process by integration from the inlet to the outlet of the exchanger. This integration giving us:

$$-\ln\left(\frac{T_{h,out} - T_{c,out}}{T_{h,in} - T_{c,in}}\right) = UA_s\left(\frac{1}{\dot{m}_h c_{p,h}} + \frac{1}{\dot{m}_c c_{p,c}}\right) \tag{2.61}$$

which by substituting the hot and cold fluid heat capacities gives [15, 16]:

$$\ln\left(\frac{T_{h,in} - T_{c,in}}{T_{h,out} - T_{c,out}}\right) = \frac{UA_s}{\dot{Q}}(T_{h,in} - T_{h,out} + T_{c,out} - T_{c,in}) \tag{2.62}$$

And from which by a final rearrangement gives us the desired expression relating the heat transfer rate taking place (within the whole HEX) to both the heat transfer surface area and the fluids' inlet and outlet temperatures as:

$$\dot{Q} = UA_s\frac{(T_{h,in} - T_{c,in}) - (T_{h,out} - T_{c,out})}{ln\left({T_{h,in}-T_{c,in}}/{T_{h,out}-T_{c,out}}\right)} \tag{2.63}$$

Showing, as it is more commonly noted, that it is possible to express the heat transfer rate taking place within a HEX in terms of HEX surface area and fluid temperatures by:

$$\dot{Q} = UA_s\Delta T_{lm} = UA_s\frac{\Delta T_1 - \Delta T_2}{ln\left({\Delta T_1}/{\Delta T_2}\right)} \tag{2.64}$$

where ΔT_{lm} is known as the LMTD, i.e. the logarithmic temperature difference between the hot and cold fluid on both sides of the HEX. In addition, owing to the properties of logarithms, the temperature differences ΔT_1 and ΔT_2 can be reversed without changing the overall LMTD value, and the LMTD expression is valid for both parallel flow and counter-flow HEXs provided we take the temperature differences for both fluids according to the following geometry:

$$\text{Parallel flow:} \begin{cases} \Delta T_1 = T_{h,in} - T_{c,in} \\ \Delta T_2 = T_{h,out} - T_{c,out} \end{cases}$$

$$\text{Counter flow:} \begin{cases} \Delta T_1 = T_{h,in} - T_{c,out} \\ \Delta T_2 = T_{h,out} - T_{c,in} \end{cases} \tag{2.65}$$

By setting the desired fluid temperatures, the LMTD method allows us to estimate the heat exchange surface area required to achieve a given heat transfer rate and for similar fluid temperatures, it can be noted that the LMTD for counter-flow HEX is always greater than that of a parallel flow exchanger. Hence, for equivalent heat transfer rates, counter-flow HEXs require smaller heat exchange surface areas compared to parallel flow HEXs, and so in practice, counter-flow HEXs are often preferred over their parallel flow counterparts.

Now it is important to note that earlier discussed LMTD expression is limited to the cases of simple parallel flow and counter-flow HEXs. However, for more complex geometries such as cross-flow HEXs or multi-pass HEXs, because it is convenient to still use a similar method of design, we can relate the LMTD to that of a counter-flow HEX by including a correction factor:

$$\text{Cross flow/multi pass:} \;\; \Delta T_{lm} = F \Delta T_{lm,Counter\,flow} \tag{2.66}$$

where F is a correction factor which varies between 0 and 1. Thus, the maximum LMTD value that can be achieved by a cross-flow or multi-pass HEX is the LMTD value that would be obtained by an equivalent counter-flow HEX with a similar heat transfer surface area. Values of correction factor F are reported in the literature for each type of HEX.

Within this section, we showed that the LMTD method allows us to size a HEX when both inlet and outlet temperatures are fixed and known. Unfortunately, in many cases the heat transfer rate and outlet temperatures of the fluids are not known and the LMTD estimation of the performance of a given HEX would require a cumbersome iterative process. Hence, a second method (commonly used in the design of HEXs) is the ε-NTU method.

2.8.2.3 Effectiveness-Number of Transfer Units (ε-NTU) Method

Developed by William Kays and Alexander London in 1955, the method considers the performance for a HEX for a given heat transfer surface area (A_s), and typically, the ε-NTU method aims at estimating the heat transfer rate taking place within the exchanger and estimating both the hot and cold fluid outlet temperatures. As the name implies, this approach is based on two dimensionless parameters: the effectiveness of the exchanger ε and the number of transfer units NTU defined as:

$$\varepsilon = \frac{\dot{Q}}{\dot{Q}_{max}} = f(NTU) \tag{2.67}$$

$$NTU = \frac{UA}{C_{min}} \tag{2.68}$$

where ε is the effectiveness of the HEX, $\dot{Q}$ is the heat transfer rate (W), and $\dot{Q}_{max}$ is the maximum heat transfer rate achievable by the HEX (W). *NTU* is the number of transfer units, U is the overall heat transfer coefficient (W/m^2 K), A is the heat transfer area (m^2), and C_{min} is the minimum heat capacity rate (W/K).

The objective of the ε-NTU technique is to directly estimate the effectiveness of the HEX ε by expressing it as a function of the number of transfer units *NTU*. And because it is relatively simple to determine the maximum achievable heat transfer rate, if the effectiveness of a HEX can be estimated, it will then be possible to predict both the heat transfer rate for the system and the fluid outlet temperatures [16, 17]. By definition, a fluid with a low heat capacity will undergo a greater change in temperature compared to a fluid with a higher heat capacity. So, the maximum achievable heat transfer rate can be obtained from cases where the fluid with minimum heat capacity reaches the temperature of the other fluid. That is:

$$\dot{Q}_{max} = C_{min}(T_{h,in} - T_{c,in}) \tag{2.69}$$

where C_{min} is the minimum heat capacity rate of the fluids employed (W/K), $T_{h,in}$ is the inlet temperature of the hot fluid (K), and $T_{c,in}$ is the inlet temperature of the cold fluid (K).

Now to estimate the number of transfer units for a given HEX, the overall heat transfer coefficient U is first required, but this can be estimated by using the electrical–thermal resistance network analogy previously discussed and by considering the conduction and forced convection thermal resistances of the HEX. To simplify the process, a number of effectiveness correlations have been developed for various standard HEX configurations, and for convenience, these equations use the capacity ratio (c), a dimensionless quantity denoting the ratio of the minimum fluid heat capacity rate to the maximum, as given by:

$$c = \frac{C_{min}}{C_{max}} \tag{2.70}$$

And based on a similar approach to the one that we used to develop the logarithmic mean temperature difference expressions, correlations for the effectiveness of several commonly used HEX configuration have been reported – some of which are reported in Table 2.16.

So by using Table 2.16, the effectiveness of a HEX can be estimated. Then by using the inlet temperatures and mass flow rates of both fluids and by estimating the maximum achievable heat transfer rate, the actual heat transfer rate taking place within the HEX can be predicted in addition to the outlet temperatures for both fluids.

2.8.3 Regenerative and Recuperative Burners

The design of burners is a wide and complex subject which, even today, is not fully understood [144]. And this is because for a complete theoretical investigation of burners many elements such as the geometry and type of burner used, the fuel combustion process, chemical reactions inside the burner, flame profile and intensity, flashback, composition of the inlet and outlet streams, radiation between surfaces,

Table 2.16 Effectiveness correlations of typical heat exchangers.

Geometry	Effectiveness relation	
	Double pipe	
Parallel flow	$\varepsilon = \frac{1-\exp[-NTU(1+c)]}{1+c}$	(2.71)
Counter-flow ($c < 1$)	$\varepsilon = \frac{1-\exp[-NTU(1-c)]}{1-c\times\exp[-NTU(1-c)]}$	(2.72)
Counter-flow ($c = 1$)	$\varepsilon = \frac{NTU}{1+NTU}$	(2.73)
	Shell and tube	
One shell pass (2, 4, … tube passes)	$\varepsilon = 2\left\{1 + c + \sqrt{1+c^2}\frac{1+\exp[-NTU\sqrt{1+c^2}]}{1-\exp[-NTU\sqrt{1+c^2}]}\right\}^{-1}$	(2.74)
n Shell passes ($2n$, $4n$, … tube passes)	$\varepsilon = \left[\left(\frac{1-\varepsilon_1\times c}{1-\varepsilon_1}\right)^n - 1\right]\left[\left(\frac{1-\varepsilon_1\times c}{1-\varepsilon_1}\right)^n - c\right]^{-1}$	(2.75)
	and ε_1 calculated for one shell pass	
	Cross-flow (single pass)	
Both fluids unmixed	$\varepsilon = 1 - \exp\left[\frac{NTU^{0.22}}{c}\{\exp[-c\times NTU^{0.78}] - 1\}\right]$	(2.76)
C_{max} mixed, C_{min} unmixed	$\varepsilon = \frac{1}{c}\times(1-\exp\{-c\times[1-\exp(-NTU)]\})$	(2.77)
C_{min} mixed, C_{max} unmixed	$\varepsilon = 1 - \exp\left\{-\frac{1}{c}\times[1-\exp(-c\times NTU)]\right\}$	(2.78)
All exchangers with $c = 0$	$\varepsilon = 1 - \exp(-NTU)$	(2.79)

surface cooling, and others factors all have an impact on the thermal performance of a burner. Usually, the design approach for burners is to firstly determine the ideal heat flux profile and then adapt the design to reach the desired heat flux. With the improvements in numerical techniques, it is tempting to use computational assistance to make the design process easier and if sizing software packages for burners are recommended, then commercial computational fluid dynamic modelling packages can be used – but limited to checking the overall energy balance of the burners. To optimise the design and performance of a complete system, the pinch point – defined as the point where the hot and cold streams are separated by the minimum temperature difference during the process – is of interest. So to reach optimal cost and energy targets, pinch point analysis consists of connecting the flows requiring heating with those requiring cooling to decrease the pinch point to its most favourable value, and if this first analysis deals with the overall design of burners, then consideration of a WHR technology for the burners requires further consideration. In burners, the total heat flux provided to the system by combustion is divided between the actual heat flux used for the process and the heat losses conveyed to the flue gases and burner surfaces is given by:

$$q''_{provided} = q''_{process} + q''_{losses} \tag{2.80}$$

where q'' is the heat flux (W/m^2). But in WHR, since the heat flux dedicated to the process remains unchanged the objective is to maximise the heat recovery from the losses rejected to the atmosphere. In burners, the two main WHR technologies are regenerative and recuperative heaters and for both the objective is to reuse the heat losses and reintroduce these to the system. By doing so, the total heat flux provided to the system then partially comprises heat recovered from the losses, so decreasing the overall amount of materials and lowering the cost of the energy required for the burner. Even so, although the objective is similar, the two WHR systems for burners require different technologies to reintroduce their waste heat.

2.8.3.1 Regenerative Burners

Regenerative burners consist of two separated burners, each with their own control, and within an industrial process the two burners are used alternately. To explain, during the operation of one burner, hot flue gas is used to heat solid heat exchange media located so as to be ready to heat the inlet of the second burner. Once the material is fully heated, the flow is switched so that the combustion air entering the second burner is preheated by this hot heat exchange media and the flue gas from the second burner is then used to heat the heat exchange media for the first burner. And although different recuperative burners technologies exist, the principle remains similar: i.e. heat is transmitted from hot flue gas to a heat exchange media by forced convection and is temporarily stored in the solid material for subsequent transient conduction. As previously noted, forced convection problems largely depend on the flow velocity and geometry and correlations for the heat transfer coefficients of forced convection consider the flow regime, turbulences, viscous forces, and solid–fluid interactions. In practice, these correlations are commonly given in terms of Nusselt number Nu_{fc} as:

$$Nu_{fc} = \frac{h_{fc}L_c}{k} = f(Re, Pr, f_{friction}) \tag{2.81}$$

$$Re = \frac{\rho v L_c}{\mu} \tag{2.82}$$

$$Pr = \frac{c_p \mu}{k} \tag{2.83}$$

where h_{fc} is the forced convection heat transfer coefficient (W/m^2 K), L_c is the characteristic length of the geometry (m), k is the fluid thermal conductivity (W/m K), Re is the Reynolds number, Pr is the Prandtl number, $f_{friction}$ is the friction coefficient, ρ is the fluid density (kg/m^3), v is the velocity of the fluid (m/s), μ is the fluid dynamic viscosity (Pa s), and c_p the fluid specific heat (J/kg K). Also, because of the transient behaviours, Newton's law of cooling is adapted as:

$$\dot{Q}(t) = h_{fc}A(T_h - T_c(t)) \tag{2.84}$$

where the heat transfer rate $\dot{Q}$ (W) is a function of time t (seconds), h_{fc} is the forced convection heat transfer coefficient (W/m^2 K), A is the heat transfer area of the solid

(m^2), and T_h is the temperature of the hot fluid (K) surrounding the solid at a temperature $T_c(t)$. As heat is transferred by forced convection from the hot flue gas to the heat transfer medium, the temperature of the media changes with time as it stores thermal energy. Between $t = 0$ and $t = t$, the total amount of heat stored (by the heat exchange media) being given by:

$$q''_{provided} = q''_{process} + q''_{losses} \tag{2.85}$$

$$Q = m_{system} c_p (T(t) - T_{t=0}) \tag{2.86}$$

where Q is the total amount of heat stored by the heat exchange media (J), m_{system} is the mass of the heat exchange media (kg), c_p is the specific heat (J/kg K), $T(t)$ is the temperature of the heat exchange media at a time $t = t$ (K), and $T_{t=0}$ is the initial temperature of the heat exchange media at a time $t = 0$. Depending on the size and thermal conductivity of the material, the temperature of the heat exchange media may be considered to be uniform or non-uniform, and when the temperature is considered to be uniform within the material, the system is denoted as 'lumped'. To determine whether the solid material is lumped or not, the Biot number (Bi) is introduced and characterises the ratio of convection heat transfer at the surface of the solid to the conduction heat transfer within the body, i.e. by:

$$Bi = \frac{h_{fc}}{k/L_c} \times \frac{\Delta T}{\Delta T} = \frac{h_{fc} L_c}{k} \tag{2.87}$$

where h_{fc} is the forced convection heat transfer coefficient (W/m^2 K), k is the solid thermal conductivity (W/m K), and L_c a characteristic length (m) given by:

$$L_c = \frac{V}{A} \tag{2.88}$$

where V is the solid volume (m^3) and A the solid external surface (m^2). When the Biot number is lower than $Bi \leq 0.1$, the temperature difference between the hot flue gas and any point within the solid material differs by less than 5% [136]. And under this condition it is generally accepted to consider the temperature distribution inside the solid heat exchange media as being uniform in all the directions. From which, if the heat exchange media is considered as lumped, its temperature $T(t)$ after being exposed to the hot flue gas from $t = 0$ to $t = t$ can be estimated by [136, 138, 144]:

$$\frac{T(t) - T_h}{T_{t=0} - T_h} = e^{-\frac{h_{fc}}{\rho c_p L_c} \times t} \tag{2.89}$$

where T_h is the temperature of the hot exhaust fluid (K), $T_{t=0}$ is the initial temperature of the heat exchange media at a time $t = 0$, h_{fc} is the forced convection heat transfer coefficient (W/m^2 K), ρ is the solid density (kg/m^3), c_p is the solid specific heat (J/kg K), L_c its characteristic length (m), and t is the time at which the temperature of the solid is calculated. However, when the temperature of the heat transfer media cannot be assumed to be uniform, then the analysis becomes more complicated and transient conduction laws inside the solid body must be applied. Now to determine the temperature at any location inside the heat exchange media, both analytical and numerical methods can be used and for most problems, the exact analytical solution is an infinite series; for which it is common to approximate the solution by only considering the first term of the series. Unfortunately, even for one-dimensional transient heat conduction with simple geometries, it is still not convenient to deal with this one-term approximation and so many constants are determined from data tables. As an example, the one-term approximations of one-dimensional transient conduction with a plane wall, a cylinder, and a sphere are:

$$\text{Plane wall:} \frac{T(x,t) - T_h}{T_{t=0} - T_h} = C_1 e^{-C_2{}^2\tau} \cos\left(\frac{C_2 x}{L}\right) \tag{2.90}$$

$$\text{Cylinder:} \frac{T(r,t) - T_h}{T_{t=0} - T_h} = C_1 e^{-C_2{}^2\tau} C_3 \left(\frac{C_2 r}{r_o}\right) \tag{2.91}$$

$$\text{Sphere:} \frac{T(r,t) - T_h}{T_{t=0} - T_h} = C_1 e^{-C_2{}^2\tau} \frac{\sin(C_2 r/r_o)}{C_2 r/r_o} \tag{2.92}$$

where x is the distance from the middle of the plane wall (m), L is half of the plane wall thickness (m), r is the radial distance to the axis of rotation (m), and r_o the outer diameter (m). The dimensionless time constant τ being defined by:

$$\begin{cases} \tau = \frac{\alpha t}{L^2} \text{ for a Plane wall} \\ \tau = \frac{\alpha t}{r_o{}^2} \text{ for a Cylinder or a Sphere} \end{cases} \tag{2.93}$$

where α is the thermal diffusivity (m^2/s), t is the time (s), L is half of the plane wall thickness (m), and r_o the outer diameter (m). The constants C_1, C_2, and C_3 depend on the Biot number and are obtained from Heisler/Gröber charts and tables are available in the literature. Helpfully, even for the simplest geometry, although we are aware that these expressions are the approximations of a series, for $\tau > 0.2$, the error made by this one-term approximation is still estimated to be less than 2%. Hence, even considering estimating the first convection coefficient, determining the Biot number for the heat transfer media, and using the above equations from the heat transfer media geometry and heat transfer area, it is still possible to estimate the time t at which the temperature of the heat transfer media will reach the temperature of the hot flue gas, and this can provide a good estimate for the time period between which the two burners should alternate.

2.8.3.2 Recuperative Burners

In recuperative burners, a HEX is placed between the outlet of the burner and its inlet to directly re-inject the heat to the combustion air before it enters the furnace. A major difference between recuperative burners and regenerative burners is therefore that recuperative burners operate in a steady-state mode whereas regenerative burners use cyclic transient conditions to alternately store energy and then release this. The implementation of a recuperative burner is similar to the sizing of HEXs previously detailed in the relevant section, and in accord with HEX theory, Jenkins and Mullinger [144], that counter-flow arrangements demonstrate better performances than other dispositions, i.e. for a similar difference in temperature between the inlets of the hot and cold streams, the relative heat transfer area needed to obtain a given rise of temperature of the fluids is smaller in the case of a counter-current flow compared to cross-flow and co-current flow geometries. So if the outlet temperature of the flue gas is a significant requirement of the recuperative burner, the LMTD method can be used to size the HEX. Alternatively, in cases where the outlet temperature of the flue gas and of that for the combustion air entering the furnace are not known, the ε-NTU method is recommended for estimating both the HEX area and the fluid temperatures. However, to do so, often requires assumptions to be made. Indeed, a common option in the design of HEXs is the use of iterative solutions that calculate scenarios based on these assumptions and later, having compared the results, to then adjust the assumptions and performed another calculation iteration. For example, we could first assume the fluid outlet temperatures and so the size of the HEX using the LMTD method. Then performing a second iteration based on the HEX size just calculated, assumptions for the fluid outlet temperatures can be modified, and this calculation process repeated until the balance between the HEX area required and the outlet temperatures of the fluids suits the designer. Lastly, mechanical and pressure drop calculations can be addressed to realise the final design of the recuperative burner.

2.8.4 Waste Heat Boilers

Typically, WHBs consist of large shell-and-tubes HEXs, with two main types of WHBs reported in the literature regarding which fluid circulates inside and which outside the tubes [145–147]. In cases where the hot exhaust gas circulates inside the tubes, the liquid within the shell surrounds the hot pipes and boils. In this scenario, a drum is generally placed on top of the waste heat boiler, linked with the shell, and where vapour accumulates. Conversely, in cases where the flue gas occupies the shell volume, the liquid circulates through the tubes and boils inside the tubes to create a two-phase flow. To design a waste heat boiler, the HEX analysis presented in Section 2.8.2 is used and in cases where the output temperature of the flue gas is fixed or assumed, the necessary heat transfer area can be estimated using the LMTD method. But, in cases where the outlet temperatures are not known, then the ε-NTU method is more suitable. However, for both methods, the overall heat transfer coefficient must be determined and in the specific case of WHBs, a correlation to estimate the boiling heat transfer coefficient must be used. Two types of boiling must also be differentiated. When a fluid is boiling without motion, boiling is called *pool* boiling – a typical example being water boiling in a pan. However, if the fluid is in motion when the phase change takes place, boiling is designated as *flow* boiling, in which case, heat transfer consists of a mixture of forced convective heat transfer together with two-phase heat transfer for the phase change. In the cases where the liquid occupies the shell volume and surround the tubes, heat is transferred inside the tubes from the hot gas to the solid tube material by forced convection and then through the tube thickness by conduction and if the liquid is not in motion, pool boiling takes place outside the tubes. Hence to estimate the overall heat transfer coefficient, a pool boiling heat transfer coefficient correlation must be used and currently, one of the most recommendable [142] pool boiling models is the *Rohsenow* [141] correlation, given by:

$$h_{boiling} = \left(\frac{q''_{boiling}}{i_{lv}}\right)^{1-r} \left[\mu_l / \sqrt{\frac{\sigma}{g(\rho_l - \rho_v)}}\right]^r \frac{c_{p,l}}{C_{sf}} Pr_l^{-s}$$

$$r = 1/3$$

$$\begin{cases} s = n = 1 & \text{for water} \\ s = n = 1.7 & \text{for other fluids} \end{cases} \tag{2.94}$$

where $q''_{boiling}$ is the heat flux per unit surface area (W/m^2), i_{lv} is the latent heat of vapourisation (J/kg), μ_l is the liquid dynamic viscosity (kg/m s), σ is the liquid surface tension (N/m), g is the gravitational acceleration (m/s^2), ρ_l and ρ_v are the density of the liquid and vapour phases, respectively, (kg/m^3), $c_{p,l}$ is the specific heat of the liquid (J/kg K), and $Pr_l = \mu c_p/k$ the Prandtl number of the liquid with k the liquid thermal conductivity (W/m K). The constant C_{sf} depends on the solid/fluid characteristics. In cases where the liquid circulates and boils inside the pipes, flow boiling occurs and again in such cases, suitable correlations must be chosen to estimate the flow boiling heat transfer coefficient. Similar to forced convection, flow boiling can be internal or external. With external flow boiling, the boiling heat transfer coefficient can be improved by the motion of the fluid and the mechanisms taking place are similar everywhere at the points of contact with the heat transfer surface. However, with internal flow boiling, the vapour cannot escape and so is forced to flow in the liquid phase. So as we move along the tube, the portion of vapour phase relative to the liquid phase increases. This is known as quality x and is defined as the ratio of the vapour mass flux to the total mass flux:

$$x = \frac{\dot{m}_v}{\dot{m}_{tot}} = \frac{\dot{m}_v}{\dot{m}_l + \dot{m}_v} \tag{2.95}$$

where $\dot{m}$ is the mass flux (kg/s) and the subscripts l, v, and *tot* referring to liquid, vapour, and total, respectively. As these quantities change inside the tube so the fluid dynamic evolves too, the two-phase flow development in a tube in addition to the evolution of the flow boiling heat transfer coefficient evolution is illustrated in Figure 2.42.

For which we can see that as the liquid enters the tube of the HEX, its temperature is below the saturation point and heat transfer takes place by liquid forced convection. As the fluid recovers thermal energy and approaches saturation conditions, small bubbles start to form near the wall where the liquid is hotter. A bubbly flow is forming and progressively increases the heat transfer coefficient. Even so, the temperature of the bulk of the liquid does not reach the saturation temperature. However, by the stirring action of bubbles forming and departing from the wall, the complete liquid volume reaches saturation and larger core bubbles develop. Then with a further increase in temperature, the bubble size increases and the bubbly flow transits to a slug flow where large bubbles occupy most of the available space and flow in unison with a volume of liquid, but once the vapour slugs are large enough to join the remaining liquid phase, it is forced to form a film near the wall and is constrained by the vapour that occupies the core of the tube. At this stage, heat transfer takes place from a hot wall to a thin liquid film which again significantly increases

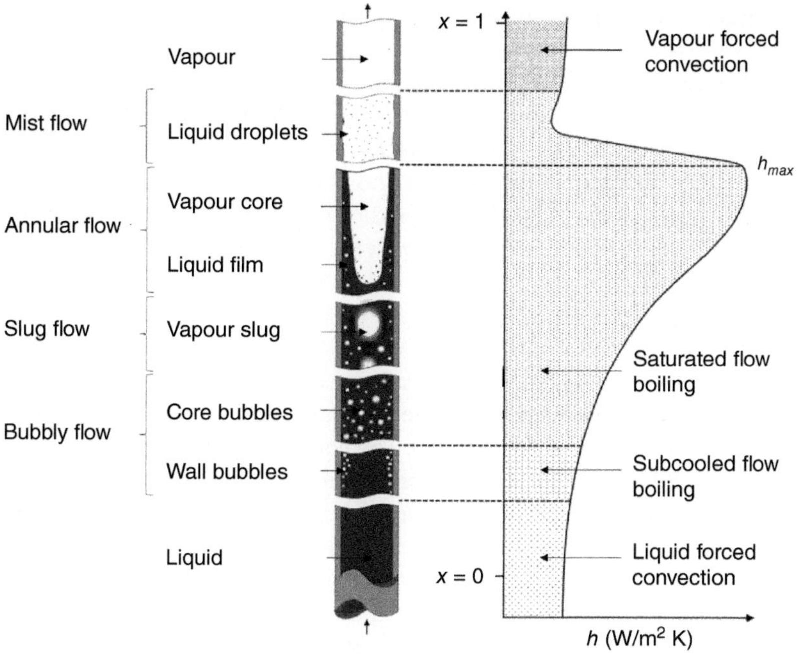

Figure 2.42 Flow regimes for forced convection boiling in a tube.

the heat transfer coefficient, and as the liquid film flows and evaporates, its thickness decreases which further enhances the heat transfer coefficient until a maximum value is obtained. Indeed, at some point, dry patches appear on the wall where evaporation of the film is no longer possible resulting in a drop in the local heat transfer coefficient because, at these points, only forced convection heat transfer between solid wall and vapour can take place. Finally, the remainder of the liquid film and droplets contained within the vapour volume evaporate and the flow quality reaches $x = 1$. At this point, only single-phase vapour forced convection heat transfer occur [136–138]. Hence, as you can imagine, estimating the flow boiling heat transfer coefficient can be laborious as it changes locally with the both type and quality of flow. For tubes where the flow type remains similar, consideration of a forced convection heat flux and then adding a pool boiling contribution can be a useful but crude first estimation. Alternatively, as different equations are made available for each flow type and are correlated in terms of fluid quality, one approximated solution is to take an average for the flow quality through the tube and to use a suitable correlation for that flow type. However, for more accurate solutions, you may need to consider an infinitesimal section of the pipe, describe the evolution of the quality x and the heat transfer coefficient on a differential volume, and then undertake an integral over the complete tube length for which that heat transfer coefficient correlation is valid.

Plus, depending on the specific characteristics of the waste heat boiler, additional design steps may also be required. For instance, some waste heat burners include an auxiliary burner unit or an afterburner to compensate for any potential insufficiency of waste heat [147]. Moreover, for some applications such as sulfur recovery units

(SRUs), the implementation of a waste heat boiler must take into account the fact that some oxidation reactions can be reversed. Specifically, as the gas passes through the waste heat boiler and is cooled, recombination reactions can take place and so change the gas composition and the heat transfer [146]. In the case of SRU units, the recombination reactions are mainly exothermic and so would create additional source of heat, but no matter whether the changes are exothermic or endothermic, changes in the flue gas composition must be predicted and are of importance in the choice of the HEX materials for a waste heat boiler.

2.8.5 Air Preheaters

With air preheaters, two technologies are typically used. With the plate type, the preheater consists of a plate-type HEX where the hot and cold fluid circulating in the HEX are separated by thin conducting plates. The equations to design a plate-type HEX are similar to those presented previously in the heat exchanger section; the LMTD method and ε-NTU method both being suitable as appropriate. However an additional characteristic of the plate-type HEX and hence the calculation of the forced convective heat transfer coefficient is that to improve the heat transfer, the plates are usually embossed or corrugated, and the type of corrugation leads to differences in the forced convective flow. Example of different plate corrugation are presented in Figure 2.43 [139].

For each type of plate, the corrugation increases the total heat transfer area of a plate, and so to consider this enhancement, a plate enlargement factor Φ is added, in the form [139]:

$$A_{tot} = \Phi \times w \times L \tag{2.96}$$

where A_{tot} is the total heat transfer area of a plate (m^2), w is the plate width (m), L is the plate length (m), and Φ is the plate area enlargement factor – usually in the range $1.15 < \Phi < 1.25$. In addition, in the design of a PHE, any additional friction

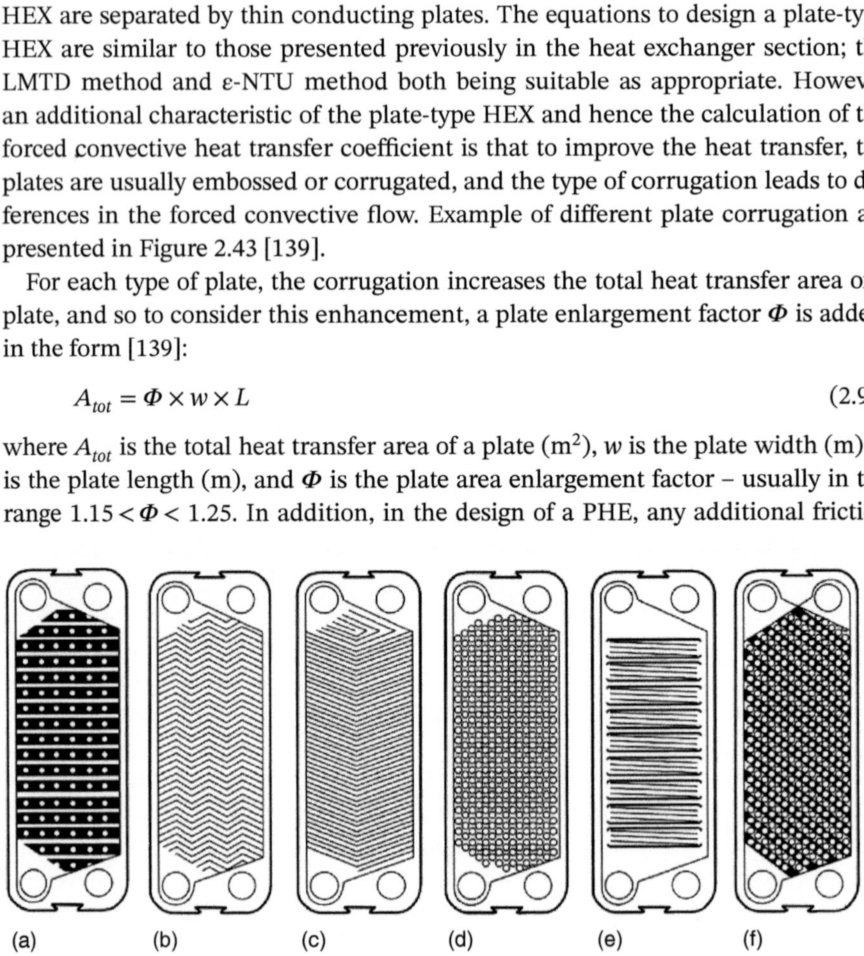

Figure 2.43 Different types of plate corrugation: (a) washboard, (b) zigzag, (c) chevron or herringbone, (d) protrusions and depressions, (e) washboard with secondary corrugation, (f) oblique washboard. Source: From Mota et al. [139].

caused by the corrugation also impacts the pressure drop and this must be balanced during the design stage.

In HPHEs, the exchanger can be analysed as two HEXs. At the heat source side, a first heat exchange takes place as the hot flue gas gives its thermal energy to the heat pipes. Then at the heat sink side, the heat pipes yields its thermal energy to the cold incoming air. Hence to size a heat pipe HEX, the LMTD method and ε-NTUs method can both be used but may require some specific adjustment. For example, to use the ε-NTU analysis method, the two sections (namely evaporator and condenser) must be analysed separately. And as heat pipe HEXs usually consist as a bank of heat pipes, to estimate the overall efficiency of the HPHE, the efficiency of one single row must first be determined. Now, in the case of a heat pipe-based HEX, since a phase change occurs inside the heat pipe, the temperature of the working fluid remains constant when heat is absorbed. Therefore, by definition, its specific heat c_p and heat capacity rate C approach infinity and as a result, the capacity ratio $C_r = C_{min}/C_{max}$ tends to zero and the effectiveness of a single row becomes [136]:

- Evaporator:

$$\varepsilon_{e,1\ row} = 1 - \exp(-NTU_e) \text{ with } NTU_e = \frac{U_e A_e}{C_e}$$

- Condenser:

$$\varepsilon_{c,1\ row} = 1 - \exp(-NTU_c) \text{ with } NTU_c = \frac{U_c A_c}{C_c} \tag{2.97}$$

where, in the above equations, the subscript e and c refer to evaporator and condenser, A represents the exchange area of one single row (m^2), C is the average heat capacity of the fluids (W/°C), and U is the overall heat transfer coefficient of the considered sections. Then to obtain the number of transfer units NTU for each section, the factor UA must be evaluated using forced convection correlations and for a second calculation, the effectiveness of the complete evaporator and complete condenser sections, with a number n_e and n_c of heat pipes rows, respectively, are determined. Again, as boiling and condensation of the working fluid inside the heat pipe takes place, the system can be studied as a boiler and condenser and the effectiveness of the total evaporator and condenser can be simplified by [137]:

- Evaporator:

$$\varepsilon_{e,n_e} = \frac{\left(\frac{1-C_{r,e}\varepsilon_{e,1row}}{1-\varepsilon_{e,1row}}\right)^{n_e} - 1}{\left(\frac{1-C_{r,e}\varepsilon_{e,1row}}{1-\varepsilon_{e,1row}}\right)^{n_e} - C_{r,e}} \xrightarrow{C_r=0} \varepsilon_{e,n_e} = 1 - (1 - \varepsilon_{e,1row})^{n_e}$$

- Condenser:

$$\varepsilon_{c,n_c} = \frac{\left(\frac{1-C_{r,c}\varepsilon_{c,1row}}{1-\varepsilon_{e,1row}}\right)^{n_c} - 1}{\left(\frac{1-C_{r,c}\varepsilon_{c,1row}}{1-\varepsilon_{c,1row}}\right)^{n_c} - C_{r,c}} \xrightarrow{C_r=0} \varepsilon_{c,n_c} = 1 - (1 - \varepsilon_{c,1row})^{n_c} \tag{2.98}$$

Finally with the overall effectiveness of the evaporator section ε_{e,n_e} and that of the condenser section ε_{c,n_c}, the total HEX effectiveness ε_{HPHE} can be determined from which fluid has the largest heat capacity rate [136, 137, 148], i.e.:

$$\text{If } C_c > C_e: \qquad \varepsilon_{HPHE} = \left(\frac{1}{\varepsilon_{e,n_e}} + \frac{C_e/C_c}{\varepsilon_{c,n_c}}\right)^{-1}$$

$$\text{If } C_e > C_c: \qquad \varepsilon_{HPHE} = \left(\frac{1}{\varepsilon_{c,n_c}} + \frac{C_c/C_e}{\varepsilon_{e,n_e}}\right)^{-1} \tag{2.99}$$

Using this process, the efficiency of the heat pipe-based HEX can be estimated, and then, as with other types of HEXs, the surface area can be determined.

For air preheaters, three main technologies are widespread: recuperators, regenerators, and RAC air preheaters. The RAC type presents similarities with heat pipe-based HEXs as this system comprises two different sections that need separate analysis. At the supply section, similar to the evaporator section for HPHE, the hot flue gas gives up its thermal energy to a fluid within the coil through a first HEX. Heat is then transported by the fluid through the coil and used to preheat the air at the second HEX. This section is designated as the exhaust section and is similar to the condenser section of heat pipe HEXs, and to make things easier, the two sections are commonly called the 'hot' HEX and 'cold' HEX. For each RAC system and depending on the application, the rate of circulation of the working fluid between the two HEXs is a critical to the design [149]. So setting the optimum rate of circulation is very important for attaining the highest possible system efficiency and minimum costs. RAC air preheaters can be symmetrical or asymmetrical, depending on whether the hot and cold HEXs are the same or not. For symmetrical RAC systems, the secondary fluid temperature distribution can be represented as [150] (Figure 2.44):

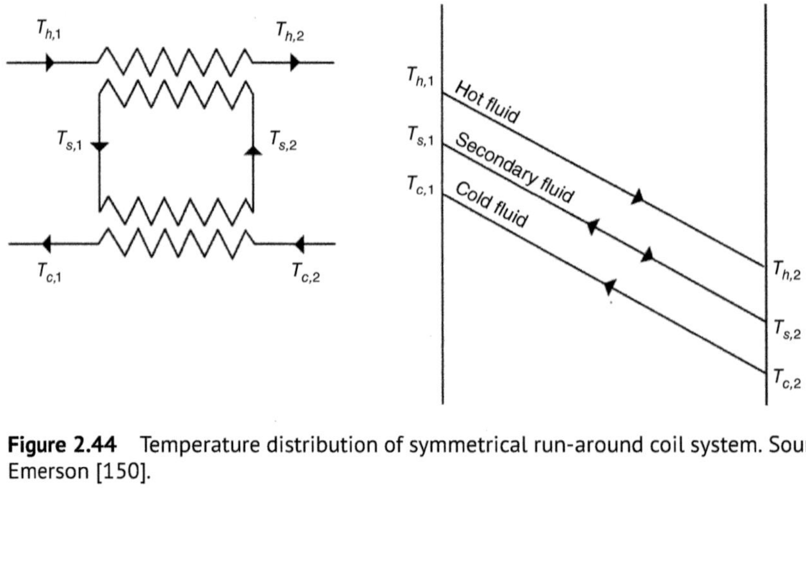

Figure 2.44 Temperature distribution of symmetrical run-around coil system. Source: From Emerson [150].

In this case (where the system is symmetric and both HEXs have the same size), the temperature of the secondary fluid circulating is given by [150]:

$$t_{s,1} = \frac{1}{2}(T_{h,1} + T_{c,1})$$
$$t_{s,2} = \frac{1}{2}(T_{h,2} + T_{c,2}) \tag{2.100}$$

where t_s is the temperature of the secondary fluid circulating (K), t_h is the temperature of the hot exhaust gas (K), and t_c the temperature of cold air to be preheated (K). Then from the previous temperatures of the secondary fluids, the mass circulation rate is related by a heat balance as:

$$\dot{m}_s = \frac{\dot{Q}}{c_{p,s}(T_{s,1} - T_{s,2})} \tag{2.101}$$

where $\dot{m}_s$ is the secondary fluid mass flow rate (kg/s), $\dot{Q}$ is the heat transfer rate (W), and $c_{p,s}$ the secondary fluid specific heat (J/kg K). For asymmetrical RAC systems, additional information regarding the two HEXs being used is required in order to estimate the optimum mass flow rate of the secondary fluid. But by balancing the cost of each HEX and their overall heat transfer coefficients, *Emerson* [150] found the optimum ratio of temperature differences β as:

$$\beta \equiv \frac{\Delta T_h}{\Delta T_h + \Delta T_c} = \frac{1}{1 + \sqrt{\left(\frac{p_c U_h}{p_h U_c}\right)}} \tag{2.102}$$

where ΔT_h and ΔT_c are the mean temperature differences in the hot and cold HEXs (K), p_h and p_c are the costs per unit area of heat transfer surface in the hot and cold HEXs (currency/m^2), and U_h and U_c are the overall heat transfer coefficients for both HEXs (W/m^2 K). Using which, the secondary fluid temperatures for an asymmetric RAC HEX can then be estimated from:

$$t_{s,1} = T_{h,1} - \beta(T_{h,1} - T_{c,1})$$
$$t_{s,2} = T_{h,2} - \beta(T_{h,2} - T_{c,2}) \tag{2.103}$$

Finally, the optimum mass transfer rate of the secondary fluid for an asymmetrical RAC air preheater can be estimated using Eq. (2.101). In addition to the HEX design methods presented earlier in this section, with RAC systems, the designer may also need to consider heat losses from the secondary fluid. Obviously, this phenomenon is amplified with increasing distance between the hot and cold HEXs and poor insulation of the secondary piping.

With recuperator air preheaters, cold air is circulating outside of a cylinder in which the hot flue gas flows. So before it can be used, thermal energy is first given to the cylinder walls and only then retransmitted to the air. However, in addition to forced convection heat transfer taking place, recuperators typically also make use of radiation to transmit heat from the hot surface to the air, and in the cases where the recuperator air preheater consists of two concentric tubes, radiative heat transfer not only takes place between the hot inner cylinder to the air circulating, but also between the outer cylinder wall and the air. Indeed, a significant part of the radiation emitted by the inner surface is received by the outer cylinder wall and scattered. Therefore, to design recuperator preheaters, a basic knowledge of radiative heat transfer is required.

Radiative bodies such as surfaces and gases are usually compared to an ideal radiative case known as the blackbody model. A blackbody is a perfect absorber and emitter of radiation and at any given temperature, the total emitted energy from the blackbody is the maximum possible. In practice, surfaces radiate less energy than black bodies and are therefore designated as grey bodies and to characterise the radiation intensity of a grey body, a total hemispherical emissivity factor (ε_s) is introduced which represents the ratio of the radiation emitted by the grey body at a given temperature to the radiation that would be emitted by a perfect blackbody at the same temperature. Now the radiative heat flux emitted by a surface per unit surface area is designated as irradiation (G) and when a surface or semi-transparent body receives a given quantity of irradiation, (say G), this radiative heat flux is divided into three components: one portion is reflected, one portion is absorbed, and the rest is transmitted. So to help describe this effect, three ratios are introduced, namely the reflectivity ρ_s, the absorptivity α_s, and the transmissivity τ_s and these are given by:

$$\rho_s = \frac{G_{ref}}{G}$$
$$\alpha_s = \frac{G_{abs}}{G}$$
$$\tau_s = \frac{G_{trs}}{G} \tag{2.104}$$

The reflectivity, absorptivity, and transmissivity of a body are always constrained between 0 and 1, and in the case of recuperator air preheaters, the air gas flowing between the cylindrical surfaces also participates in the absorption and emission process because it contains asymmetric molecules such as H_2O and CO_2 [151, 152]. Hence, the reflected, absorbed, and transmitted irradiation from the gas can be schematised as per Figure 2.45.

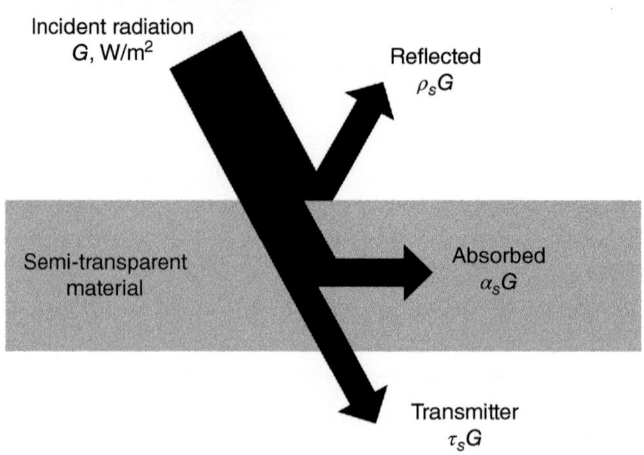

Figure 2.45 Reflection, absorption, and transmission of radiation by a semi-transparent material.

In practice, the incident radiation that a material receives (such as the gas in the case of air preheater) comes from different sources, different directions, and with different wavelengths. Accordingly, the reflectivity, absorptivity, and transmissivity of the body can vary locally, but for simplicity, these characteristics are usually averaged on a whole body consideration and where if these properties are independent of the wavelength of the received radiation, the body is designated 'grey'.

When radiative heat transfer occurs between two surfaces, the orientation of the two surfaces influences the transfer and the radiation properties. To consider this effect, a view factor ($F_{i\rightarrow j}$) is introduced which represents the fraction of the radiation leaving a surface i that then strikes a surface j. So, in the general case, the view factor of a surface 1 to a surface 2 is given by [137]:

$$F_{1\rightarrow 2} = \frac{1}{A_1}\int_{A_1}\int_{A_2}\frac{\cos\theta_1\cos\theta_2}{\pi r^2}dA_1 dA_2 \tag{2.105}$$

where r is the distance between the surfaces dA_1 and dA_2 (m), and θ_1 and θ_2 are the angles between the normals to their respective surfaces and the line connecting dA_1 and dA_2. Note that in the case of cylinders and concave surfaces, a surface can have a view factor relative to itself. Also, because the view factor can be considered from the viewpoint of both surfaces A_1 and A_2, a reciprocity for view factors is also observed as:

$$A_1F_{1\rightarrow 2} = A_2F_{2\rightarrow 1} \tag{2.106}$$

If a surface completely surrounds the other, the view factor is unity. However, most of the time, the view factor between surfaces and bodies must be estimated and estimating the view factor between surfaces and between the bodies implied in the heat transfer process can be difficult. Correlations and equations reported in the literature for the simplest cases may be used or adapted and so for cylindrical recuperator's air preheaters, it is possible to consider these as two cylinders between which the air flows and is preheated [136–138].

Using an electrical analogy, it is often convenient to write the radiative heat transfer rate as the ratio of a difference in potential to a radiation resistance. To do so, the radiosity (J_i) is introduced and which represents the total radiative energy leaving a surface i per unit area per unit of time. Using this analogy, we can consider that the radiosity from a surface consists of both the emitted radiative energy and the reflected radiative energy from that surface, and so, based on the concept of view factor and radiosity, the radiative heat transfer rate from a surface i to a surface j can be written as:

$$\dot{Q}_{i\rightarrow j} = \frac{J_i - J_j}{R_{i\rightarrow j}} \tag{2.107}$$

where $\dot{Q}_{i\rightarrow j}$ is the radiative heat transfer rate from surface 'i' to surface j (W), J_i and J_j are the respective radiosities (W/m^2), and $R_{i\rightarrow j}$ is the space radiative resistance (m^{-2}) defined by:

$$R_{i\rightarrow j} = \frac{1}{A_iF_{i\rightarrow j}} \tag{2.108}$$

and where A_i is the surface i (m^2) and $F_{i\rightarrow j}$ the view factor from surface i to j. For the situation where several surfaces or bodies exchange radiative heat transfer, the problem can be solved using standard electrical network calculation employing the previously described potentials and resistances. So for surfaces with a specific heat transfer rate, the net heat transfer rate $\dot{Q}_i$ from a surface i to a number N of surfaces can be given by [136]:

$$\dot{Q}_i = A_i \sum_{j=1}^{N} F_{i\rightarrow j}(J_i - J_j) \tag{2.109}$$

Whereas for surfaces with specific temperatures, this temperature T_i can be obtained from:

$$\sigma_{sb} T_i^{\,4} = J_i + \frac{1-\varepsilon_{s,i}}{\varepsilon_{s,i}} \sum_{j=1}^{N} F_{i\rightarrow j}(J_i - J_j) \tag{2.110}$$

where $\sigma_{sb} \approx 5.67\times10^{-8}$ W/m^2 K^4 and is the Stefan–Boltzmann constant, and $\varepsilon_{s,i}$ is the surface emissivity. Hence using the above equations for all the bodies and surfaces implied in a radiative heat transfer between N number of surfaces and so N radiosity transfers, these can then be solved for that system. For gas and gas mixture especially, the emissivity has been reported in existing literature and data tables.

Also, we need to remember that for regenerators and rotary regenerators, the system design is based on transient conduction principles. Therefore, transient conduction laws described in the regenerative burner section need to be applied.

2.8.6 Heat Recovery Steam Generator

In HRSGs, energy from hot exhaust gas coming from burners is recovered in order to produce steam – the hot gas successively passing through three exchangers, namely the superheater, the evaporator, and the economiser. As for the water line, water circulates in the opposite direction, and to better understand the heat transfer process taking place within the water circuit, the temperature–enthalpy diagram of the process is illustrated in Figure 2.46.

Pressurised water liquid is first supplied by a feed pump to the economiser where it is heated but remains below the saturation point (State A to B). The warm water liquid is then introduced to the evaporator where it reaches its saturation point, after which boiling takes place and vapour is created. During this process a phase change takes place and the energy provided by the HRSG is completely used for the mass transfer from the liquid phase to the vapour phase. Note also that during the phase change the temperature of the system remains constant (State B to C). The liquid and gaseous phases are then separated in a drum and the vapour is sent to the superheater where it is heated further – this stage being mainly used to bring the vapour to the desired conditions to match user requirement and, in the case where steam is used to drive a turbine, maximise the efficiency for electricity production. The presented stages are common to all HRSGs. However, depending on the application, different options can be added to a steam generator for further thermal treatment of the steam. For instance, following initial use by a process, some

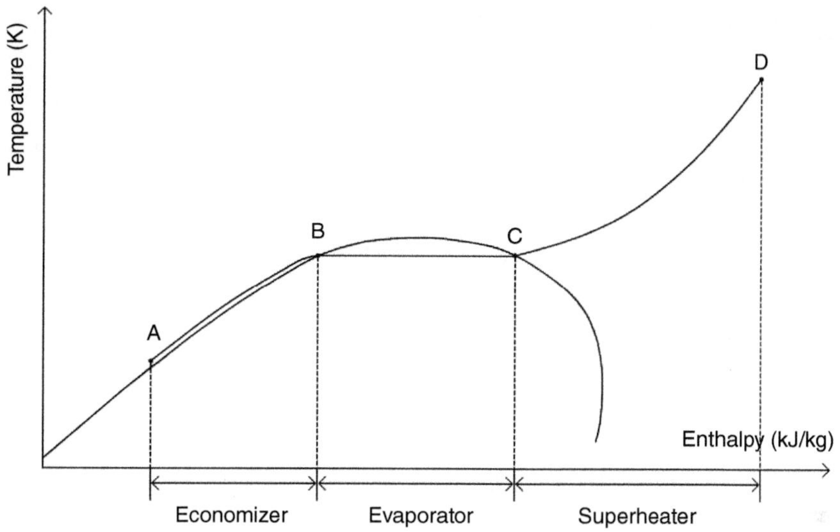

Figure 2.46 Temperature–enthalpy diagram of water during a heat recovery steam generator process.

HRSGs may also include a reheater to reheat and so reuse the steam. In addition, depending on the number of drums in the boiler, the HRSG can be a single-, dual-, or triple-pressure type, the advantage of adding several drums in a HRSG being to change the pressure in each line, which then changes the boiling point of the water. Hence, in a triple-pressure-type HRSG, a high-pressure, intermediate-pressure, and low-pressure sections are used to increase the overall performance.

For design purposes, the pressure chosen is highly significant and is connected to the thermal cycle operated during the process. So to design a HRSG, several design steps are usually recommended. Firstly, the need for the steam, its application, and thus the requirements expected of the steam must be identified. Then, depending upon the use for the steam, the quality, temperature, and pressure needed from the HRSG may differ and so impact the sizing of the equipment. In such considerations, both short-term and long-term requirements must be anticipated and it is common practice to over produce and allow excess steam to be created to ensure that the minimum amount of steam required can always be supplied. Nevertheless, even where accurate predictions are aimed at such generalised requirements, whenever possible, designs should remain relatively simple so as to limit the cost of an installation [153].

The design stages can generally be divided into preliminary design, detailed design, and final design. During the preliminary design, the objective is to rapidly provide the first concept for the system without involving numerous resources. The mass and material balance of the water and steam circuit providing a first idea for the required heat transfer rates. Where applicable, the designer may also extrapolate the new design from existing systems which then allows for a better estimate for the size, performance, and cost for a new HRSG system. Then for detailed design, the

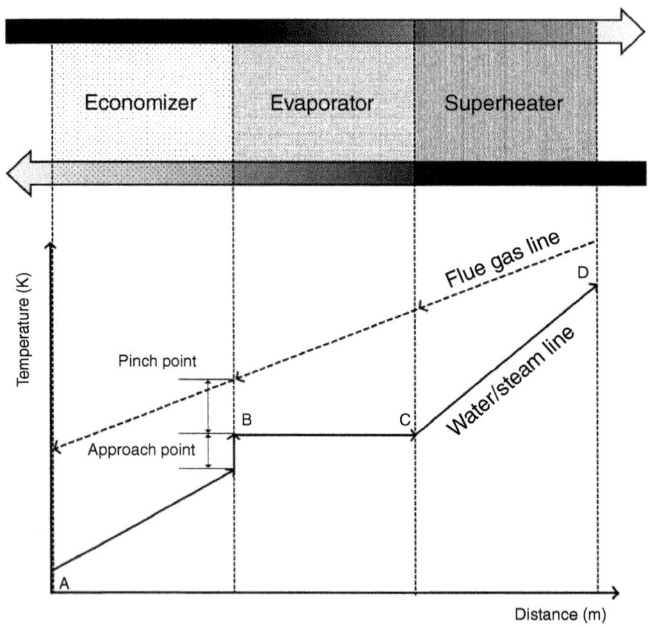

Figure 2.47 Schematic analysis of a HRSG and temperature profile.

first step is to better characterise the steam, the hot exhaust fluid, and the working environment. The composition and temperature range of the exhaust gas is crucial as the chemical particles within the gas may affect the performance of the HEX and react with the materials chosen. However, at this stage will also aim to design the heat transfer surfaces, optimise the surface area, and improve the compactness of the system whilst still reaching the desired performance. The stage is crucial and is a key challenge in the design of an installation, and as a first approach, the economiser, evaporator, and superheater can be considered as three counter-current HEXs operating in series. A schematic analysis of a HRSG and its temperature profile are presented in Figure 2.47.

The temperature profiles of the flue gas line and of the water/steam line are linked by the heat balance equation:

$$\dot{Q} = \dot{Q}_h = \dot{Q}_c = \dot{m}_h(i_{h,in} - i_{h,out}) = \dot{m}_c(i_{c,out} - i_{c,in}) \tag{2.111}$$

where $\dot{Q}$ is the heat transfer rate (W), $\dot{m}_h$ and $\dot{m}_c$ are the hot and cold fluid mass flow rates (kg/s), and i the enthalpy (kJ/kg). In HRSGs, two parameters are of significant importance: The pinch point and the approach point and these two points can be measured at the evaporator section. At the exit of the economiser, the water is in its liquid stage at a temperature lower than the saturation temperature and the difference between this temperature and the saturation temperature of water is called the approach point. The approach point therefore describes the amount of energy required by the evaporator before the water liquid phase turns to steam, and, not surprisingly, designers typically try to reduce the approach point as much as possible whilst still maintaining a safe margin to ensure that the water remains liquid within the circuit before the evaporator. On the other hand, the pinch point is the temperature difference between the flue gas leaving the evaporator and the saturation temperature and this point represents the location where the design is most constrained. To limit costs and maximise the performance of a HRSG, pinch and approach point are normally in the range of 10–15 °C [59]. So in practice, the pinch and approach points are often the starting points for thermal calculations and it is not uncommon to first set and assume both the pinch and approach points and only then estimate the temperatures and size of the system.

Now since these systems are typical HEXs, the LMTD and ε-NTU method are both central in the thermal design of HRSGs, but as the outlet temperatures of the streams are not known, the area calculation is typically estimated using the ε-NTUs technique with the equation:

$$A = \frac{NTU \times C_{min}}{U} \tag{2.112}$$

where A is the heat transfer area (m^2), NTU is the number of transfer units, C_{min} is the minimum heat capacity rate (W/K), and U the overall heat transfer coefficient (W/m^2 K).

To use the ε-NTU method, the overall heat transfer coefficient of the HEX is required, but this can be determined using an appropriate equivalent thermal resistance network and suitable conduction, convection, and radiation heat transfer correlations. However, within the evaporator, boiling takes place and so two-phase heat transfer must be considered for the prediction of the HRSG performance. To estimate the pool boiling heat transfer coefficient, the pool boiling model by *Rohsenow* [141] is recommended. Once the overall heat transfer coefficient of the HEXs is estimated, the outlet temperatures of both fluids and the size of the HEX can be estimated using the ε-NTUs method.

Optimum system geometries and arrangement will also be studied using heat transfer analysis, using different simulations and considering several scenarios. Finally, a control for the steam quality and temperature may also be relevant depending on the desired steam requirements. The overheating of steam and its cooling using an attemperator is a common practice to reach the expected steam conditions (the attemperator sprays water droplets into the superheated steam to ensure that the downstream, mixed, steam temperature will not adversely affect downstream components). A second step is generally the fluid mechanical design where the designer selects the number of drums, the pumps, fans, and blowers. And a critical choice in the design of HRSGs is the number of drums and therefore the number of pressurised water lines. Depending on the number of drums, a HRSG can be single, dual, or triple pressure type. Alternatively and conversely to studying the number of drums and hence the assisted circulation of water within their pressurised sections, the designer may also consider using natural circulation to reduce the cost of the installation. Whereas assisted circulation HRSGs mainly have a vertical gas path, for natural circulation, the exhaust gas path is generally horizontal. Therefore, depending on the available space, the implementation of the steam generator that requires a high vertical chimney or a long horizontal ducting may not be practical. Whereas although natural circulation HRSGs have the advantages of being more economical, use less energy and more reliable and safer in earthquakes zones, their start-up times from cold are usually longer and greater maintenance is required. Also in the selection of fans and blowers, pressure drop most be considered and a similar analysis should be undertaken for the water and steam lines to estimate the pressure drop and head required for their pumps. Last but not least, within this stage the two-phase flow must be considered in relation to sizing and this may also require an analysis of the natural convection streams relevant to protect zones where steam can naturally accumulate. Perhaps it is obvious, but it needs to be stated that in all cases of equipment selection, standard sizes are advised to reduce the cost.

In the design of HRSG, different types of evaporators can be selected. The D-frame evaporator is the most compact form of evaporator and is therefore best suited for small turbines or for restricted areas as opposed to O-frame evaporators which are advised for larger gas flows. Alternatively, the I-frame evaporator is relatively economical and adaptable which has made them very popular. In addition, it is common to choose the number of rows in the tube bundles and the orientation of parallel modules if the I-frame evaporator is constructed from several modules. In the mechanical design stage, tube materials and their thickness are of interest – the tube materials and size usually being determined according to their operating temperature and pressure range within the system. However, if tubes need to be sized according to their structural or mechanical requirements, a balance must be found because thicker tubes deteriorate the heat transfer performances. In addition, tube supports are also sized according to a stress analysis and the loads that need to be supported. Furthermore, regarding the design of supports and enclosures, environmental factors such as wind, seismic risk, and soil condition must be included. Headers and drums are also sized during the mechanical design stage but in addition, the composition of the exhaust gas and its possible interaction with any chosen solid materials must be considered.

Indeed, during the heat transfer process, particles from the exhaust gas can accumulate, decrease the heat transfer performance, and, in the worst cases, damage the material. In particular, attention must be made to ensure that the design keeps the flue gas above its dew point to prevent condensation taking place within the HEX because condensation and formation of liquid from an exhaust gas can lead to the appearance of acids that would threaten the complete installation.

Lastly, in the final design step, all the previous steps need to be reviewed and detailed further, manufacturing drawings are generated, and the final design is tested in simulations according to their different working conditions. Amongst others, the extent of emission levels need to be checked.

2.8.7 Heat Pumps

In heat pumps, a thermodynamic cycle is used to transport energy from one location to another. This is achieved by the means of a working fluid travelling from a hot exchanger where heat is recovered to a cold exchanger where the heat is released, and in vapour compression heat pumps, the working fluid used to transport this heat exists in two different phases. At the evaporator HEX, the liquid phase of the working fluid evaporates and turns to vapour, but conversely, at the condenser, the vapour condenses back to the liquid phase. In the context of WHR, the evaporator HEX needs to be placed within the hot exhaust stream from which the thermal energy is to be recovered. Whereas the location of the condenser HEX depends on the end use for this recovered heat. For example, in cases where the cold HEX is used to preheat an air stream, the vapour compression heat pump system looks similar to the RAC air preheater system previously discussed, but heat pumps are widely used due to the higher efficiencies explained by the nature their two phases of working fluid flow. Indeed, the evaporation and condensation heat transfer coefficients are significantly higher than those of forced convection. Also, another major difference for heat pumps is their use of a compressor and an expansion valve that provides a thermodynamic treatment of the working fluid and assures circulation of the system fluid. So to design vapour compression heat pumps, the thermodynamic cycle on which the system rely should be studied, a schematised explanation of which is presented in Figure 2.48.

At the evaporator HEX, the liquid phase of the working fluid absorbs heat from the flue gas and transforms to saturated vapour. Even so, this vapour remains at

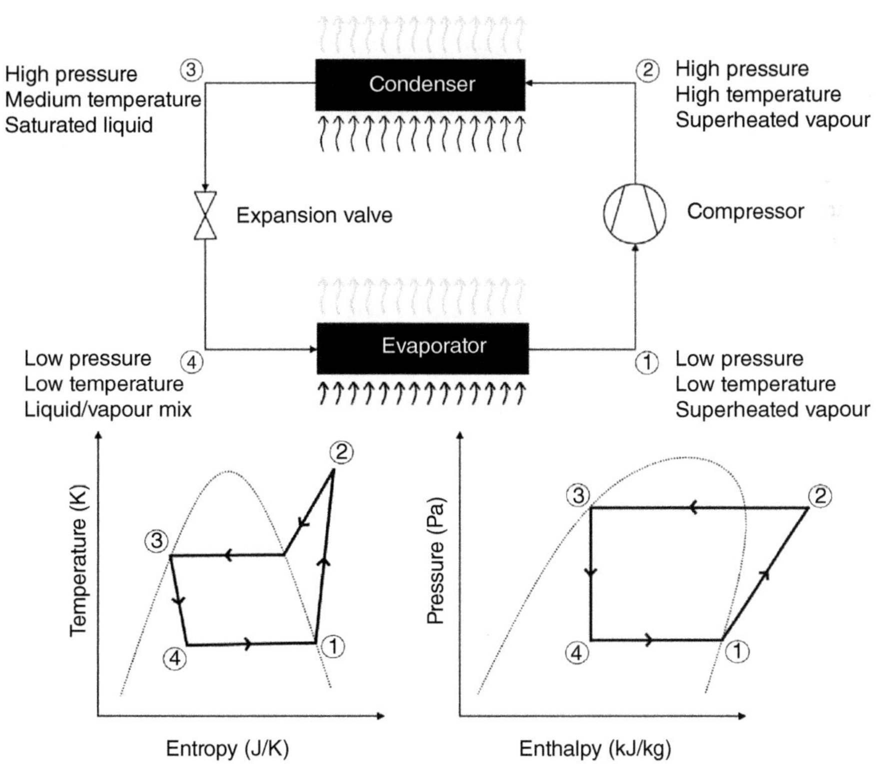

Figure 2.48 Thermodynamic principle of heat pumps.

a relatively low temperature and pressure (State 1). Then to ensure circulation of the working fluid and improve heat transfer, a compressor increases the vapour pressure and thus the temperature. At this point electrical power is supplied to the system and the work undertaken by the compressor is conveyed as additional energy. At the outlet from the compressor, the high-pressure vapour is superheated and reached the highest temperature of the cycle (State 2). Then, a high-pressure line leads the vapour to the condenser where the waste heat is released by condensation of the working fluid inside another HEX. The working fluid remains at a high pressure, but its temperature decreases as thermal energy is transferred. Leaving the condenser, the working fluid exists in a liquid phase at high pressure and medium temperature (State 3), and finally, the working fluid moves from the high-pressure line to the low-pressure line by passing through the expansion valve. In doing so, the decrease of pressure brings a decrease of temperature. This can be seen on the temperature–entropy diagram and during this transition (State 3 to 4), a portion of liquid may turn to vapour. Hence, after the expansion valve, the working fluid is a liquid/vapour mixture at low pressure and low temperature (State 4).

From the described thermodynamic principles of heat pumps, several design limits can be identified. To begin with, at the evaporator, the saturation temperature of the working fluid at low pressure must be lower than the flue gas temperature for evaporation to take place. Similarly, at the condenser, the pressure delivered by the compressor must be set so that the condensation temperature of the refrigerant at this high pressure is higher than the temperature of the cold fluid going through the condenser [154]. In addition, during the transition from high-pressure liquid to low pressure within the expansion valve, vapour can appear which, if in significant amounts at this stage, can reduce the performance of the heat pump because a part of the working fluid will already exist in the vapour phase before going through the evaporator. To reduce this, heat pumps can be designed so that they sub-cool the liquid after the condenser and by doing so, reduce pressure of the working fluid within the expansion valve and so eliminate vapour formation allowing the entire liquid volume to then evaporate within the hot HEX [154].

To estimate the performance of a heat pump, an important efficiency measure for heat pumps is the COP. This coefficient represents the ratio of the useful output energy to the energy supplied to the system in the form of work. The COP is given by:

$$\text{COP} = \frac{\text{useful output}}{\text{power input}} = \frac{\dot{Q}_{condenser}}{W_{provided}} \tag{2.113}$$

where $\dot{Q}_{cooling}$ is the heat transfer rate recovered at the condenser (W) and $W_{provided}$ is the work provided to the system in the form of electrical power (W). For a heat pump system to be economically viable, the COP should be higher than 2.5 [154], and in the case of an ideal thermodynamic cycle (as presented in Figure 2.48.), the COP can be expressed according the enthalpy of the working fluid in its different states as:

$$\text{COP} = \frac{i_2 - i_3}{i_2 - i_1} \tag{2.114}$$

where i is the enthalpy (kJ/kg), and the subscripts 1, 2, 3 refer to the working fluid states. Similarly, the evaporator cooling capacity ($\dot{q}_{evap}$) is defined by [155]:

$$\dot{q}_{evap} = i_1 - i_4 \tag{2.115}$$

the heat released at the condenser ($\dot{q}_{cooling}$) by [155]:

$$\dot{q}_{condenser} = i_1 - i_4 \tag{2.116}$$

and the work provided by the compressor ($w_{provided}$) as [155]:

$$w_{provided} = i_2 - i_1 \tag{2.117}$$

The enthalpy noted can be read from the phase diagram of the working fluid selected. To check the equilibrium of the system at steady state, the amount of energy given to the system must be equal to the energy leaving it – for which the following equation is appropriate:

$$\dot{Q}_{evaporator} + W_{provided} = \dot{Q}_{condenser} \tag{2.118}$$

where $\dot{Q}_{evaporator}$ is the heat transfer rate from the hot flue gas to the working fluid at the evaporator (W), $W_{provided}$ is the work provided to the system by the compressor (W), and $\dot{Q}_{condenser}$ the heat transfer rate that is recovered at the condenser (W).

However, in addition to the HEX theories used to size the evaporator and condenser HEXs, several other design characteristics of heat pumps are also important to develop a heat pump system – the most important component of which is the compressor, and this needs to be appropriately chosen according to the pressure and temperature needed at the condenser. The objective is to ensure that the condensation temperature of the refrigerant is higher than the inlet temperature of the fluid used to recover heat. In addition, the choice of the compressor must be balanced against the relatively long life of heat pump systems and their operating conditions of noise, vibration, and the frequency of necessary maintenance [154]. The compressor is very decisive in the performance of a heat pump and is the element with the largest impact on the COP figure. In heat pumps, contrary to many refrigeration systems, the electric motor does not need to be oversized as it is usually underloaded [154]. Then the choice of the refrigerant is usually made by considering its suitability for the condensing pressure of the working fluid and the temperature of the fluid used to recover the thermal energy. Finally the piping should aim at reduce the flow resistance.

2.8.8 Direct Electrical Conversion Device

Direct electrical conversion devices create electricity directly from a difference of temperature applied to their surface and in the context of WHR, hot flue gas can be used to heat one side of a direct electrical device which then produces electricity. Direct electrical conversion devices use four different technologies for this electricity generation: thermoelectric, piezoelectric, thermionic, and thermophotovoltaic (TPV), and since the design of such systems depend on their operation, these four categories are discussed separately in the following paragraphs:

With thermoelectric generation, a device made from semiconductor material generates electricity when its surfaces are exposed to different temperatures. Using the effect noted by Seebeck, heated electrons travel to the cooler side of the material and thereby generate a current. The current density (J_{el}) from a module is a function of the gradient of temperature applied and is given by:

$$J_{el} = \sigma_{el}(-\nabla V + E_{emf}) \tag{2.119}$$

$$E_{emf} = -S\nabla \mathrm{T} \tag{2.120}$$

where σ_{el} is the local electrical conductivity (S/m), V is the local voltage (V), E_{emf} is the electromagnetic field, S is the Seebeck coefficient (V/K), and T the temperature (K). To provide a measure of thermoelectric performance for these generators, a thermoelectric figure of merit (zT) has been introduced as given by:

$$zT = \frac{S^2 \sigma_{el} T}{k} \tag{2.121}$$

where k is the thermal conductivity of the TEG material (W/m K). Then from this thermoelectric figure of merit, the efficiency of the TEM (η_{TEG}) can be defined as [156–158]:

$$\eta_{TEG} = \left[\frac{(T_h - T_c)}{T_h}\right]\left[\frac{\sqrt{1+zT}-1}{\sqrt{1+zT}+\frac{T_c}{T_h}}\right] \tag{2.122}$$

where T_h is the temperature of the hot side of the TEG (K), T_c is the temperature of the cold side of the TEG (K), and zT the thermoelectric figure of merit. In the design of TEG, we need to be aware that these modules have relatively low efficiencies (5% [6]) and a high conversion resistance in the range of 1.5 °C/W [157, 158]. Therefore, the temperature gradients imposed on TEGs need to be relatively high. In this regard, the choice of material must also consider the difficult mechanical and thermal conditions in which these modules operate, and so the stresses and fatigue induced within the materials. Hence to improve their power output, TEGs are generally placed in series, and since the heat sink and heat source are usually separated, it is common to use high thermal conductivity devices to transport the thermal energy from the heat source to the TEG. For instance, heat pipes can be used to recover waste heat from a flue gas and convey this energy directly to a TEG module [157].

In piezoelectric power generation, vibration of a material (such as a membrane) is used and converted to electricity [159], and in the context of WHR, the waste heat uses a thermoacoustic–piezoelectric (TAP) device to create the vibrations from the temperature gradient [160]. This device consists of five sections: the heat cavity, the stack, a resonator tube, a Helmholtz resonator cavity, and the piezoelectric diaphragm. When heat is applied to the heat cavity, a temperature gradient is generated within the stack which releases acoustic waves. These acoustic waves are then amplified by the resonator and used to create electricity from the piezoelectric section. The governing piezoelectric equation for the coupling between the acoustic and electrical fields is given by [160]:

$$\begin{Bmatrix} \frac{x_{center}(i\omega)}{q_{elec}(i\omega)} \end{Bmatrix} = \begin{bmatrix} Y_{11}(i\omega) & Y_{12}(i\omega) \\ Y_{21}(i\omega) & Y_{22}(i\omega) \end{bmatrix} \begin{Bmatrix} \frac{\Delta P(i\omega)}{V(i\omega)} \end{Bmatrix} \tag{2.123}$$

where x_{center} is the centre point deflection depending on the piezo-diagram frequency (m), ω is the frequency of the oscillation (Hz), q_{elec} is the charge (C), Y is the admittance (S), ΔP is the pressure across the element (Pa), V is the voltage (V), Y_{11} is the acoustic admittance, Y_{22} is the electrical admittance, and Y_{12} and Y_{21} are the coupled electroacoustic admittances. It should be noted that piezoelectric power generation from WHR systems is still at an early stage and are currently being developed to provide better conversion performances. This means that piezoelectric power generators can only be profitable by using high-power generation capacity piezoelectric material, but high reliability and good stability are also necessary to decrease maintenance costs and allow for easy control of the piezoelectric power generator [161]. Moreover, the resonance frequency of the self-charging assembly needs to be studied and adapted to fit the excitation frequency and so maximise the piezoelectric power output. When doing so, whilst the material must produce high piezoelectric strain, its flexibility must also be considered. Indeed, due to its constant exposure to vibration, the material is subjected to constant mechanical forces and so its behaviour with regard to mechanical fatigue is yet another important design criterion.

Thermionic generators consist of a hot cathode and a cold anode, respectively, called the emitter and the collector. When a heat source is applied to the emitter, a flow of electrons is generated which move from the emitter to the collector in contact with a heat sink. Simplistically, thermionic generators can be considered as a diode that will conduct and produce current in one direction, but only when waste heat is received. These devices can therefore be considered as thermal engines where the source of power is the temperature difference between the cathode and the anode and where the efficiency of generator (η) is always less than:

$$\eta < 1 - \frac{T_a}{T_c} \tag{2.124}$$

where T_a and T_c are the anode and the cathode temperatures (K), respectively. In practice however, radiative heat transfer can also take place between the cathode and anode which reduces the thermionic generation efficiency. In which case, taking this radiative loss in account, the maximum realisable efficiency of a thermionic generator (η_{max}) can be obtained from [162]:

$$\eta_{max} = \frac{j_c(e_c - e_a)}{j_c e_c + P_{rad}} = \eta_{carnot} / \left(1 + \frac{P_{rad}}{j_c e_c}\right) \tag{2.125}$$

$$\eta_{carnot} = 1 - \frac{e_a}{e_c} \tag{2.126}$$

where j_c is the cathode current density (A/m^2), e_c and e_a are the cathode and anode potentials (V) and P_{rad} is the dissipated radiation power (W). The losses due to radiation can be difficult to estimate but will never exceed the power loss given by the Stefan–Boltzmann law from:

$$P_r \leq \varepsilon_{s,c} \sigma_{sb} T_c^{\ 4} \tag{2.127}$$

where $\varepsilon_{s,c}$ is the total emissivity of the cathode surface, $\sigma_{sb} \approx 5.67 \times 10^{-8}$ W/m^2 K^4 is the Stefan–Boltzmann constant, and T_c the cathode surface temperature (K).

Similar to solar panels, TPV generators rely on the production of electricity from radiation. So to create the necessary radiation from waste heat, an emitter is necessary which then transforms the temperature gradient into a radiative source for use by the photovoltaic module to produce electricity. For the design of TPV generators, radiation theory is used, the exchanged power $\dot{Q}$ from the emitter to the photovoltaic cell being given by [163]:

$$\dot{Q} = \int_{\lambda_{min}}^{\lambda_{max}} \dot{Q}(\lambda)d\lambda = \int_{\lambda_{min}}^{\lambda_{max}} \frac{E(\lambda, T_{emitter}) - E(\lambda, T_{PV})}{\frac{1-\varepsilon_{s,emitter}(\lambda)}{\varepsilon_{s,emitter}(\lambda)} + \frac{1}{F_{emitter\to PV}} + \frac{1-\varepsilon_{s,PV}(\lambda)}{\varepsilon_{s,PV}(\lambda)}} d\lambda \tag{2.128}$$

where λ is the radiation wavelength (m), $E(\lambda,T)$ is the spectral blackbody radiated power (W) from a surface at a temperature T (K), ε_s is the surface emissivity, and $F_{emmiter\to PV}$ the view factor between the emitter and the photovoltaic cell, and which can be obtained from [164, 165]:

$$\begin{aligned} F_{emitter\to PV} = \frac{2}{\pi\overline{x}\overline{y}} \Bigg\{ & \ln \left[\frac{(1+\overline{x}^2)(1+\overline{y}^2)}{1+\overline{x}^2+\overline{y}^2} \right]^{1/2} + \overline{x}(1+\overline{y}^2)^{1/2} \\ & \arctan\left(\frac{\overline{x}}{(1+\overline{y}^2)^{1/2}} \right) + \overline{y}(1+\overline{x}^2)^{1/2} \arctan\left(\frac{\overline{y}}{(1+\overline{x}^2)^{1/2}} \right) \\ & -\overline{x}\arctan\overline{x} - \overline{y}\arctan\overline{y} \Bigg\} \end{aligned} \tag{2.129}$$

where x, y is the width and length of the emitter (m), D is the distance between the emitter, and the PV cell (m) and the normalised parameters $\overline{x}$ and $\overline{y}$ are defined as follows:

$$\overline{x} = \frac{x}{D} \tag{2.130}$$

$$\overline{y} = \frac{y}{D} \tag{2.131}$$

In practice, direct electrical conversion devices are not widely used due to their low efficiencies [6], and this is mainly explained by the numerous energy transformations that involve different loss factors. Furthermore, these systems require high temperature differences in order to produce a significant quantity of electrical energy.

2.8.9 Thermal Storage

The design of TES is simpler than many WHR systems because theoretically the mechanisms are relatively simple – the main challenges are more of a technical nature of reaching optimal performance by the choice of suitable materials. The selection requiring compromises between mechanical strength, storage and thermal properties of the enclosure, the thermal conductivity and heat capacity of the heat exchange media, the operating temperatures and pressures, and the like. And even for the best systems, during any TES process, losses occur during both charging and discharging phases. As such, the total efficiency η_{tot} of TES can be considered from the product of the charging and discharging efficiencies, i.e. as:

$$\eta_{tot} = \eta_{charging} \times \eta_{discharging} = \frac{\overline{Q}_{stored}}{\overline{Q}_{supplied}} \times \frac{\overline{Q}_{delivered}}{\overline{Q}_{stored}} \tag{2.132}$$

where $\overline{Q}_{stored}$ is the thermal energy stored during the charging process (kWh), $\overline{Q}_{supplied}$ is the thermal energy supplied to the TES during charging (kWh), and $\overline{Q}_{delivered}$ is the thermal energy recovered from the TES during the discharge process (kWh). Note that Eq. (2.133) does not consider the internal losses of energy between the charging and discharging process. Moreover, as electrical energy is used during the charging and discharging processes, it seems relevant to also include this usage in the estimation of the TES performances. In which case a COP can be introduced and defined as [166]:

$$COP = \frac{\overline{Q}_{th,delivered}}{\overline{Q}_{e,charging} + \overline{Q}_{e,discharging}} \tag{2.133}$$

where $\overline{Q}_{th,delivered}$ is the thermal energy delivered during the discharge process (kWh), and $\overline{Q}_{e,charging}$ and $\overline{Q}_{e,discharging}$ is the electrical energy provided to the TES during the charging and discharging operations (kWh). However, since different TES systems exist, the design method may change accordingly using suitable equations to describe the heat transfers taking place. In the case of sensible heat storage (widely used), a fixed mass of heat exchange media such as a liquid is heated by the heat source and this increases its internal temperature without phase change. In which case this is simply described as the heating of a fixed mass by the following equation:

$$Q = m_{hxmedia} c_{p,hxmedia} (T_f - T_i) \tag{2.134}$$

where Q is the thermal energy provided to the constant mass of heat exchange media during the heating (J), $m_{hxmedia}$ is the mass of the heat exchange media (kg), $c_{p,hxmedia}$ its specific heat (J/kg K), and T_f and T_i are the final and initial temperatures of the heat exchange media during the charging process (K).

With latent heat storage, the energy received by the TES is used to produce a phase change, and since during the process the temperature of the medium remains constant, the energy stored is related to the system latent heat by:

$$Q = m_{hxmedia} i_{pc} \tag{2.135}$$

where $m_{hxmedia}$ is the mass of the heat exchange media (kg) and i_{pc} is the phase change latent heat (J/kg). For latent heat TES, the phase change transition can be from solid to liquid or from liquid to vapour. With the introduction of PCMs, it is now common for latent heat TES to be combined with sensible heat storage and in such cases, the heat exchange media is typically below the phase change point when energy is first stored. So the thermal energy is initially used to bring the heat exchange media to the phase change temperature and then used to operate the phase change. Once the heat exchange media has changed its phase, sensible heating within the new phase can also take place. In this case, the energy stored to the system Q is linked to both the sensible and latent heat transfers by:

$$Q = m_{hxmedia} \int_{T_i}^{T_{pc}} c_{p,phase1}(T)dT + m_{hxmedia} i_{pc} + m_{hxmedia} \int_{T_{pc}}^{T_f} c_{p,phase2}(T)dT \quad (2.136)$$

where $m_{hxmedia}$, is the mass of the heat exchange media (kg), T_i is the initial temperature of the heat exchange media (K), T_{pc} is the phase change temperature of the heat exchange media (K), T_f is the final temperature of the heat exchange media (K), $c_{p,phase1}(T)$ is the specific heat of the first phase of the heat exchange media depending on the temperature (J/kg K), $c_{p,phase2}$ (T) is the specific heat of the second phase of the heat exchange media depending on the temperature (J/kg K), and i_{pc} the phase change latent heat (J/kg). Based on this, for design purposes, the mass of PCM required for the TES system can be obtained from [167]:

$$m_{hxmedia} = \frac{Q}{\int_{T_i}^{T_{pc}} c_{p,phase1}(T)dT + i_{pc} + \int_{T_{pc}}^{T_f} c_{p,phase2}(T)dT} \quad (2.137)$$

where Q, is the thermal energy given to the TES system during the charging (J). With the development of computational methods, simulation-based optimisation of TES are becoming prominent [168], and by considering different scenarios depending on the climate and operating condition of the system, design choices such as the type of insulation, the thickness of the PCM, and the heating and cooling demands have all been studied and optimised. A particular interest for using such simulations is the possibility of studying and estimating system behaviour over time and so evaluate its performance whilst embedded into a process.

References

1 Incropera, F.P., Prescott, P.J., and Voelkel, D.D. (1985). Hybrid systems for furnace waste heat recovery: I. Use of a radiation recuperator with a Rankine cycle. *J. Heat Recover. Syst.* 5: 321–330. https://doi.org/10.1016/0198-7593(85)90006-2.

2 Thekdi, A. and Nimbalkar, S.U. (2014). Industrial Waste Heat Recovery – Potential Applications, Available Technologies and Crosscutting R&D Opportunities. https://doi.org/10.2172/1185778.

3 Beltrán, R.A.P. (2017). *Study of Radiative Recuperators in High Temperature Furnaces*: 112.

4 Ilgaz, Y. (2006). Thermal energy equipment: waste heat recovery. *Energy Effic. Guid. Ind. Asia* 1: 1–18.

5 Mullinger, P. and Jenkins, B. (2008). *Industrial and Process Furnaces*, 1e, vol. 7506. Butterworth-Heinemann: Architectural Press.

6 Jouhara, H., Khordehgah, N., Almahmoud, S. et al. (2018). Waste heat recovery technologies and applications. *Therm. Sci. Eng. Prog.* 6: 268–289. https://doi.org/10.1016/J.TSEP.2018.04.017.

7 Johnson, I., Choate, W.T., and Davidson A. (2008). Waste Heat Recovery: Technology Opportunities in the US Industry – prepared by Bcs, Incorporated. *US Dep. Energy – Ind. Technol. Progr.*: 1–112.

8 Morris, A. (2015). Improving energy Effieciency with recuperative and regenerative burners. *Ind. Heat.*

9 Reay, D.A. (1981). A review of gas-gas heat recovery systems. *J. Heat Recover. Syst.* 1: 3–41. https://doi.org/10.1016/0198-7593(81)90003-5.

10 The National Productivity Council of India. (2006). *United Nations Environment Programme, Waste Heat Recovery.*

11 Jasper GmbH. (n.d.). EcoReg, Rotary bed regenerator.

12 Henning, B. and Jasper, R. (2004). Multimelter – The new generation of aluminium melting furnaces. Heat Process.

13 Schlesinger, M.E. (2013). *Aluminium Recycling*. New Edition. Taylor & Francis Inc.

14 Hassan, M.I. and Al Kindi, R. (2014). Feasibility study of regenerative burners in aluminum holding furnaces. *J. Miner.* 66: 1603–1611. https://doi.org/10.1007/s11837-014-1110-0.

15 Costa, D.S. and Whipple, D.F. (2003). The use of regenerative burners systems to improve aluminium melting efficiencies. 5th World Congress on Aluminium (20 March 2003). Interall Publishing, Modena, Italy.

16 Osaka Gas. (n.d.). Regenerative Burner System.

17 Suzukawa, Y., Mori, I., and Fukushima, S. (2002). Eco-friendly regenerative burner heating system. *SEAISI Q (South East Asia Iron Steel Institute)* 31: 23–32.

18 Voyer, R. and Caron, F. (n.d.). Energy efficiency of casthouse furnaces. *Light Met.* 2010: 651–655.

19 Schalles D. Regenerative Burners – Are They Worth It? 2004.

20 Zajemska, M., Poskart, A., and Musiał, D. (2015). The kinetics of nitrogen oxides formation in the flame gas. *Econ. Environ. Stud.* 15: 445–460.

21 Toolan, S. (2012). *Heat Recovery for Commercial Buildings*. Dublin Institute of Technology.

22 Levenspiel, O. (2014). *Engineering Flow and Heat Exchange*, 3e. https://doi.org/10.1007/978-1-4899-7454-9.

23 Miao, Q., You, S., Zheng, W. et al. (2017). A grey-box dynamic model of plate heat exchangers used in an urban heating system. *Energies* 10 https://doi.org/10.3390/en10091398.

24 Pandya, N.S., Shah, H., Molana, M., and Tiwari, A.K. (2020). Heat transfer enhancement with nanofluids in plate heat exchangers: a comprehensive review. *Eur. J. Mech. - B/Fluids* 81: 173–190. https://doi.org/10.1016/j.euromechflu.2020.02.004.

25 Picón-Núñez, M., Canizalez-Dávalos, L., Martínez-Rodríguez, G., and Polley, G.T. (2007). Shortcut design approach for spiral heat exchangers. *Food Bioprod. Process.* 85: 322–327. https://doi.org/10.1205/fbp07073.

26 Jouhara, H., Chauhan, A., Nannou, T. et al. (2017). Heat pipe based systems - advances and applications. *Energy* 128: 729–754. https://doi.org/10.1016/j.energy.2017.04.028.

27 Johnson I, Choate W. (n.d.) Waste Heat Recovery Technology and Opportunities in U.S. Washington D.C: U.S Department of Energy.

28 Turner, W.C. (2009). *Mechanical Engineering Handbook*. London: Taylor & Francis Ltd.

29 J.+G (n.d.). *Waste Heat Boilers – New Energy from Old Energy*. Schwetzingen: Jünger+Gräter GmbH.

30 TLV. (n.d.). Waste Heat Recovery.

31 Ganapathy, V. (n.d.). *Steam Generators and Waste Heat Boilers for Process and Plant Engineers*. Taylor & Francis Group: CRC Press.

32 Eriksen VL, Schroeder JE. *Heat Recovery Steam Generator Technology* n.d.

33 King, M.J., Davenport, W.G., and Moats, M.S. (n.d.). *Sulfuric Acid Manufacture*. Elsevier.

34 Primasonics. (2020). Acoustic Cleaning Technologies.

35 Storm, K. (n.d.). *Industrial Process Plant Construction Estimating and Man-Hour Analysis*. Elsevier.

36 Viessmann. Waste Heat Boiler 2015.

37 Muench, R.K., Knauss, D.T., and Purne, G. (n.d.). *A Study of Waste-Heat-Boiler Size and Performance of a Conceptual Marine Cogas System*. Bethesda, Maryland: Department of the Navy.

38 Hatami, M., Ganji, D., and Gorji-Bandpy, M. (n.d.). A review of different heat exchangers designs for increasing the diesel exhaust waste heat recovery. *Renewable Sustainable Energy Rev.* 37: 168–181.

39 Liu, X., Gong, G., and YiWua, H.L. (n.d.). Thermal performance analysis of Brayton cycle with waste heat recovery boiler for diesel engines of offshore oil production facilities. *Appl. Therm. Eng.*: 320–328.

40 Lagzi, I., Mészáros, R., and Györgyi Gelybó, Á.L. (n.d.). *Atmospheric Chemistry*. Eötvös Loránd University.

41 England G. Development of fine particulate emission factors and speciation profiles for oil and gas-fired combustion systems n.d.

42 Kim, D.H., Park, T., and Lee, C.-E. (n.d.). Heat recovery boilers with water spray: part 2: parametric analysis and optimization of design specifications. *Therm. Sci. Eng. Prog.*

43 Sasikuma, S., Srikanth, S., and Ravikumar, B. (n.d.). Failure of evaporator tubes initiated by lamellar tearing during the commissioning of a waste heat recovery boile. *Eng. Fail. Anal.* 14: 262–278.

44 Kawahara, Y. (n.d.). High temperature corrosion mechanisms and effect of alloying elements for materials used in waste incineration environment. *Corros. Sci.* 44: 223–245.

45 Chena, G., Zhang, N., Maa, W. et al. (n.d.). Investigation of chloride deposit formation in a 24 MWe waste to energy plant. *Fuel* 140: 317–327.

46 Rongju, S., Weiqiang, W., Yan, L. et al. (n.d.). Root cause analysis of stress corrosion at tube-to-tubesheet joints of a waste heat boiler. *Eng. Fail. Anal.* 45: 398–405.

47 Leferink, R. and Huijbregts, W. (n.d.). Nitrate stress corrosion cracking in waste heat recovery boilers. *Anti-Corros. Methods Mater.* 49: 118–126.

48 Srikanth, S., Ravikumar, B., Das, K.S. et al. (n.d.). Analysis of failures in boiler tubes due to fireside corrosion in a waste heat recovery boiler. *Eng. Fail. Anal.* 10: 59–66.

49 Ebara, R., Tanaka, F., and Kawasaki, M. (n.d.). Sulfuric acid dew point corrosion in waste heat boiler tube for copper smelting furnace. *Eng. Fail. Anal.* 33: 29–36.

50 Ding, Q., Tang, X.-F., and Yang, Z.-G. (2017). Failure analysis on abnormal corrosion of economizer tubes in a waste heat boiler. *Eng. Fail. Anal.* 73: 129–138. https://doi.org/10.1016/j.engfailanal.2016.12.011.

51 Jouhara, H., Khordehgah, N., Almahmoud, S. et al. (2018). Waste heat recovery technologies and applications. *Therm. Sci. Eng. Prog.* 6 https://doi.org/10.1016/j.tsep.2018.04.017.

52 Jouhara, H. and Olabi, A.G. (2018). Editorial: industrial waste heat recovery. *Energy* 160: 1–2. https://doi.org/10.1016/J.ENERGY.2018.07.013.

53 Gupta, U.S., Mishra, S.K., and Bohra, M.A. (2013). New approach to waste heat and pressure energy systems. *Int. J. Res. Mech. Eng. Technol.* 4: 2249–5762.

54 Johnson, L., Choate, W., Davidson, A. (2008). US Department of Energy, Office of Scientific and Technical Information.

55 Reddy, B.V., Ramkiran, G., Ashok Kumar, K., and Nag, P.K. (2002). Second law analysis of a waste heat recovery steam generator. *Int. J. Heat Mass Transfer* 45: 1807–1814. https://doi.org/10.1016/S0017-9310(01)00293-9.

56 din, A.A.S.E., Elhosseini, M.A., and Arafat Ali, H. (2018). Modelling and practical studying of heat recovery steam generator (HRSG) drum dynamics and approach point effect on control valves. *Ain Shams Eng. J.* 9: 3187–3196. https://doi.org/10.1016/j.asej.2018.06.004.

57 Krishnan, V.V. (2020). Thermodynamics and energy engineering. In: *Design and Operation of Solid Oxide Fuel Cells*, 43–84. Elsevier https://doi.org/10.1016/b978-0-12-815253-9.00002-1.

58 Thekdi, A.C. (2014). Industrial Waste Heat Recovery: Potential Applications, Available Technologies and Crosscutting R&D Opportunities.

59 Ahmed, A., Khodary Esmaeil, K., Irfan, M.A., and Al-Mufadi, F.A. (2018). Design methodology of heat recovery steam generator in electric utility for waste heat recovery. *Int. J. Low-Carbon Technol.* 13 (4): 369–379. https://doi.org/10.1093/ijlct/cty045.

60 Nag, P.K. and De, S. (1997). Design and operation of a heat recovery steam generator with minimum irreversibility. *Appl. Therm. Eng.* 17: 385–391. https://doi.org/10.1016/s1359-4311(96)00033-6.

61 Franco, C.A. (2005). Thermodynamic optimization of the recovery in combined power plants. *Int. J. Appl. Thermodyn.* 4: 43–52.

62 Ganapathy, V. (1966). Heat recovery steam generators: understand the basics. *Chem. Eng. Prog.*: 32–45.

63 Theotokatos, G., Rentizelas, A., Guan, C., and Ancic, I. (2020). Waste heat recovery steam systems techno-economic and environmental investigation for ocean-going vessels considering actual operating profiles. *J. Cleaner Prod.* 267: 121837. https://doi.org/10.1016/j.jclepro.2020.121837.

64 Du, W., Li, J., and Yuan, B. (2020). Dynamic characteristics analysis of a once-through heat recovery steam generator. *Appl. Therm. Eng.* 173 https://doi.org/10.1016/j.applthermaleng.2020.115155.

65 Charisiadis, C. (2018). Brine Zero Liquid Discharge (ZLD) Fundamentals and Design; A guide to the basic conceptualization of the ZLD/MLD process design and the relative technologies involved. Researchgate. doi:https://doi.org/10.13140/RG.2.2.19645.31205.

66 Palen JW. *Heat Exchangers, Vaporizers, and Condensers*. 2015. doi:https://doi.org/10.1002/9781118985960.meh409.

67 Eriksen, V.L. (2017). *Heat Recovery Steam Generator Technology* | ScienceDirect, vol. 1. Woodhead Publishing.

68 Roslyakov, P.V., Pleshanov, K.A., and Sterkhov, K.V. (2014). A study of natural circulation in the evaporator of a horizontal-tube heat recovery steam generator. *Therm. Eng.* (English Transl Teploenerg) 61: 465–472. https://doi.org/10.1134/S004060151407009X.

69 Wardle, T.J. (2000). Creep-Rupture Assessment of Superheater Tubes Using Nondestructive Oxide Thickness Measurements:1–6.

70 Nordin, A. and Majid, M.A.A. (2016). Parametric study on the effects of pinch and approach points on heat recovery steam generator performance at a district cooling system. *J. Mech. Eng. Sci.* 10: 2134–2144. https://doi.org/10.15282/jmes.10.2.2016.17.0201.

71 Leelananda, R. and Suen, K.O. Influence of liquid receiver on the performance of reversible heat pumps using refrigerant mixtures. *Int. J. Refrig.* 27 (1): 53–62. ISSN 01407007 2004;doi:https://doi.org/10.1016/S0140-7007(03)00092-6.

72 Grassi W. *Heat Pumps: Fundamentals and Applications*. 2018. doi:https://doi.org/10.1007/978-3-319-62199-9.

73 Carroll, P., Chesser, M., and Lyons, P. (2020). Air source heat pumps field studies: a systematic literature review. *Renewable Sustainable Energy Rev.* 134: 110275. https://doi.org/10.1016/j.rser.2020.110275.

74 Mustafa Omer, A. (2008). Ground-source heat pumps systems and applications. *Renewable Sustainable Energy Rev.* 12: 344–371. https://doi.org/10.1016/j.rser.2006.10.003.

75 Zhao, Y., Dunn, A., Lin, J., and Shi, D. (2019). Chapter 13, Photothermal effect of nanomaterials for efficient energy applications. In: *Chen Environmental and Energy Applications XBT-NN for B*, editors. Micro Nano Technol., (ed. X. Wang), 415–434. Elsevier https://doi.org/10.1016/B978-0-12-814497-8.00013-8.

76 Suppes, G.J. and Storvick, T.S.B.T. (2007). Sustainable nuclear power- a volume in sustainable world. *Sustain. World*, Burlington: Academic Press: 383–393. https://doi.org/10.1016/B978-012370602-7/50031-4.

77 Cleveland, C.J. (ed.) (2006). *Encyclopedia of Energy*, vol. 38, 712. Elsevier (six volumes, 5376 pages, ISBN 0-12-176480-X, £1,300). Energy Build – ENERG BLDG https://doi.org/10.1016/j.enbuild.2006.02.001.

78 Trasatti, S. (1999). 1799–1999: Alessandro Volta's 'electric pile': two hundred years, but it doesn't seem like it. *J. Electroanal. Chem.* 460: 1–4. https://doi.org/10.1016/S0022-0728(98)00302-7.

79 Luo, D., Wang, R., Yu, W., and Zhou, W. (2020). A novel optimization method for thermoelectric module used in waste heat recovery. *Energy Convers. Manage.* 209: 112645. https://doi.org/10.1016/j.enconman.2020.112645.

80 Kraemer, D., Jie, Q., McEnaney, K. et al. (2016). Concentrating solar thermoelectric generators with a peak efficiency of 7.4%. *Nat. Energy* 1: 16153. https://doi.org/10.1038/nenergy.2016.153.

81 Ishaq, H., Islam, S., Dincer, I., and Yilbas, B.S. (2020). Development and performance investigation of a biomass gasification based integrated system with thermoelectric generators. *J. Cleaner Prod.* 256: 120625. https://doi.org/10.1016/j.jclepro.2020.120625.

82 Sargolzaeiaval, Y., Padmanabhan Ramesh, V., Neumann, T.V. et al. (2020). Flexible thermoelectric generators for body heat harvesting – enhanced device performance using high thermal conductivity elastomer encapsulation on liquid metal interconnects. *Appl. Energy* 262: 114370. https://doi.org/10.1016/j.apenergy.2019.114370.

83 Aravind, B., Khandelwal, B., Ramakrishna, P.A., and Kumar, S. (2020). Towards the development of a high power density, high efficiency, micro power generator. *Appl. Energy* 261: 114386. https://doi.org/10.1016/j.apenergy.2019.114386.

84 Ostroverkhova O. *Handbook of Organic Materials for Electronic and Photonic Devices* n.d.

85 Naito, M., Yokoyama, T., Hosokawa, K., and Nogi, K.B. (2018). *Nanoparticle Technology HandbookNanoparticle Technology Handbook*, 3e (eds. M. Naito, T. Yokoyama and K. Hosokawa), Nogi KBT-NTH, 845–877. Elsevier https://doi.org/10.1016/B978-0-444-64110-6.18001-1.

86 Buschow, K.H.J. (2001). *Encyclopedia of Materials: Science and Technology*. Elsevier.

87 Memon, S. and Tahir, K.N. (2018). Experimental and analytical simulation analyses on the electrical performance of thermoelectric generator modules for direct and concentrated quartz-halogen heat harvesting. *Energies* 11 https://doi.org/10.3390/en11123315.

88 Chen, A. and Wright, P. (2012). Medical applications of Thermoelectrics. In: *Modules, Systems, and Applications in Thermoelectrics*, 21–26. CRC Press https://doi.org/10.1201/b11892-30.

89 Yadali Jamaloei, B. and Asghari, K. (2015). The joule-Thomson effect in petroleum fields: I. well testing, multilateral/slanted Wells, hydrate formation, and drilling/completion/production operations. *Energy Sources, Part A Recover Util. Environ. Eff.* 37: 217–224. https://doi.org/10.1080/15567036.2010.551258.

90 Koster, G., Huijben, M., and Rijnders, G. (2015). *Epitaxial Growth of Complex Metal Oxides*. Elsevier Ltd.

91 Demirel, Y. and Sandler, S.I. (2004). Nonequilibrium thermodynamics in engineering and science. *J. Phys. Chem. B* 108: 31–43. https://doi.org/10.1021/jp030405g.

92 Meseguer, J., Pérez-Grande, I., and Sanz-Andrés, A. (2012). *Spacecraft Thermal Control.* Woodhead Publishing https://doi.org/10.1533/9780857096081.175.

93 Hendricks, T.C.W.T. (2006). [2]. Engineering scoping study of thermoelectric generator Systems for Industrial Waste Heat Recovery. United States. Department of Energy. Office of Energy efficiency and. *Renewable Energy.*

94 Deepak, K., Pattanaik, M.S., and Ramanujan, R.V. (2019). Figure of merit and improved performance of a hybrid thermomagnetic oscillator. *Appl. Energy* 256: 113917. https://doi.org/10.1016/j.apenergy.2019.113917.

95 Ohtani, B. (2011). Chapter 10, Photocatalysis by inorganic solid materials: revisiting its definition, concepts, and experimental procedures. In: *Inorganic Photochemistry*, vol. 63 (ed. R. van Eldik), Stochel GBT-A in IC, 395–430. Academic Press https://doi.org/10.1016/B978-0-12-385904-4.00001-9.

96 Janardhanam, V., Yun, H.-J., Jyothi, I. et al. (2019). Fermi-level depinning in metal/Ge interface using oxygen plasma treatment. *Appl. Surf. Sci.* 463: 91–95. https://doi.org/10.1016/j.apsusc.2018.08.187.

97 Massaguer, E., Massaguer, A., Montoro, L., and Gonzalez, J.R. (2014). Development and validation of a new TRNSYS type for the simulation of thermoelectric generators. *Appl. Energy* 134: 65–74. https://doi.org/10.1016/J.APENERGY.2014.08.010.

98 Orr, B., Taglieri, J., Ding, L.C., and Akbarzadeh, A. (2016). Validating an alternative method to predict thermoelectric generator performance. *Energy Convers. Manage.* 116: 134–141. https://doi.org/10.1016/j.enconman.2016.02.074.

99 Wee, D. (2011). Analysis of thermoelectric energy conversion efficiency with linear and nonlinear temperature dependence in material properties. *Energy Convers. Manage.* 52: 3383–3390. https://doi.org/10.1016/j.enconman.2011.07.004.

100 Chen, Z.-G., Han, G., Yang, L. et al. (2012). Nanostructured thermoelectric materials: current research and future challenge. *Prog. Nat. Sci. Mater. Int.* 22: 535–549. https://doi.org/10.1016/j.pnsc.2012.11.011.

101 Joshi, G., Lee, H., Lan, Y. et al. (2008). Enhanced thermoelectric figure-of-merit in nanostructured p-type silicon germanium bulk alloys. *Nano Lett.* 8: 4670–4674. https://doi.org/10.1021/nl8026795.

102 Wang, J., Cao, P., Li, X. et al. (2019). Experimental study on the influence of Peltier effect on the output performance of thermoelectric generator and deviation of maximum power point. *Energy Convers. Manage.* 200: 112074. https://doi.org/10.1016/j.enconman.2019.112074.

103 EASE. Thermal Storage Position Paper. EASE 2017.

104 Commission Staff Working Document (2017). Energy storage – the role of electricity. *Eur. Comm.*: 1–25.

105 EASE-EERA. (n.d.) Joint EASE/EERA recommendations for a European Energy Storage Technology Development Roadmap towards 2030. EASE-EERA.

106 Fallahi, A., Guldentops, G., Tao, M. et al. (2017). Review on solid-solid phase change materials for thermal energy storage: molecular structure and thermal properties. *Appl. Therm. Eng.* 127: 1427–1441. https://doi.org/10.1016/j.applthermaleng.2017.08.161.

107 Sarbu, I. and Sebarchievici, C. (2018). A comprehensive review of thermal energy storage. *Sustain* 10 https://doi.org/10.3390/su10010191.

108 EASE-EERA. Joint EASE/EERA recommendations for a European Energy Storage Technology Development Roadmap 2017 Update. 2017.

109 Koohi-Fayegh, S. and Rosen, M.A. (2020). A review of energy storage types, applications and recent developments. *J. Energy Storage* 27: 101047. https://doi.org/10.1016/j.est.2019.101047.

110 Mehling, H. and Cabeza, L.F. (2008). *Heat and Cold Storage with PCM*. Berlin, Heidelberg: Springer Berlin Heidelberg https://doi.org/10.1007/978-3-540-68557-9.

111 Elias, C.N. and Stathopoulos, V.N. (2019). A comprehensive review of recent advances in materials aspects of phase change materials in thermal energy storage. *Energy Procedia* 161: 385–394. https://doi.org/10.1016/j.egypro.2019.02.101.

112 Rathore, K.P.S. and Shukla, S.K. (2019). Potential of macroencapsulated PCM for thermal energy storage in buildings : a comprehensive review. *Constr. Build. Mater.* 225: 723–744. https://doi.org/10.1016/j.conbuildmat.2019.07.221.

113 Guelpa, E. and Verda, V. (2019). Thermal energy storage in district heating and cooling systems: a review. *Appl. Energy* 252: 113474. https://doi.org/10.1016/j.apenergy.2019.113474.

114 Rathore, P.K.S., Shukla, S.K., and Gupta, N.K. (2020). Potential of microencapsulated PCM for energy savings in buildings: a critical review. *Sustain. Cities Soc.* 53: 101884. https://doi.org/10.1016/j.scs.2019.101884.

115 Badenhorst, H. (2018). A review of the application of carbon materials in solar thermal energy storage. *Sol. Energy* 192: 35–68. https://doi.org/10.1016/j.solener.2018.01.062.

116 Gasia, J., Miró, L., and Cabeza, L.F. (2017). Review on system and materials requirements for high temperature thermal energy storage. Part 1: general requirements. *Renewable Sustainable Energy Rev.* 75: 1320–1338. https://doi.org/10.1016/j.rser.2016.11.119.

117 Li, G. (2015). Energy and exergy performance assessments for latent heat thermal energy storage systems. *Renewable Sustainable Energy Rev.* 51: 926–954. https://doi.org/10.1016/j.rser.2015.06.052.

118 Sunku Prasad, J., Muthukumar, P., Desai, F. et al. (2019). A critical review of high-temperature reversible thermochemical energy storage systems. *Appl. Energy* 254: 113733. https://doi.org/10.1016/j.apenergy.2019.113733.

119 Li, Q., Li, C., Du, Z. et al. (2019). A review of performance investigation and enhancement of shell and tube thermal energy storage device containing molten salt based phase change materials for medium and high temperature applications. *Appl. Energy* 255: 113806. https://doi.org/10.1016/j.apenergy.2019.113806.

120 Magendran, S.S., Khan, F.S.A., Khan, A. et al. (2019). Synthesis of organic phase change materials (PCM) for energy storage applications: a review. *Nano-Struct. Nano-Objects* 20: 100399. https://doi.org/10.1016/j.nanoso.2019.100399.

121 Alva, G., Lin, Y., Liu, L., and Fang, G. (2017). Synthesis, characterization and applications of microencapsulated phase change materials in thermal energy storage: a review. *Energy Build.* 144: 276–294. https://doi.org/10.1016/j.enbuild.2017.03.063.

122 Umair, M.M., Zhang, Y., Iqbal, K. et al. (2019). Novel strategies and supporting materials applied to shape-stabilize organic phase change materials for thermal energy storage–a review. *Appl. Energy* 235: 846–873. https://doi.org/10.1016/j.apenergy.2018.11.017.

123 Li, D., Wu, Y., Wang, B. et al. (2020). Optical and thermal performance of glazing units containing PCM in buildings: a review. *Constr. Build. Mater.* 233: 117327. https://doi.org/10.1016/j.conbuildmat.2019.117327.

124 Al-Waeli, A.H.A., Kazem, H.A., Chaichan, M.T., and Sopian, K. (2019). Experimental investigation of using nano-PCM/nanofluid on a photovoltaic thermal system (PVT): technical and economic study. *Therm. Sci. Eng. Prog.* 11: 213–230. https://doi.org/10.1016/j.tsep.2019.04.002.

125 Wu, S., Yan, T., Kuai, Z., and Pan, W. (2019). Thermal conductivity enhancement on phase change materials for thermal energy storage: a review. *Energy Storage Mater.* 12 https://doi.org/10.3390/en12010075.

126 Qureshi, Z.A., Ali, H.M., and Khushnood, S. (2018). Recent advances on thermal conductivity enhancement of phase change materials for energy storage system: a review. *Int. J. Heat Mass Transfer* 127: 838–856. https://doi.org/10.1016/j.ijheatmasstransfer.2018.08.049.

127 Mahdi, J.M., Lohrasbi, S., and Nsofor, E.C. (2019). Hybrid heat transfer enhancement for latent-heat thermal energy storage systems: a review. *Int. J. Heat Mass Transfer* 137: 630–649. https://doi.org/10.1016/j.ijheatmasstransfer.2019.03.111.

128 Reyes, A., Henríquez-Vargas, L., Vásquez, J. et al. (2020). Analysis of a laboratory scale thermal energy accumulator using two-phases heterogeneous paraffin wax-water mixtures. *Renewable Energy* 145: 41–51. https://doi.org/10.1016/j.renene.2019.06.007.

129 Gao, H., Wang, J., Chen, X. et al. (2018). Nanoconfinement effects on thermal properties of nanoporous shape-stabilized composite PCMs: a review. *Nano Energy* 53: 769–797. https://doi.org/10.1016/j.nanoen.2018.09.007.

130 Qu, Y., Wang, S., Tian, Y., and Zhou, D. (2019). Comprehensive evaluation of paraffin-HDPE shape stabilized PCM with hybrid carbon nano-additives. *Appl. Therm. Eng.* 163 https://doi.org/10.1016/j.applthermaleng.2019.114404.

131 Al-Waeli, A.H.A., Kazem, H.A., Yousif, J.H. et al. (2020). Mathematical and neural network modeling for predicting and analyzing of nanofluid-nano PCM photovoltaic thermal systems performance. *Renewable Energy* 145: 963–980. https://doi.org/10.1016/j.renene.2019.06.099.

132 Huang, X., Zhu, C., Lin, Y., and Fang, G. (2019). Thermal properties and applications of microencapsulated PCM for thermal energy storage: a review. *Appl. Therm. Eng.* 147: 841–855. https://doi.org/10.1016/j.applthermaleng.2018.11.007.

133 Rathore, P.K.S. and Shukla, S.K. (2019). Potential of macroencapsulated PCM for thermal energy storage in buildings: a comprehensive review. *Constr. Build. Mater.* 225: 723–744. https://doi.org/10.1016/j.conbuildmat.2019.07.221.

134 Alehosseini, E. and Jafari, S.M. (2019). Micro/nano-encapsulated phase change materials (PCMs) as emerging materials for the food industry. *Trends Food Sci. Technol.* 91: 116–128. https://doi.org/10.1016/j.tifs.2019.07.003.

135 Wang, C., Feng, L., Li, W. et al. (2012). Shape-stabilized phase change materials based on polyethylene glycol/porous carbon composite: the influence of the pore structure of the carbon materials. *Sol. Energy Mater. Sol. Cells* 105: 21–26. https://doi.org/10.1016/J.SOLMAT.2012.05.031.

136 Cengel, Y.A. (2003). *Heat Transfer: A Practical Approach*, 785–841. New York: McGraw-Hill https://doi.org/10.1017/CBO9781107415324.004.

137 Incropera, F.P., DeWitt, D.P., Bergman, T.L., and Lavine, A.S. (2007). Fundamentals of heat and mass transfer. *Fundam. Heat Mass Transfer* 997 https://doi.org/10.1016/j.applthermaleng.2011.03.022.

138 Rohsenow, W.M., Hartnett, J.R., Cho, Y.I. et al. (eds.) (1998). *Handbook of Heat Transfer*, 3e. McGraw-Hill.

139 Mota, F.A.S. (2015). Chapter 7, modeling and Design of Plate Heat Exchanger. In: *Heat transfer Studies and Applications* (ed. E.P. Carvalho), 166–167. Rijeka: IntechOpen https://doi.org/10.5772/60885.

140 Nusselt, W. (1916). The condensation of steam on cooled surfaces. *VDI-Zs* 60: 541.

141 Rohsenow, W.M. (1952). A method of correlating heat-transfer data for surface boiling of liquids.

142 Guichet, V., Almahmoud, S., and Jouhara, H. (2019). Nucleate pool boiling heat transfer in wickless heat pipes (two-phase closed thermosyphons): a critical review of correlations. *Therm. Sci. Eng. Prog.* 13 https://doi.org/10.1016/j.tsep.2019.100384.

143 Guichet, V. and Jouhara, H. (2019). Condensation, evaporation and boiling of falling films in wickless heat pipes (two-phase closed thermosyphons): a critical review of correlations. *Int. J. Thermofluids* 100001 https://doi.org/10.1016/J.IJFT.2019.100001.

144 Jenkins, B. and Mullinger, P. (2011). *Industrial and Process Furnaces: Principles, Design and Operation*. Elsevier.

145 Shao, H., Zhao, Q., Liang, Z., and Wang, Y. (2018). Numerical investigation on a separated structure shell-and-tube waste heat boiler based on experiment. *Int. J. Heat Mass Transfer* 117: 1006–1018. https://doi.org/10.1016/j.ijheatmasstransfer.2017.09.127.

146 Manenti G, Papasidero D, Manenti F, Bozzano G, Pierucci S. Design of SRU thermal reactor and waste heat boiler considering recombination reactions. Procedia Eng., vol. 42, Elsevier Ltd; 2012, p. 376–83. doi:10.1016/j.proeng.2012.07.429.

147 Turner, W.C. and Doty, S. (2007). *Energy management handbook*. Inc.: The Fairmont Press.

148 Ramos, J., Chong, A., and Jouhara, H. (2016). Experimental and numerical investigation of a cross flow air-to-water heat pipe-based heat exchanger used in waste heat recovery. *Int. J. Heat Mass Transfer* 102: 1267–1281. https://doi.org/10.1016/j.ijheatmasstransfer.2016.06.100.

149 Emerson, W.H. (1984). Making the most of run-around coil systems. *J. Heat Recover. Syst.* 4: 265–270. https://doi.org/10.1016/0198-7593(84)90065-1.

150 Emerson, W.H. (1983). Designing run-around coil systems. *J. Heat Recover. Syst.* 3: 305–309. https://doi.org/10.1016/0198-7593(83)90011-5.

151 Sharma, H. and Kumar, A. (2012). Varun, Performance analysis of metallic concentric tube recuperator in parallel flow arrangement. *Int. J. Heat Mass Transfer* 55: 7760–7771. https://doi.org/10.1016/j.ijheatmasstransfer.2012.07.084.

152 Modest, M.F. (1993). *Radiative Heat Transfer*. Hightstown, NJ: McGraw-Hill. Inc.

153 Basu, P., Kefa, C., and Jestin, L. (2012). *Boilers and Burners: Design and Theory*. Springer Science & Business Media.

154 Von Cube, H.L. and Steimle, F. (2013). *Heat Pump Technology*. Elsevier.

155 Perko J, Dugec V, Topic D, Sljivac D, Kovac Z. Calculation and design of the heat pumps. 3rd Int. youth Conf. Energ., Leiria: 2011, p. 1–7.

156 Sivaprahasam, D. (2018). Automotive waste heat recovery by thermoelectric generator technology, chapter 9. In: (ed. S. Harish). Rijeka: IntechOpen https://doi.org/10.5772/intechopen.75443.

157 Remeli, M.F., Date, A., Orr, B. et al. (2016). Experimental investigation of combined heat recovery and power generation using a heat pipe assisted thermoelectric generator system. *Energy Convers. Manage.* 111: 147–157. https://doi.org/10.1016/j.enconman.2015.12.032.

158 Remeli, M.F., Tan, L., Date, A. et al. (2015). Simultaneous power generation and heat recovery using a heat pipe assisted thermoelectric generator system. *Energy Convers. Manage.* 91: 110–119. https://doi.org/10.1016/j.enconman.2014.12.001.

159 Gambier, P., Anton, S.R., Kong, N. et al. (2012). Piezoelectric, solar and thermal energy harvesting for hybrid low-power generator systems with thin-film batteries. *Meas. Sci. Technol.* 23 https://doi.org/10.1088/0957-0233/23/1/015101.

160 Smoker, J., Nouh, M., Aldraihem, O., and Baz, A. (2012). Energy harvesting from a standing wave thermoacoustic-piezoelectric resonator. *J. Appl. Phys.* 111 https://doi.org/10.1063/1.4712630.

161 Sun, X. (2017). An overview on piezoelectric power generation system for electricity generation. *J. Power Energy Eng.* 05: 11–18. https://doi.org/10.4236/jpee.2017.52002.

162 Cayless, M.A. (1961). Thermionic generation of electricity. *Br. J. Appl. Phys.* 12.

163 Chen, M., Yan, H., Zhou, P., and Chen, X. (2020). Performance analysis of solar thermophotovoltaic system with selective absorber/emitter. *J. Quant. Spectrosc. Radiat. Transfer* 253.

164 Meseguer, J., Pérez-Grande, I., and Sanz-Andrés, A. (2012). Thermal radiation heat transfer. *Spacecr. Therm. Control*: 73–86.

165 Tao, W.Q. and Sparrow, E.M. (1985). Ambiguities related to the calculation of radiant heat exchange between a pair. *Int. J. Heat Mass Transfer* 28: 1786–1787.

166 Piché, P., Haillot, D., Gibout, S. et al. (2020). Design, construction and analysis of a thermal energy storage system adapted to greenhouse cultivation in isolated northern communities. *Sol. Energy* 204: 90–105. https://doi.org/10.1016/j.solener.2020.04.008.

167 Hossain, M.S., Pandey, A.K., Rahim, N.A. et al. (2020). Self-cleaning assisted photovoltaic system with thermal energy storage: design and performance evaluation. *Sol. Energy* 206: 487–498. https://doi.org/10.1016/j.solener.2020.06.031.

168 Lin, W., Ma, Z., McDowell, C. et al. (2020). Optimal design of a thermal energy storage system using phase change materials for a net-zero energy Solar Decathlon house. *Energy Build.* 208: 109626. https://doi.org/10.1016/j.enbuild.2019.109626.

3
Low-Temperature Applications

The commonly applied definition of *low-temperature* applications are those technologies and processes that operate below 0 °C, and common examples are those that operate in the cryogenic domain and some refrigeration applications. Of relevance to the technologies of waste heat recovery, such usage is widespread and is involved in many industrial processes such as the cryogenic preservation for the food and healthcare industries, the refrigeration of produce, and heating, ventilation, and air conditioning (HVAC) systems for chiller rooms.

Unlike high-temperature applications, which commonly focus on the efficiency of burner technology and those technologies that emit high temperature, corrosive, and particulate-heavy exhaust gases, the cryogenic spectrum of waste heat recovery presents a different set of issues such as a change in material properties, which can increase brittleness and so lead to subsequent failure of a technology. This then significantly changes the compatibility of different materials for their application within this temperature spectrum.

3.1 Refrigeration

The application of refrigeration for commercial refrigerator units used in supermarket has been briefly discussed in Chapter 4. In terms of 'refrigeration', the temperature range is typically defined from 0 to 10 °C whereas 'freezing' is defined from −20 °C and below, and this identifies a developing region of sub-zero cooling. The application of this (sub-zero cooling) has been studied, and Liu et al. [1] discussed the development of sub-zero refrigerators for the preservation of fruit and vegetables, and the composition of such a system typically consists of multiple technologies as discussed in Chapter 2. A key part being the addition of heat pumps and heat exchangers. Wu et al. [2] investigated the role of ammonia in sub-zero refrigerators and in heat pumps. This work studied ammonia because of its low freezing point which then eliminates the risk of crystallisation, a potential problem for the role of refrigeration in which a coolant fluid is typically circulated through a condenser and absorber to remove heat from each respective process. A schematic for which is shown in Figure 3.1.

Waste Heat Recovery in Process Industries, First Edition. Hussam Jouhara.

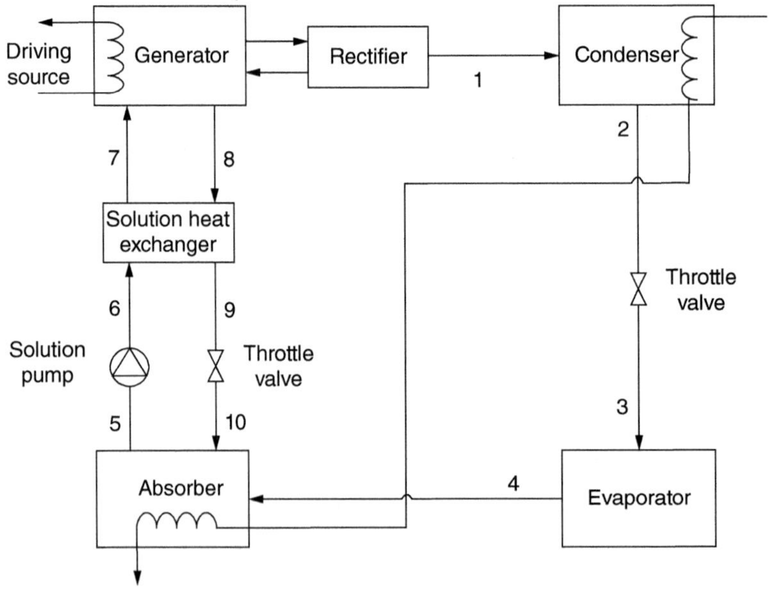

Figure 3.1 Ammonia-based refrigeration cycle – NH_3–H_2O absorption heat pump. Source: Wu et al. [2].

The application of ammonia itself is limited due to handling and safety concerns which outweigh its benefits as a refrigerant. However, significant studies have been conducted to improve the properties and offer safer handling of ammonia by introducing salts, turning the solution into a binary mixture. A similar study also has been conducted by Singh and Dasgupta [3], who investigated the use of carbon dioxide within a refrigeration process in the dairy processing industry, highlighting the operation of a low-grade waste heat recovery in an existing warm climate, which typically relies on evaporative cooling to maintain products below a maximum of 4 °C. The proposed system, shown in Figure 3.2, allows the technology to be combined with ammonia whilst eliminating the role of an evaporative cooler; the operation allowing the carbon dioxide evaporator side to absorb heat from the ammonia cycle and the boiler section to heat groundwater at a constant temperature. By utilising the proposed system, the temperature of the final product remained stable and approximately 37 kW was recovered, but in addition, 5000 l of water which would have been used for evaporative cooling was eliminated.

The use of carbon dioxide with heat pumps highlights research into necessary alternative refrigerants as many are currently being phased out due to their high global warming potential (GWP), and studies conducted by Polzot et al. [4] investigated the energy benefit of using a carbon dioxide-based heat pump system within a commercial refrigeration unit as shown within Figure 3.3. The presented system features a water loop heat pump, where electrical pumps are used to control the thermal zones of the system. The refrigeration cabinet has two operating programmes: a heating mode and a cooling mode. The heating operation allowing cooled air to pass through the system instead of water, whilst the cooling scenario is typically activated

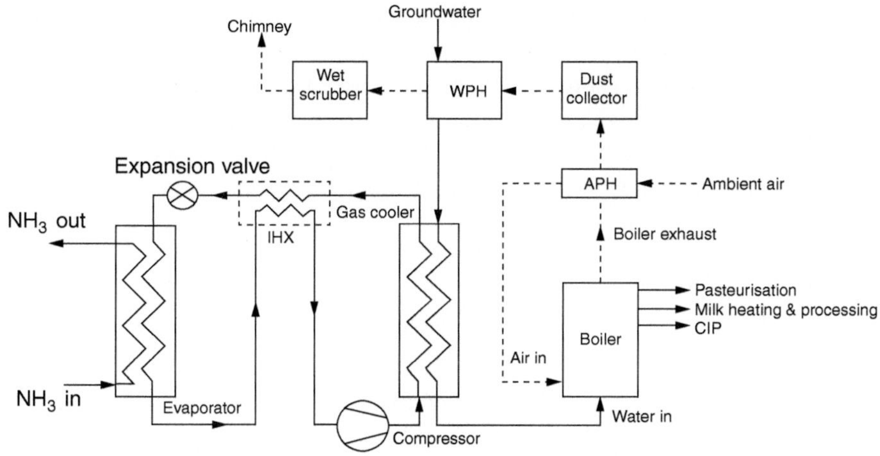

Figure 3.2 CO_2 heat pump system used within a dairy application. Source: Singh and Dasgupta [3].

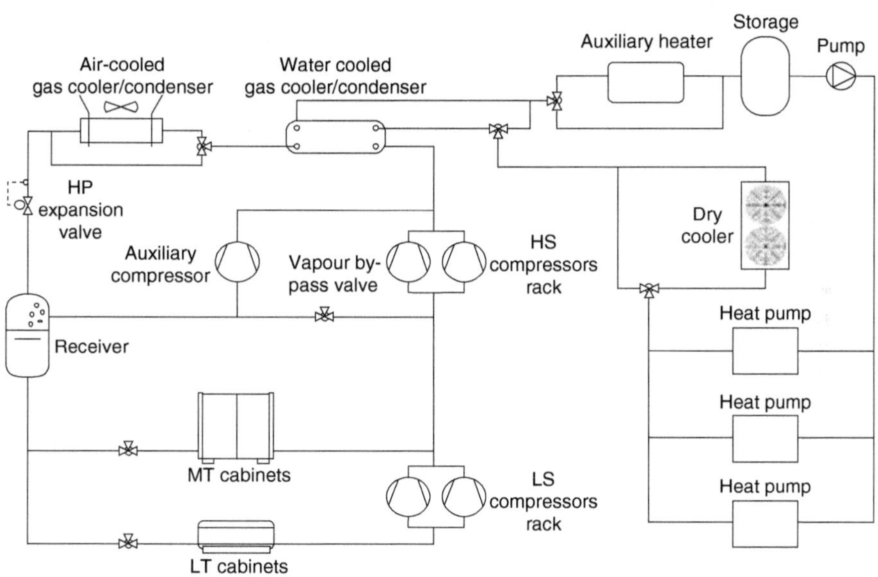

Figure 3.3 Integrated CO_2 heat pump system within a commercial refrigerator. Source: Polzot et al. [4].

when the heat available in the heat pump system is low, so allowing the water temperature to increase, and a dry cooler on the water loop to reject this heat to ambient. For this study an alternative refrigerant, R1234ZE (E), was employed (reflective of a commercial unit) and fed through the system at a constant mass flow rate. The outcome of the study highlighted an improved operation by the use of the CO_2 heat pump system, with approximately 9.4% of energy savings being recovered, and lower operational costs compared to traditional refrigeration cabinets.

To maximise the efficiency of traditional refrigeration cabinets, the use of binary heat pump fluids in absorption systems has also been investigated by Salehi and Yari [5], who investigated ammonia/water, ammonia/sodium thiocyanate, and ammonia/lithium nitrate within the three different systems shown in Figure 3.4. These systems involved three test rigs, the first being a conventional heat pump system (system 1), which featured two compressors, an evaporator, a condenser, a heat exchanger, and a separation vessel. The liquid leaving the condenser outlet is passed through a separation vessel, allowing the isolation of vapour and liquid – the vapour subsequently being passed through a compressor to enter a condenser, where the heat rejected is transferred through a heat exchanger to provide help meet the building heating load. In comparison, for system 2 both compressors are removed, and an absorption section is added where the heat is absorbed by a refrigeration evaporator. In system 3, both compressors are again removed along with the separation vessel. The condensed liquid from the condenser splits into two streams, whereby they enter into the evaporators. The vapour leaving the first evaporator is then pressurised using the vapour-driven ejector. However, the vapour from the second evaporator is pressurised using the solution-driven ejector. Therefore, the condenser supplies the heating load and the second evaporator supplies the refrigeration load. This enables the solution flowing through the generator to be heated by any heat source such as solar energy, industrial waste heat, and others.

The study tested each binary mixture under a set of fixed conditions, with a traditional system acting as a comparative baseline. The outcome from the study showed that the mixture of ammonia and water resulted in a low coefficient of performance (COP) for all three systems, and that both ammonia/sodium thiocyanate and ammonia/lithium nitrate performed equally on system 2. System 3 however showed the main superiority of ammonia/lithium nitrate in that the fluid recorded a higher COP and improved evaporator temperatures. The study also reported an average power generation efficiency of 39.8%, but with the utilisation of ammonia/lithium nitrate, this value increased by 7.5%.

The use of ammonia was also been investigated by Palm [6], who used ammonia within low-capacity refrigeration systems. The study primarily focused on the applicability of ammonia within different heat pump applications and configurations, and made comparisons between ammonia and existing refrigerants. The system developed by Palm [6] (as shown in Figure 3.5) allows feed water to pass through the condenser, and then separated, where a small amount of the flow will pass through a desuperheater, whilst the water exiting the system will return back into the water distribution system. To ensure correct mixing of the water tank, a valve was installed within the tank. On the evaporator side of the system, the loop contains a refrigerant which can be heated. The study was investigated at different loads ranging from 3.7 to 7.1 kW on the cooling side – the feed water remaining constant. The outcome of the study highlighted the success of the system noting the effects of using minute amounts of ammonia in small-scale systems. In addition, the study also extended the system evaporation temperature range, from −13° to 2 °C.

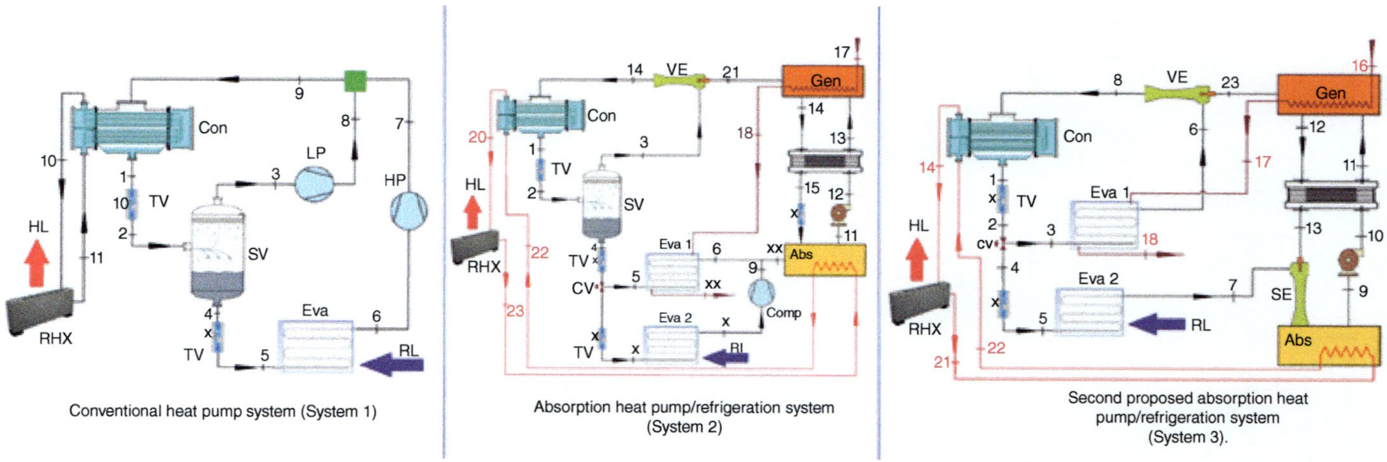

Figure 3.4 Proposed test rig. Salehi and Yari [5].

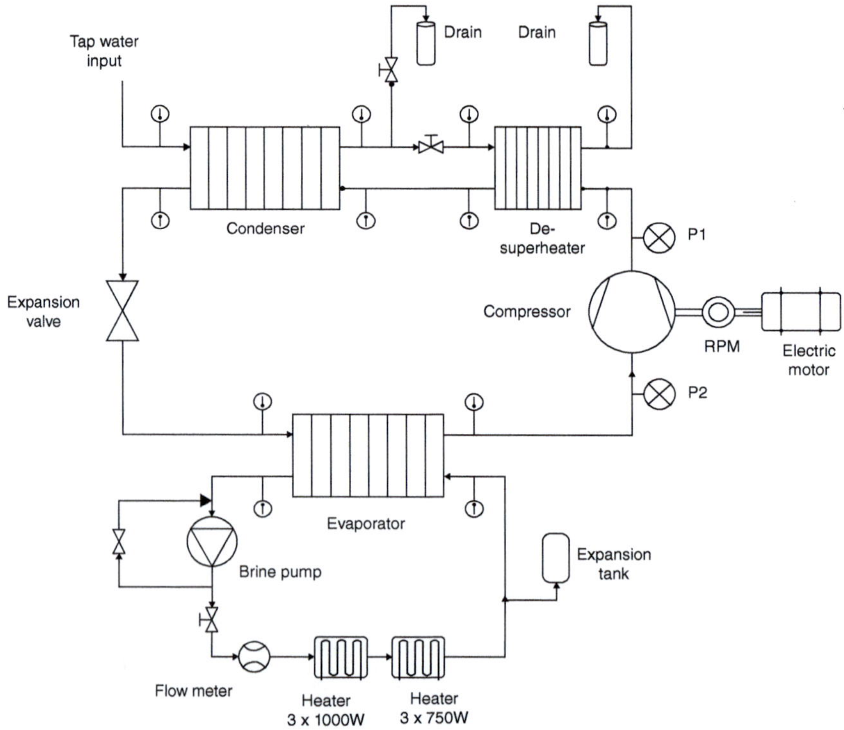

Figure 3.5 Proposed ammonia low-capacity heat pump system. Source: Palm [6].

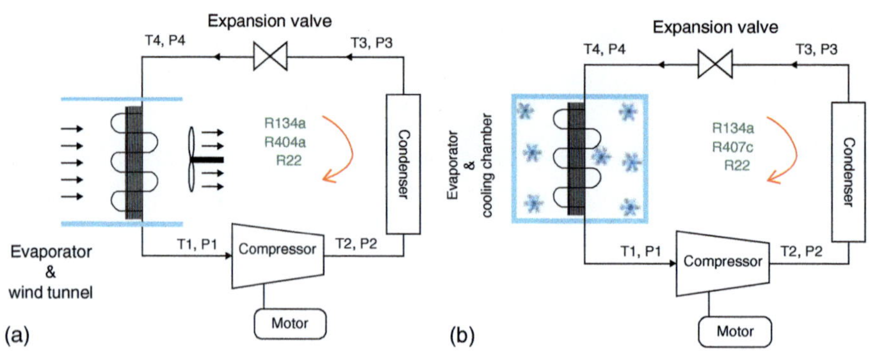

Figure 3.6 (a) Test rig utilising a heat pump and (b) a freezer operation. Source: Afshari et al. [6].

Similar comparative studies of different refrigerants and heat pumps within a refrigeration process have also been conducted by Afshari et al. [7]. In this study, R134a, R407c, R22 and R404a were each investigated in two modes of operation, and Figure 3.6 shows the two operating modes used. Figure 3.6a highlights including the heat pump and Figure 3.6b highlighting the operation with a freezer. The system operates on a constant airflow which acts as the heat source and water as the heat sink (secondary fluid), and the system also features a reciprocating compressor to recirculate the refrigerant. In comparison between the two modes of operation, the heat pump exhibits stable steady-state performance in terms of power consumption and evaporator temperature, unlike the freezer which showed a variable evaporator temperature.

The application of waste heat technologies within the low-temperature domain is not limited to traditional heat exchangers and heat pump networks, thermoelectric generators (TEGs) has also been applied within the refrigeration field but on a smaller scale. For example, Rodríguez et al. [8] investigated the role of TEG modules within ice makers, as shown in Figure 3.7. The operation allows absorption from the cold side of a Peltier module, which then cools and freezes the water in contact with it – ice forming on aluminium trays which are in contact with the cold side of the module. In testing, the ice production increased by 84% when 2 V was applied to the module

Similarly, a study conducted by Söylemez et al. [9] investigated the application of a TEG module within a domestic refrigerator as shown in Figure 3.8. The system consisted of a Peltier module sandwiched between two heat exchangers, and the study considered the addition of simulated food blocks and how they cooled. The outcome demonstrated not only a significantly faster chilling process but also lower temperature limit was reached compared to traditional refrigeration units.

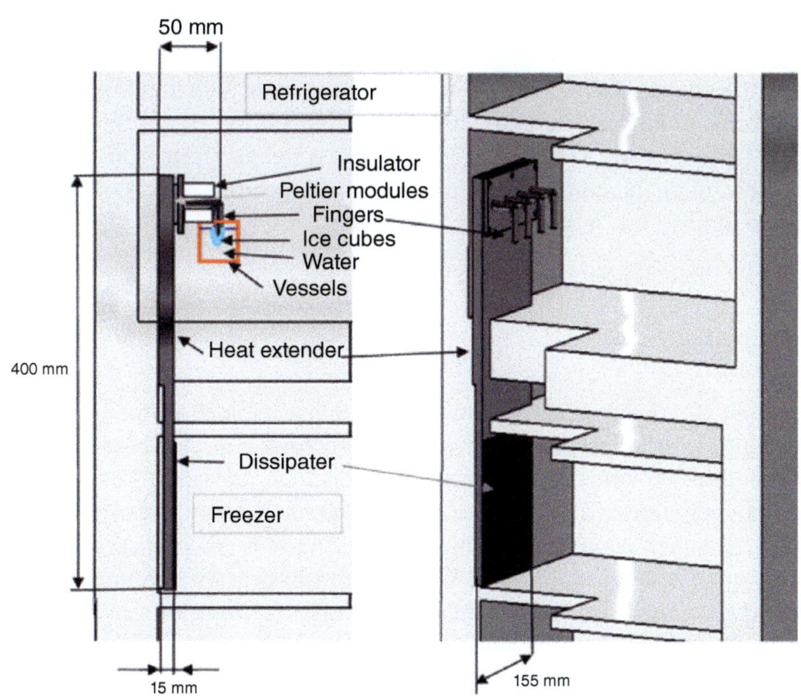

Figure 3.7 TEG application within an ice maker. Source: Rodríguez et al. [8].

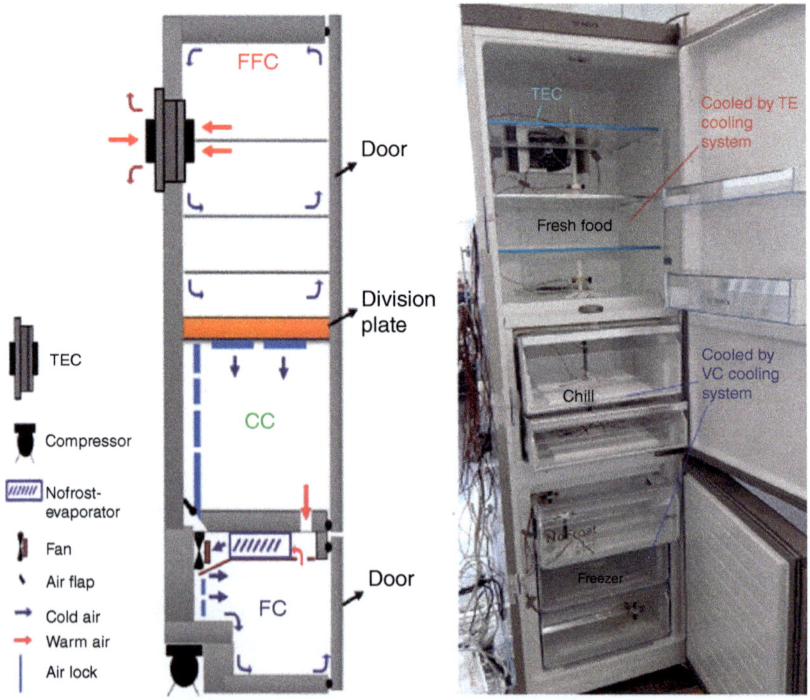

Figure 3.8 Proposed TEG domestic refrigerator. Source: Söylemez et al. [9].

To improve the efficiency and application of TEG modules, Min and Rowe [10] developed multiple prototype TEG refrigerators. The outcome of their study noted that although TEG units were easily able to maintain a temperature of 5 °C, their electrical consumption, in comparison to traditional refrigerators, was 30% higher and gave rise to significantly larger maintenance costs.

3.2 Cryogenics

The development of low-temperature and cryogenic heat pipes has been investigated, with multiple working fluids and operations being considered from both experimental and computational perspectives. For example, Long and Zhang [11] looked into the operation of a cryogenic thermosyphon at different fill ratios and inclinations. The system comprised a single copper thermosyphon charged with nitrogen, fed from a reservoir tank filled with liquid nitrogen, the thermosyphon and reservoir then being linked by a condenser section. To simulate a heat load, the evaporator was attached to a heating wire. The test exposed a problem. The lack of controls to normalise the evaporator temperature resulted in a consistent fluctuation, resulting in constant on and off operation, leading to an unsteady heat transfer rate and hence fluctuating condenser temperatures. So when the filling ratio was increased, the increasingly unsteady evaporator temperatures led to a drying out of the thermosyphon at the higher filling ratios.

The authors also developed a system of a single heat pipe charged with liquid helium encased with a copper shell [12]. Similar to the nitrogen-charged heat pipe, the evaporator also contained a heated wire to simulate its heat load. The condenser system further contained a cryocooler to act as a heat sink. Unlike the nitrogen configuration, the system contained two radiation shields to minimise radiative effects, and the initial tests highlighted a minimum radiative effect of 1 W during start up. The authors also developed a relationship between the filling ratio and the risk of drying out, with drying out occurring at 0.7 W and a 100% filling ratio.

As others have shown, helium is an optimum working fluid, and is often used to create conditions close to absolute zero with cryogenic heat pipes and cooling systems being applied at CERN for the cooling of particle colliders. The loop is cooled by liquid helium, but its storage and containment are extremely difficult, hence using helium within a thermosyphon could be problematic owing to such leakages, thereby altering the filling ratio. In respect of other cryogenic fluids, liquid nitrogen has been investigated by Bolozdynya et al. [13], and again the basis of their test was a copper thermosyphon with an electrical heater wire applied to the evaporator, the condenser heat sink comprising a liquid nitrogen-cooled head. The test considered the filling ratio and its effect on the subsequent cooling rate with the addition of a heat load, and displayed a poor cooling rate of 0.6 K/min with a filling ratio of 32%, but with a smaller filling ratio of 6.5% the test showed a cooling rate at 3 K/min (under identical conditions).

3.2.1 Loop Heat Pipe

A loop heat pipe (LHP) is based on a closed loop circuit containing both liquid and vapour phases simultaneously. The operation begins with a compensation chamber which contains the working fluid and a wick structure to ensure that there is always working fluid within the fluid chamber of the heat pipe. A section of the LHP leading from the compensation chamber houses an evaporator section, where the heat input is supplied. The vapour then travels through the adiabatic section of the LHP to the condenser section where it is transformed back into liquid, before finally travelling back to the fluid chamber. Loop heat pipes are widely applied to cryogenic systems, and Bai et al. [14] investigated cryogenic LHPs charged with nitrogen for the cooling of space exploration devices. The heat pipe employs the same principle (as shown in Figure 3.9), with the addition of secondary auxiliary loops – these secondary loops contain additional evaporator and condenser sections to initiate supercritical operation. In the study, an electrical heater was used to supply the thermal load to both the auxiliary evaporator and the main evaporator. Other, similar applications for space technologies have used LHPs. For example, Guo et al. [16] investigated a methane-based cryogenic LHP for thermal regulation within an infrared detection system, which was tested to study conditions such as the performance under supercritical conditions and its heat transport capacity. The research introduced an innovative compact system with an integrated wick structure within both

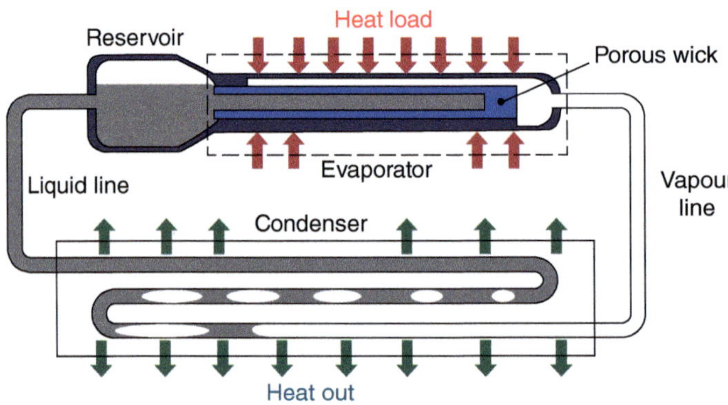

Figure 3.9 Loop heat pipe operation. Source: Yan et al. [15].

evaporator and compensation chambers, thereby eliminating the risk of dry out. The condenser consists of a primary and secondary coil wrapped around a cryocooler acting as a heat sink. The research's aim was to study supercritical heat pipe operation at ambient temperatures, and although successful during start-up, allowing the evaporator to drop the operating range of the working fluid, once the auxiliary heat load was reduced to 1 W in the cycling process the system failed to detect the heat load due to parasitic heat load from the surroundings.

Multiple studies on cryogenic LHPs have noted their potential to dry out and their failure to reach supercritical operation during start up, but in addition, the importance of selecting the correct working fluid. Yan et al. [15] studied various fluids to determine an optimum for cryogenic LHPs, and their initial tests were conducted using liquid nitrogen at different filling ratios and filling pressures. The system contains both a primary and a secondary evaporation loop – where the secondary evaporator loop is connected to a reservoir tank. With traditional closed LHPs, one of the main problems is the difficulty in maintaining a constant level of working fluid within the compensation chamber, so the study suggested an improvement of introducing another chamber to mitigate dry out and heat pipe failure and the second loop reducing the risk of failure during supercritical conditions. In this way, the secondary loop circulates fluid through the system whilst the evaporator reaches a suitable operating range for the working fluid. The relationship between the secondary loop and the filling ratio was widely considered within the study, and because of the improved circulation of fluid, it was found that the need for excess fluid only hindered the heat pipe performance, leading to an increase in saturation pressure; in as much as, the change in fill pressure (with constant fill ratio), highlighted an increase in saturation pressure, which led to an increase in resistance and a decrease in heat pipe performance.

Another variant is the pulsating heat pipe (PHP) that embodies a closed loop circuit, but one which relies on variations in both vapour bubbles and liquid slugs within the system, the temperature and pressure differentials between the evaporator and condenser sections creating a pulsating motion, allowing a liquid slug to pass through the system as shown in Figure 3.10.

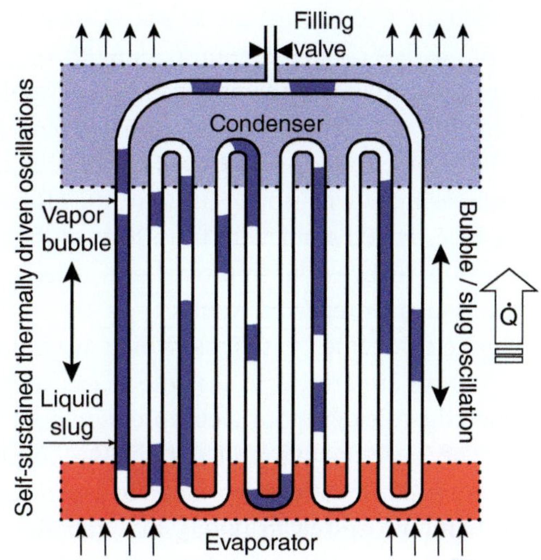

Figure 3.10 Schematic of pulsating heat pipe. Source: Liang et al. [17].

The system typically operates without a wick, making it a viable option for cryogenic heat pipes, but the system performance is highly dependent on the number of turns. Liang et al. [17] investigated the performance of an oscillating heat pipe charged with neon as the working fluid. The system operated from room temperature to the critical point of the working fluid and was designed to investigate the cool down period of the heat pipe. A heat load was applied to the evaporator and a cooling plate was fixed to the condenser section. The system showed that its performance was dependent on the rate of liquid neon entering the system because the cooling rate took nine hours from 223 to 31 K. It is worth noting that the system experienced significant pressure changes – indicating dry out – which also occurred in other PHPs when charged with a cryogenic fluid. For example, the effects of dry out were also experienced by Xu et al. [18] in their study to investigate the operation of a PHP charged with liquid nitrogen. The evaporator section was connected to a heating wire wrapped around the breadth of the system and the electrical input to the heater was increased accordingly to observe the thermal performance. The PHP was charged with liquid nitrogen, but increases in evaporator temperature led to the dry out of the PHP. Xu et al. [18] investigated a PHP charged with helium for ultra-fast cooling applications, in which a cryocooler was attached to the condenser and a heater wire was used for the evaporator to apply the heat load to the system. Prior to operation, a pre-cool function was applied to allow the PHP to reach the desired operating temperature, the helium PHP then allowed the cooling down from room temperature to 7 K in eight hours. Initially the configuration was aligned vertically and highlighted the effect of gravity, allowing the fluid to reach the evaporative system and surpass the operating temperature. However, when the test was repeated horizontally, no major impact was noted. Other variations and configurations do, nonetheless, play a key role in the operation, with the number of turns increasing the resistance of the system, and Barrak et al. [19] investigated different variations of PHPs with a selection of working fluids. The study involved six PHPs charged with butane and refrigerant R134a. A heater was applied through the evaporator section, with a coolant stream passing through the condenser. The location of the heater and cooler varied with a single or double contact with a variable power supply and the six PHPs were tested with both contact positions and electrical loads. The results demonstrated an improved performance for a double contact evaporator with an increasing power input. Similarly, the effects of horizontal and vertical positions were investigated along with the effects of thermal resistance, and for this research, the results reflected an improved and more reliable vertical operation against the horizontal, over repeated results.

Unlike cryogenic heat pipes, which are mainly limited to research applications, heat pumps are critical in the role of gas separation, and Fu and Gundersun [20] defined the role of vapour recompression heat pumps within the air separation process. The basic operation of the distillation process allows air to be drawn into a column, and nitrogen and oxygen are separated into their distinct forms; both gases then being processed to liquify in a reboiler, compressor, and a condenser combination. In a traditional system, a significant amount of heat loss occurs within the processing of impurities and compression, with an estimated 40% of heat being wasted to ambient. As an improvement, Figure 3.11 highlights an optimised liquification unit, the air entering the distillation cycle, where the flow is compressed and purified. The gas is then subcooled to its dew point (within a heat exchanger) and is then passed to a distillation column where it is cooled and expanded. This method also allows both gas streams to pass through a heat exchanger to liquify their impurities, which can be handled in a separate process (dashed line). The streams leaving the heat exchangers are further purified and cooled until both gases are liquefied. This proposed operation allows nitrogen to be used as a refrigerant with the heat pump, the outcome of this producing a 5% saving in energy consumption.

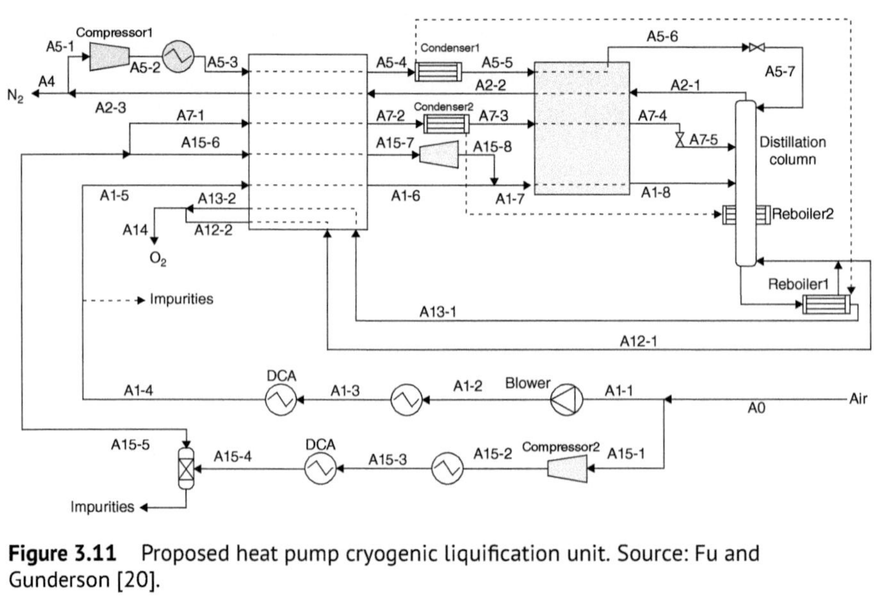

Figure 3.11 Proposed heat pump cryogenic liquification unit. Source: Fu and Gunderson [20].

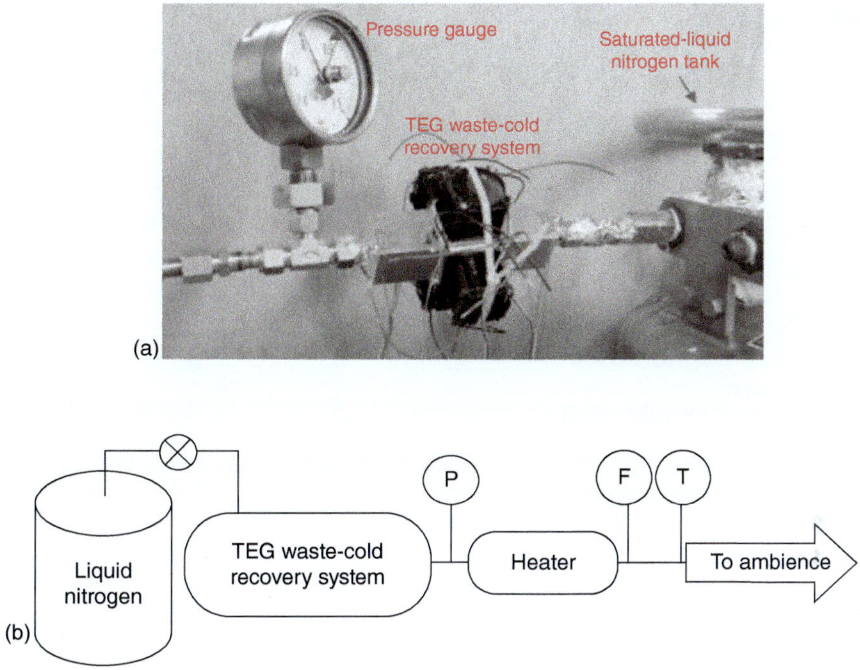

Figure 3.12 (a) Experimental set up of a TEG module attached to a nitrogen exhaust line and (b) schematic highlighting the operation of the system. Source: Weng et al. [21].

Considerations regarding impurities and vented nitrogen has also been studied to maximise the utilisation of the wasted thermal energy, and a study conducted by Weng et al. [21] investigated the applicability of TEG modules for cold recovery from vented cryogenic gases. The schematic highlighted in Figure 3.12 shows the installation of a TEG cold recovery unit on the outlet of a liquid nitrogen tank whilst in operation. As the tank separates liquid nitrogen from gaseous nitrogen, the gaseous phase is vented and passes through a TEG module to recover some cold energy. The outcome of the study showing a recovery of 0.93 W when the outlet flow was 3.7 g/s.

The utilisation of cold energy has also been investigated by Ge et al. [22], their research included the development of a cold storage method as shown in Figure 3.13. The unit is installed at the outlet of the tank where it is primarily gaseous nitrogen – the operation being similar to the study conducted by Weng et al. [21], but with one key difference – the introduction of a pre-cooling tank which contains the liquid nitrogen to ensure that liquid reaches the TEG module, the gas being re-liquified (if necessary) before passing to the TEG module. The study was conducted at various flow rates ranging from 40 to 100 l/min at a constant pressure and ambient temperature, and the outcome of the study showed that the addition of a TEG for this application only produced an efficiency of 0.03–0.035%, with only 20 W being recovered.

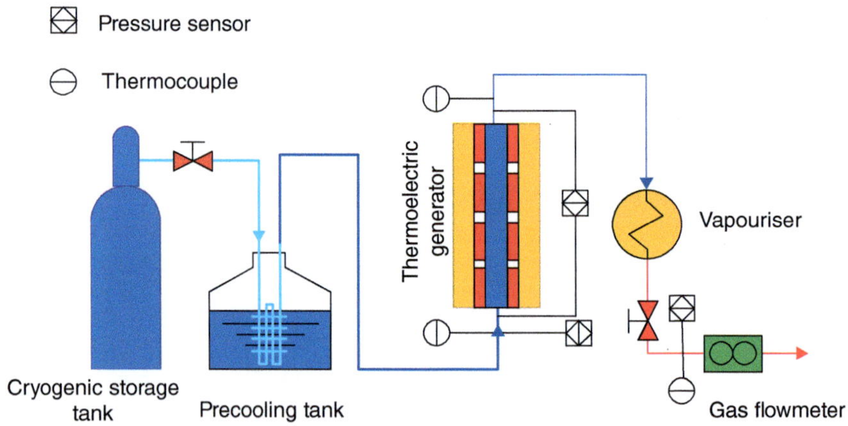

Figure 3.13 Schematic highlighting the application of TEG to recover energy from a liquid nitrogen line. Source: Ge et al. [22].

3.3 HVAC

The application possibilities of waste heat recovery technologies within the ventilation industry are vast, with multiple opportunities to combine technologies to optimise the amount of waste heat recovered from various buildings forms, ranging from residential homes to hospitals and office blocks.

One such example is the application of waste heat recovery technologies by Sukarno et al. [23], who investigated the role of heat pipe heat exchangers (HPHE) to reduce energy consumption and mitigate the spread of infectious disease for a hospital HVAC systems. The proposed system, shown in Figure 3.14, allows infectious air to be drawn from the room and act as a heat sink. At the same time,

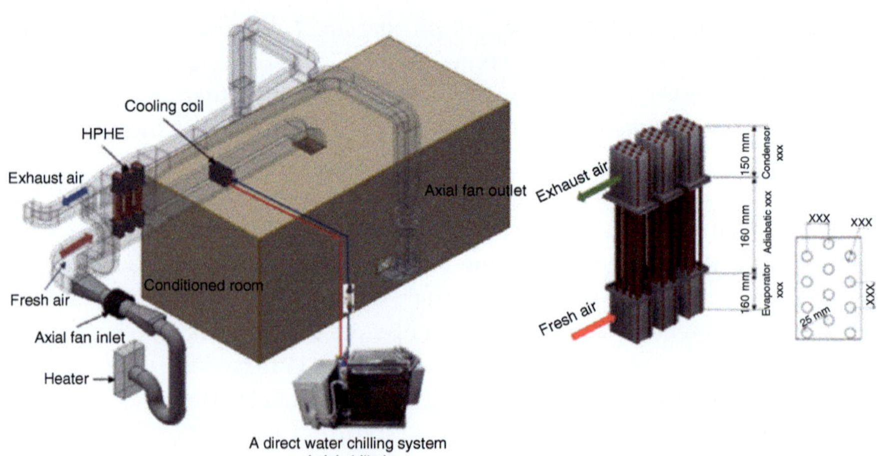

Figure 3.14 HPHE developed for a hospital application to maximise the energy efficiency and infection. Source: Sukarno et al. [23].

the evaporator section allows cooler clean air to be heated and reintroduced into the room. The system was tested to determine its energy efficiency at various inlet velocities, the outcome of the study highlighting heat recoveries ranging from 82 to 767 W depending on the inlet velocity.

The use of heat pipes has also been investigated by Jouhara and Meskimmon [24] and Jouhara and Ezzudin [25], the developed systems involved the utilisation of a wraparound loop heat pipe (WLHP), used in combination with a cooling coil as shown in Figure 3.15. The principle behind the system reverses the temperature difference across the heat pipe, which then contributes towards the dehumidification process and maximises the potential of the cooling coil, allowing the WLHP to act as a precooler for incoming air.

The use of heat exchangers has also been investigated by Eidan et al. [26] to compare the performance of HPHEs as a dehumidification process in subtropical climates. The study involved the development of a climate-controlled chamber to maintain a constant operating condition for the installed HVAC system as shown in Figure 3.15. The tests involved the gradual increase in relative humidity ranging from 15% to 100%, the outcome from the study highlighting a significant decrease in

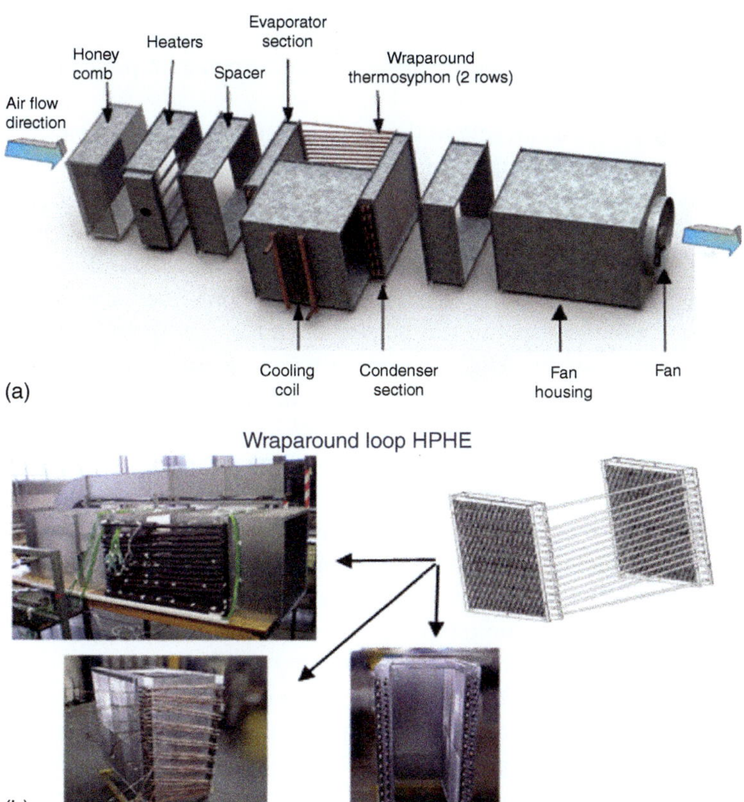

Figure 3.15 (a) Schematic highlighting the operation of WLHP and (b) developed WLHP prototype. Source: (a) Modified from Jouhara and Meskimonn [24].

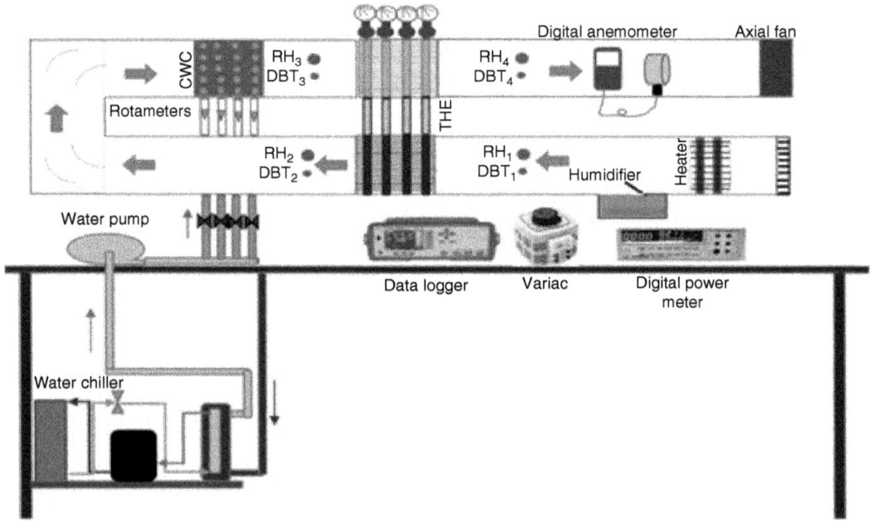

Figure 3.16 Subtropical HVAC system tested at a variable humidity. Source: Eidan et al. [26].

the dehumidification process as a result of utilising a cooling coil with pre-cooling from a HPHE (Figure 3.16).

Heat pumps also play a significant role in the development of HVAC systems to maximise all available heat sources from unconventional locations. For instance, a study conducted by Lyu et al. [27] developed HVAC system utilising geothermal energy. The plant involves the addition of a heat pump and an air-to-water heat exchanger to heat clean incoming air, but with additional channels to allow for pipes to be embedded within the walls. When the unit is cooling, the water from the condenser section of the heat pump can be circulated through these pipes to cool and remove heat stored in the walls. Conversely, when the unit is in its heating mode, the evaporator section from the heat pump can introduce warm water throughout the pipe-embedded walls. Studied under ambient conditions, the system showed significant energy savings of approximately 29%, and reduced the reliance on air-conditioning by 34% (Figure 3.17).

Similar work on utilising geothermal energy was also investigated by Magraner et al. [28], where a coupled geothermal heat pump system was developed in transient system simulation tool (TRNSYS) as shown in Figure 3.18. The network consists of a reversible water-to-water heat pump (optimised using a blend of propane and refrigerant) and a borehole heat exchanger. The heat pump operates in heating and cooling modes by (respectively) absorbing heat from the ground or rejecting heat into it, and the performance was studied for a series of estimations regarding occupancy and weather for a building complying with current building regulations – the outcome from all of which showed an improved energy performance of 15% compared to a generic house.

The application of geothermal hybrid systems has also been considered by Villarino et al. [29], for use within a multi-storey office building in Spain. The schematic,

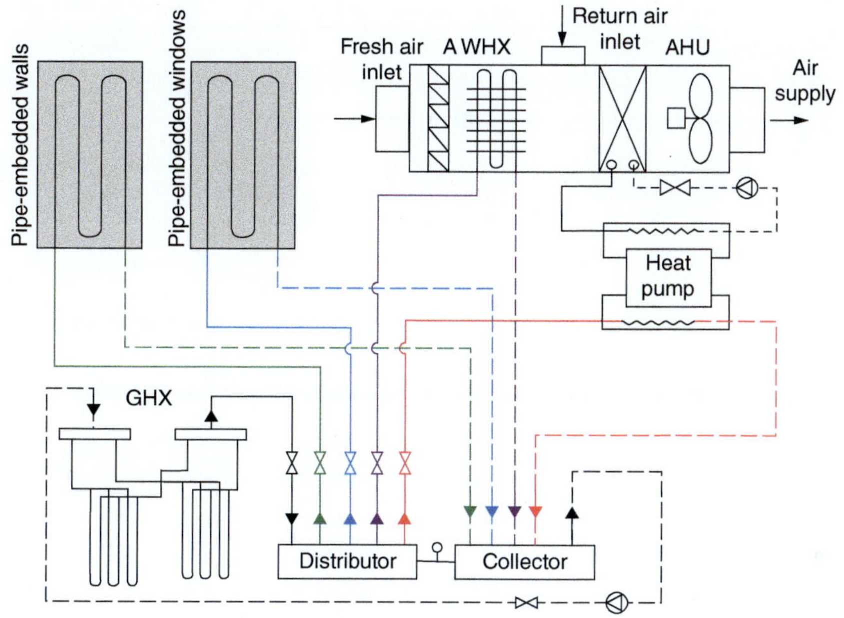

Figure 3.17 Geothermal HVAC system featuring pipe-embedded walls for heating and cooling applications. Source: Lyu et al. [27].

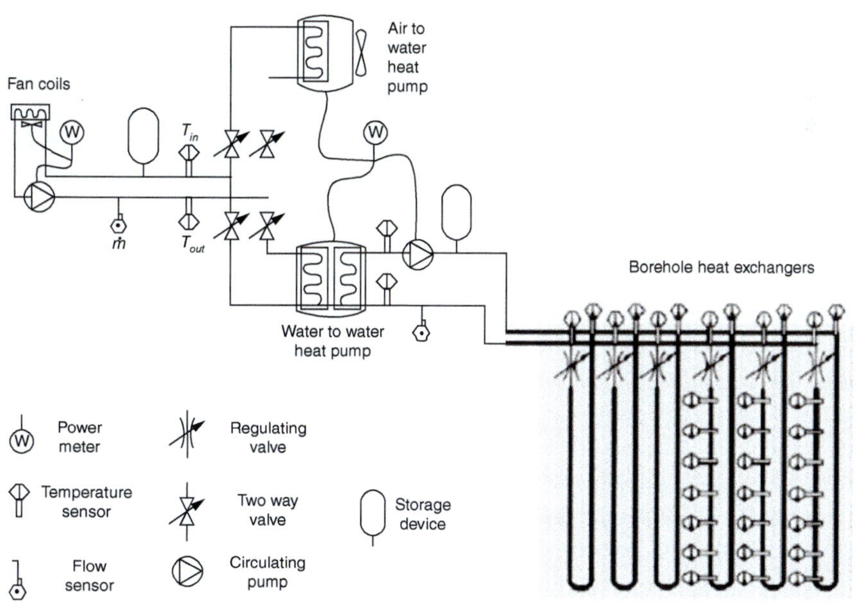

Figure 3.18 Developed geothermal heat pump network. Source: Magraner et al. [28].

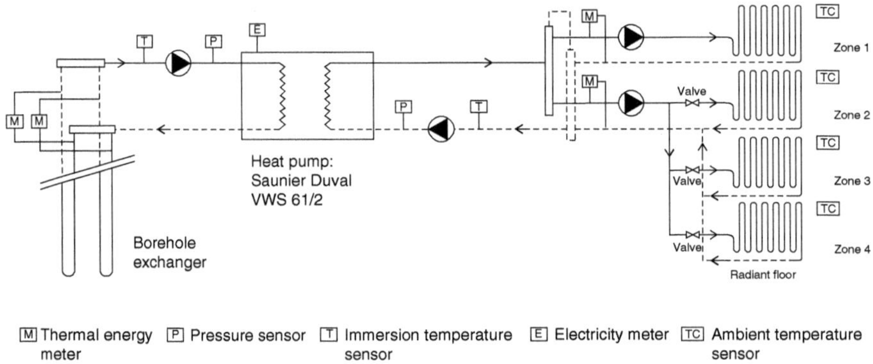

Figure 3.19 Geothermal system developed for a multi-storey office building located in Spain. Source: Villarino et al. [29].

shown in Figure 3.19, consists of a borehole heat exchanger connected to a heat pump (which will allow for heating and cooling modes) feeding a radiant floor system. The system was tested in summer conditions in July for one week, and the results indicated that the geothermal heat pump system could provide a room temperature of 21.7 °C in cooling mode and 24.8 °C in heating mode.

By way of performance improvement and to maximise the available waste heat potential, the application of phase change material (PCM) to HVAC systems has also been investigated. For example, Maccarini et al. [30] investigated the application of a PCM heat exchanger combined with a reversible air-to-water pump in a commercial building. Traditionally, PCM heat exchanger systems function by allowing simultaneous heating and cooling via a two-way loop, in which water is fed into the system and kept at a constant of 22 °C throughout all zones in the building regardless of their individual heating or cooling requirements. The advantage of the system presented within this study highlights the potential for waste heat to be transferred to cooler zones, so reducing the building's reliance on its heating system(s). In addition, the use of the PCM allows for heat to be stored and discharged to a cold-water source in the evening, and by such operation, saved approximately 6.3 kWh/m^2 compared to a traditional configuration. As a further application, PCM's have also been studied in the form of PCM integrated walls, one such study conducted by Kishore et al. [31], investigated the potential of the technology within different climates within the United States (US). The schematic highlighted in Figure 3.20 shows a cross section of the proposed PCM integrated walls, the principle behind this being the charging of the PCM during daylight hours, and discharging during the evening. The investigation of the PCM showed that during the peak summer months, the PCM was 100% melted, hence highlighting that the PCM was not functional during these times. However, during cooler months the PCM showed a significant but varied heat gain across all five locations. Locations based on desert climates showed minimal heat gain values of under 4% but mountainous regions registered a heat gain of 47%.

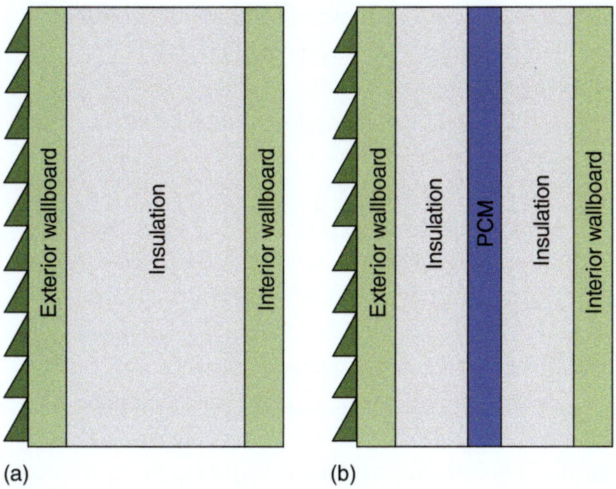

Figure 3.20 (a) Regular wall and (b) PCM integrated wall [29]. Source: Kishore et al. [31].

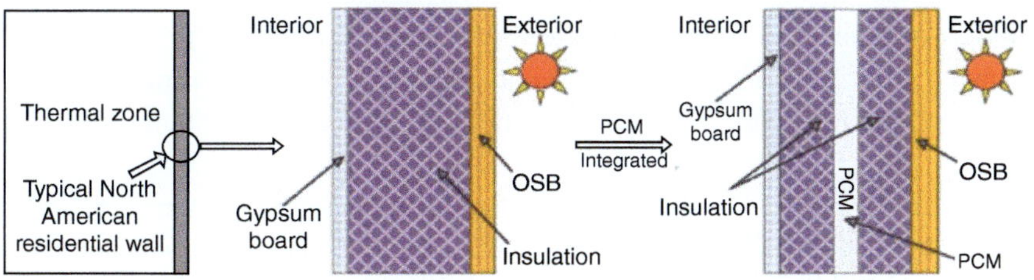

Figure 3.21 Thin integrated PCM walls. Source: Li et al. [32].

A similar study was conducted by Li et al. [32], who also investigated thin PCM-reinforced walls in a multitude of climate zones within the US ranging from deserts to high-altitude regions. Similar to the study conducted by Kishore et al. [31], the developed schematic shown in Figure 3.21 also features a thinner PCM sandwiched between layers of insulation.

The study investigated the thickness of the PCM throughout the different zones and months, and the study indicated that increasing the PCM layer to 15 mm also increased the building efficiency (due to the increased thermal storage capacity) without altering the melt point. In addition, the study showed that when the PCM was applied in hot climates and deserts – where the temperature exceeds 30 °C – the charging of the PCM was able to maintain comfortable room temperatures.

References

1 Liu, D.K., Xu, C.C., Guo, C.X., and Zhang, X.X. (2020). Sub-zero temperature preservation of fruits and vegetables: a review. *J. Food Eng.* 275: 109881. https://doi.org/10.1016/j.jfoodeng.2019.109881.

2 Wu, W., Wang, B., Shi, W., and Li, X. (2014). An overview of ammonia-based absorption chillers and heat pumps. *Renewable Sustainable Energy Rev.* 31: 681–707. https://doi.org/10.1016/j.rser.2013.12.021.

3 Singh, S. and Dasgupta, M.S. (2017). CO2 heat pump for waste heat recovery and utilization in dairy industry with ammonia based refrigeration. *Int. J. Refrig* 78: 108–120. https://doi.org/10.1016/j.ijrefrig.2017.03.009.

4 Polzot, A., Dipasquale, C., D'Agaro, P., and Cortella, G. (2017). Energy benefit assessment of a water loop heat pump system integrated with a CO_2 commercial refrigeration unit. *Energy Procedia* 123, Elsevier Ltd: 36–45. https://doi.org/10.1016/j.egypro.2017.07.282.

5 Salehi, S. and Yari, M. (2019). Exergoeconomic assessment of two novel absorption-ejection heat pumps for the purposes of supermarkets simultaneous heating and refrigeration using $NaSCN/NH_3$, $LiNO_3/NH_3$ and H_2O/NH_3 as working pairs. *Int. J. Refrig* 101: 178–195. https://doi.org/10.1016/j.ijrefrig.2019.03.029.

6 Palm, B. (2008). Ammonia in low capacity refrigeration and heat pump systems. *Int. J. Refrig* 31: 709–715. https://doi.org/10.1016/j.ijrefrig.2007.12.006.

7 Afshari, F., Comakli, O., Karagoz, S., and Zavaragh, H.G. (2018). A thermodynamic comparison between heat pump and refrigeration device using several refrigerants. *Energy Build.* 168: 272–283. https://doi.org/10.1016/j.enbuild.2018.03.037.

8 Rodríguez, A., Vián, J.G., and Astrain, D. (2009). Development and experimental validation of a computational model in order to simulate ice cube production in a thermoelectric ice-maker. *Appl. Therm. Eng.* 29: 2961–2969. https://doi.org/10.1016/j.applthermaleng.2009.03.005.

9 Söylemez, E., Alpman, E., Onat, A., and Hartomacıoğlu, S. (2021). CFD analysis for predicting cooling time of a domestic refrigerator with thermoelectric cooling system. *Int. J. Refrig* 123: 138–149. https://doi.org/10.1016/j.ijrefrig.2020.11.012.

10 Min, G. and Rowe, D.M. (2006). Experimental evaluation of prototype thermoelectric domestic-refrigerators. *Appl. Energy* 83: 133–152. https://doi.org/10.1016/j.apenergy.2005.01.002.

11 Long, Z.Q. and Zhang, P. (2012). Impact of cooling condition and filling ratio on heat transfer limit of cryogenic thermosyphon. *Cryogenics* 52: 66–76. https://doi.org/10.1016/J.CRYOGENICS.2011.11.004.

12 Long, Z.Q. and Zhang, P. (2013). Experimental investigation of the heat transfer characteristics of a helium cryogenic thermosyphon. *Cryogenics* 57: 95–103. https://doi.org/10.1016/J.CRYOGENICS.2013.04.006.

13 Bolozdynya, A.I., Dmitrenko, V.V., Efremenko, Y.V. et al. (2015). The two-phase closed tubular cryogenic thermosyphon. *Int. J. Heat Mass Transfer* 80: 159–162. https://doi.org/10.1016/J.IJHEATMASSTRANSFER.2014.09.001.

14 Bai, L., Lin, G., Zhang, H. et al. (2012). Experimental study of a nitrogen-charged cryogenic loop heat pipe. *Cryogenics* 52: 557–563. https://doi.org/10.1016/J.CRYOGENICS.2012.07.005.

15 Yan, T., Zhao, Y., Liang, J., and Liu, F. (2013). Investigation on optimal working fluid inventory of a cryogenic loop heat pipe. *Int. J. Heat Mass Transfer* 66: 334–337. https://doi.org/10.1016/J.IJHEATMASSTRANSFER.2013.07.043.

16 Guo, Y., Lin, G., Zhang, H., and Miao, J. (2018). Investigation on thermal behaviours of a methane charged cryogenic loop heat pipe. *Energy* 157: 516–525. https://doi.org/10.1016/J.ENERGY.2018.05.133.

17 Liang, Q., Li, Y., and Wang, Q. (2017). Experimental investigation on the performance of a neon cryogenic oscillating heat pipe. *Cryogenics* 84: 7–12. https://doi.org/10.1016/J.CRYOGENICS.2017.03.004.

18 Xu, D., Li, L., and Liu, H. (2016). Experimental investigation on the thermal performance of helium based cryogenic pulsating heat pipe. *Exp. Therm Fluid Sci.* 70: 61–68. https://doi.org/10.1016/J.EXPTHERMFLUSCI.2015.08.024.

19 Barrak, A.S., Saleh, A.A.M., and Naji, Z.H. (2019). An experimental study of using water, methanol, and binary fluids in oscillating heat pipe heat exchanger. *Eng. Sci. Technol. Int. J.* https://doi.org/10.1016/J.JESTCH.2019.05.010.

20 Fu, C. and Gundersen, T. (2013). Recuperative vapor recompression heat pumps in cryogenic air separation processes. *Energy* 59: 708–718. https://doi.org/10.1016/j.energy.2013.06.055.

21 Weng, C.C., Lin, M.C., and Huang, M.J. (2016). A waste cold recovery from the exhausted cryogenic nitrogen by using thermoelectric power generator. *Energy* 103: 385–396. https://doi.org/10.1016/j.energy.2016.02.146.

22 Ge, M., Li, Z., Wang, Y. et al. (2021). Experimental study on thermoelectric power generation based on cryogenic liquid cold energy. *Energy* 220: 119746. https://doi.org/10.1016/j.energy.2020.119746.

23 Sukarno, R., Putra, N., Hakim, I.I. et al. (2021). Utilizing heat pipe heat exchanger to reduce the energy consumption of airborne infection isolation hospital room HVAC system. *J. Build. Eng.* 35: 102116. https://doi.org/10.1016/j.jobe.2020.102116.

24 Jouhara, H. and Meskimmon, R. (2010). Experimental investigation of wraparound loop heat pipe heat exchanger used in energy efficient air handling units. *Energy* 35: 4592–4599. https://doi.org/10.1016/j.energy.2010.03.056.

25 Jouhara, H. and Ezzuddin, H. (2013). Thermal performance characteristics of a wraparound loop heat pipe (WLHP) charged with R134A. *Energy* https://doi.org/10.1016/j.energy.2012.10.016.

26 Eidan, A.A., Najim, S.E., and Jalil, J.M. (2017). An experimental and a numerical investigation of HVAC system using thermosyphon heat exchangers for sub-tropical climates. *Appl. Therm. Eng.* 114: 693–703. https://doi.org/10.1016/j.applthermaleng.2016.12.027.

27 Lyu, W., Li, X., Yan, S., and Jiang, S. (2020). Utilizing shallow geothermal energy to develop an energy efficient HVAC system. *Renewable Energy* 147: 672–682. https://doi.org/10.1016/j.renene.2019.09.032.

28 Magraner, T., Montero, Á., Quilis, S., and Urchueguía, J.F. (2010). Comparison between design and actual energy performance of a HVAC-ground coupled heat pump system in cooling and heating operation. *Energy Build.* 42: 1394–1401. https://doi.org/10.1016/j.enbuild.2010.03.008.

29 Villarino, J.I., Villarino, A., and Fernández, F.Á. (2017). Experimental and modelling analysis of an office building HVAC system based in a ground-coupled heat pump and radiant floor. *Appl. Energy* 190: 1020–1028. https://doi.org/10.1016/j.apenergy.2016.12.152.

30 Maccarini, A., Hultmark, G., Bergsøe, N.C., and Afshari, A. (2018). Free cooling potential of a PCM-based heat exchanger coupled with a novel HVAC system for simultaneous heating and cooling of buildings. *Sustainable Cities Soc.* 42: 384–395. https://doi.org/10.1016/j.scs.2018.06.016.

31 Kishore, N. and Ramteke, R.R. (2016). Forced convective heat transfer from spheres to Newtonian fluids in steady axisymmetric flow regime with velocity slip at fluid–solid interface. *Int. J. Therm. Sci.* 105: 206–217.

32 Li, M., Cao, Q., Pan, H. et al. (2021). Effect of melting point on thermodynamics of thin PCM reinforced residential frame walls in different climate zones. *Appl. Therm. Eng.* 188: 116615. https://doi.org/10.1016/j.applthermaleng.2021.116615.

4

Medium-Temperature Applications

A significant amount of our daily essentials is reliant on industrial processes – from the food we buy in supermarkets to the cars we drive, but this also results in an increase in fossil fuel dependency and an increase in the emissions generated. Often on an industrial scale, these generated emissions contain harmful greenhouse gases and valuable waste heat – which has the potential to be applied in other processes. The multiple technologies mentioned in Chapter 2 reflect the potential for waste heat recovery and the associated reduction in emissions and economic benefits. This chapter reflects the application of waste heat technologies in the medium-temperature range (0–400 °C).

4.1 Food Industry

Energy efficiency in the food-processing sector is becoming an increasingly important concern. Because not only are the costs of both electricity and fossil fuel resources dramatically increasing, we also have the challenging targets for lowering greenhouse gas emissions (as set out in the Climate Change Act 2008), and so waste heat recovery is essential in improving industrial energy efficiency. The estimated recoverable heat that is lost annually in the United Kingdom alone is 11.4 TWh, of which a quarter comes from the food- and beverage-processing industry. Utilisation of this heat, which can be described as a green, carbon-neutral energy source may therefore be used to reduce fossil fuel consumption and reduce greenhouse gas emissions and so increase the cost-effectiveness of a process.

Table 4.1 indicates the best-case theoretical scenarios in which all of the available waste heat is recovered to replace heating duties traditionally provided by natural gas.

4.1.1 Energy Use in the Industry

As mentioned earlier, this sector accounts for around 25% of industrial energy usage within the United Kingdom, around 42 TWh/year, and this energy usage can be broken down according to various processes as shown Figure 4.1. From this figure, we can see that energy use (within the industry) is dominated by low-temperature

Waste Heat Recovery in Process Industries, First Edition. Hussam Jouhara.

Table 4.1 Potential cost and emissions savings per year due to recovery of industrial waste heat.

	Units of waste energy (kWh/year)	Cost per unit gas [1] (£/kWh)	Greenhouse gas emissions per unit of gas [2]	Potential cost saving per year (£m/year)	Potential emissions saving per year (tCO_2eq/year)
Industry total	11.4×10^9	0.025	0.1836	£285 m	2 093 040
Food and drinks sector	2.8×10^9	0.025	0.1836	£70	514 080

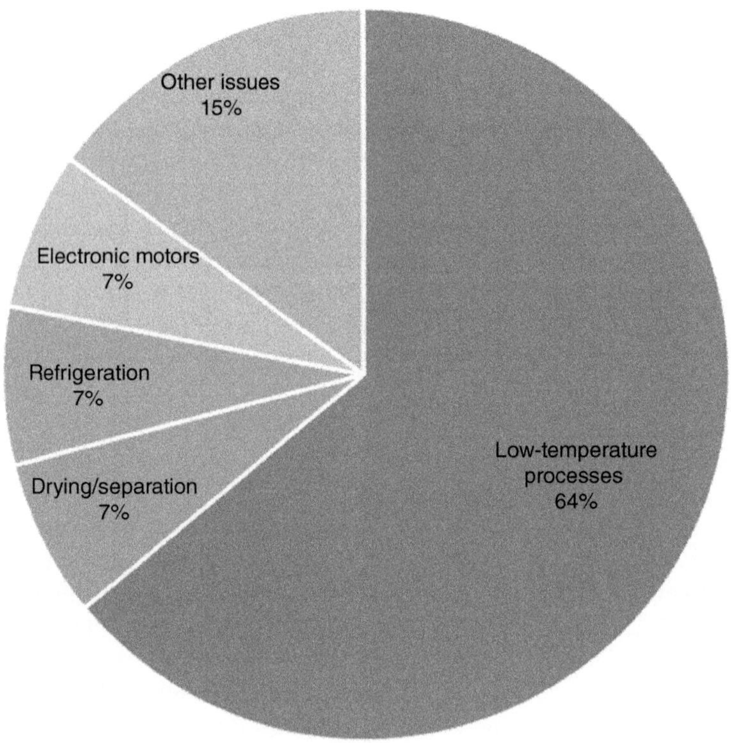

Figure 4.1 Breakdown of sector energy usage (Climate Change Act).

processes (including cooking food in ovens/fryers, etc.), with the only other high intensity thermal processes associated with drying/separation and refrigeration.

4.1.2 Case Study 1: Heat Recovery Potential of the Crisps Manufacturing Process

In the production phase of crisps manufacturing: natural gas, fryer foul gas – consisting primarily of fat (optional in some processes), combustion air, and water (used to regulate the temperature of the combustion chamber) are all burnt within the combustion chamber at a temperature of about 472 °C. After combustion, the hot waste stream from the combustion chamber flows into the heat exchanger where it is used to heat the cooking oil from 154 to 170.2 °C.

In certain industrial facilities, once the waste stream is discharged from the heat exchanger part of it is filtered back to the combustion chamber [3], whilst the remainder is emitted to the atmosphere via an exhaust pipe. However, in other processing facilities, the waste stream is directly emitted into the atmosphere without any waste heat recovery/recycling methods being employed at all, and since crisp production requires huge amounts of energy, it might be presumed that utilisation of waste heat during the production process would be of the greatest priority; especially as there are multiple waste heat recovery possibilities from the fryer section of the crisps manufacturing plant. For example:

- Using the waste heat from the foul gas exiting the fryer to drive an organic Rankine cycle (ORC) system for power generation – In the crisps manufacturing process, the foul gas from the fryer is mostly hot vapourised cooking oil at a temperature of about 120 °C with a mass flow rate of about 3.172 kg/s and mass fraction of 0.606 of $H_2O_{(g)}$, 0.299 of N_2, 0.095 of O_2, and negligible CO_2
- Using the waste heat from the waste gas stream from the heat exchanger to drive an ORC system for power generation – The waste stream from the heat exchanger to the exhaust stack is at a temperature of approximately 164 °C, with an approximate mass flow rate of 10.509 kg/s with mass fraction of 0.411 of $H_2O_{(g)}$, 0.050 of CO_2, 0.506 of N_2 and 0.033 of O_2.

4.1.3 Case Study 2: Temperature and Energy Performance of Open Refrigerated Display Cabinets Using Heat Pipe Shelves

One of the biggest users of electricity in the United Kingdom are the supermarket food stores. Based on the distribution area of the store, the trading procedures, the shop layout, the shopping operation, and the equipment used in the store to conserve and show food items, their energy usage is measured at about 2000 kWh/m^2/year [4]; although we need to be aware that cooling systems only consume about half of this energy, the other support facilities, such as the air-conditioning, utilise the remainder.

Vertical open-style display cabinets are the most popular type of refrigerators and are found in supermarkets and department stores worldwide. These feature a recirculating air barrier – which acts as an aero-thermodynamic barrier between the cabinet interior and the outside world – but also enables food items to be seen without physical restriction. In a typical cabinet configuration, the heat circulation between foodstuffs and shelves is comparatively limited, which means that food items will cool inefficiently in places where the products need to have a heavy interaction with the shelves. Furthermore, during any defrost periods, they also exhibit large temperature fluctuations and major temperature rises.

In compliance with ISO 23953-2:2015, the UK government specifies that all food items shall be stored at a temperature below 5 °C [5], but unfortunately, current cabinet prototypes seem incapable of holding minimum temperatures without inducing localised freezing in the cabinet. Public Group surveys have found that 70% of foodstuffs in supermarket display cabinets are kept at temperatures above 8 °C, while 60% of items are kept at or above 10 °C [6]. It is therefore essential to revisit these refrigeration systems.

Refrigeration functions by preventing or at least retarding deterioration, spoilage, and development of bacteria within a product. Hence, the overall objective of any cooling system is not just to maintain the acceptability of food goods, but also (in regards to the colour, smell, taste, and vitamins content) to maintain food items at their highest quality, because any observable changes in these variables reduces a product's consumer value and so contributes to economic loss.

In several industries, heat pipe-based heat exchangers are used as heat recovery and energy-saving devices [7, 8] and because of the high concentrations of heat they can transfer even with relatively small temperature gradients, heat pipes are also sometimes known as thermal superconductors. In their basic form, heat pipes are referred to as 'thermosiphons', and their work depends on gravity, where heat is transmitted only from the lower end of the pipe to the higher end. A heat pipe is a hermetically sealed tube partially filled with a small amount of working fluid, which then exists in both the liquid and the vapour phases, and is key to its operation. At the higher temperature 'evaporator' section of the heat pipe, the contained working fluid in the liquid phase evaporates and flows at high speed into the cooler end of the pipe. This vapour then condenses (within the 'condenser' section) and gives up its heat. Finally, by gravitational effect, the liquid operating fluid returns to the vessel's evaporator.

A series of straight heat pipes may be viewed therefore as a heat recovery unit connected to form one system – where high thermal conductivity, passive and reliable operation, consistent temperature distribution, affordable costs, and no conventional external pumping systems exchangers – are all to its advantage. Relevant to such an application, Jouhara et al. [9] have developed a novel technology that incorporates an actively cooled heat pipe-based refrigerator shelf, the shelves being designed so that food items can be placed on the upper surface of the flat heat pipe which are made from aluminium materials filed with their working fluid of ammonia (NH_3). Figure 4.2 illustrates the structure of the heat pipe-based refrigerator shelf.

This novel adaption of heat pipes to shelving integrates three additional heat transfer processes into the current cooling system, (as can be seen from Figure 4.3): (i) heat transfer due to conduction from the top surface of the heat pipe from food items positioned on it, (ii) heat transfer by natural convection between the food and the base surface of a second shelf placed directly above the food, and (iii) heat transfer due to radiation which helps transfer heat from the warmer food items to the colder surfaces of the heat pipe shelves.

Case studies employed two identical merchandised open display cabinets for the purpose of experimental analysis and comparison. One case was equipped with its

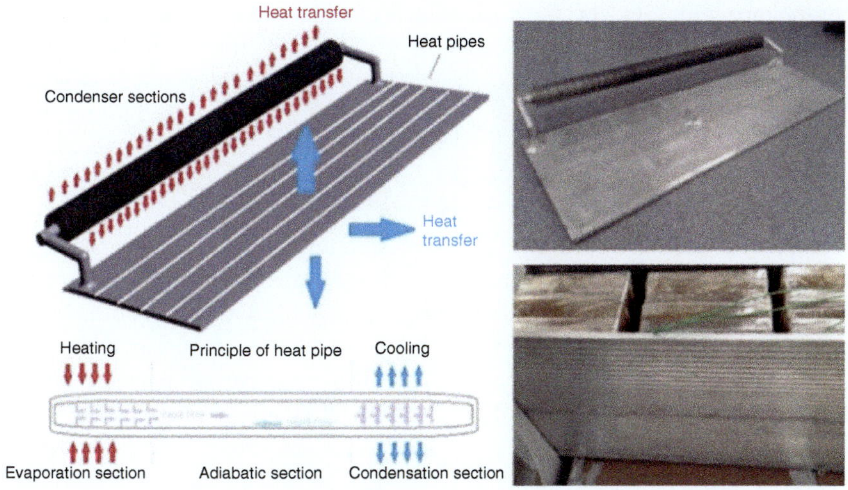

Figure 4.2 Heat pipe-based refrigerator shelf.

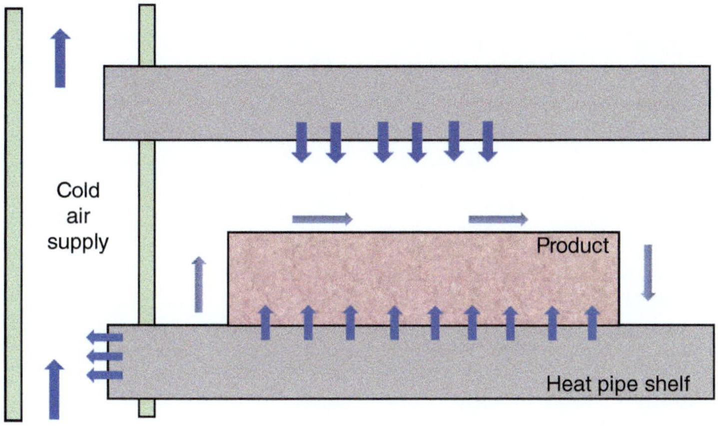

Figure 4.3 Heat transfer mechanics of the heat pipe shelf.

standard shelving whilst the second was fitted with the advanced actively cooled shelf, but both cabinets were then sited inside an environmental-controlled research chamber. The position of the products and the estimated shelf life for each product used in the tests are shown in Figure 4.4.

The data acquisition (DAQ) system was linked to a total of 28 k-type thermocouples for quantifying and tracking the primary temperatures of the products, the traditional shelf and the actively cooled shelf surface temperatures, the cabin back panel temperature, and the ambient air temperature, and in order to record the energy usage for the cabinets and the temperature of the food products, experiments were carried out for 24 hours. As illustrated in Figure 4.5, products in the front row of the shelves were directly measured with thermocouples. For products in touch with the shelf in the back row, in each front column, in the back row of the shelves, and those

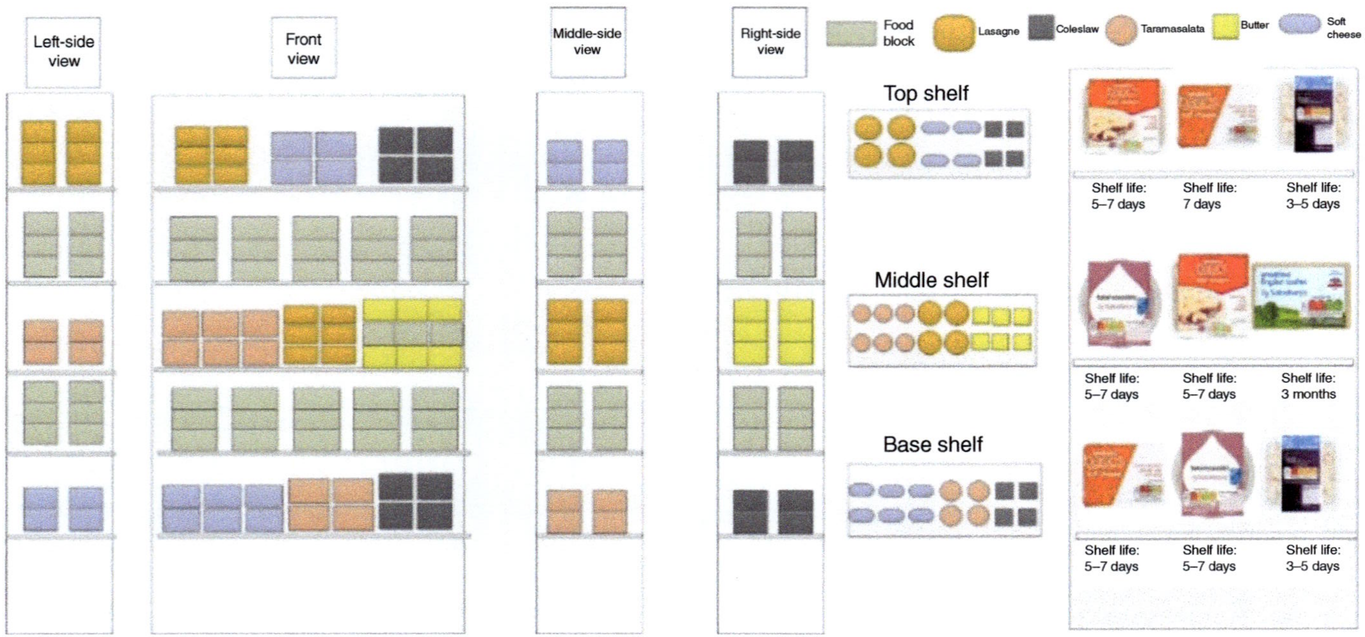

Figure 4.4 Cabinets' food layout.

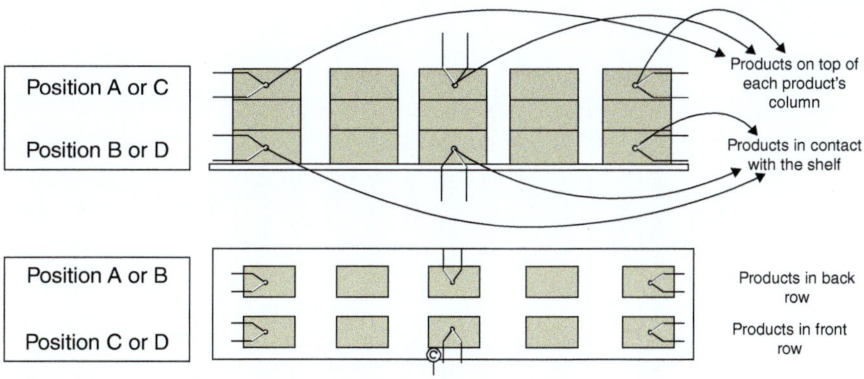

Figure 4.5 Different positions of food products with thermocouples within the cabinet.

on the top of each product column (as seen in Figure 4.5), they were measured by thermocouples placed in the following positions: (i) on top of each product's column, (ii) in contact with the shelf, (iii) in the back row, (iv) in the front row.

Table 4.2 shows the reflective results for the purpose of demonstrating the temperature distributions for the experiment. Dependent on their location within the cabinet, the lowest and highest temperatures of all food items were recorded. With the exception of the food items located on the top shelf of the system, most other food items located on a heat pipe shelf demonstrated lower core temperatures compared to their equivalent product sited on a traditional shelf. Furthermore, it was noted that for the heat pipe shelves, food items in positions B and D, i.e. those in direct contact with the shelf, experienced the lowest temperatures. However, and unexpectedly, the food items located in point D for the middle and bottom traditional shelves were subject to hot spots – which were not anticipated.

It can be seen that the average core temperatures of the products placed on the top heat pipe shelf fluctuate less compared to the same products on traditional shelving, but also, the tests showed that for conventional shelves, the lowest core temperatures occurred with back-row products (positions A and B).

Overall, the experimental findings indicated that heat pipe shelves were able to obtain a more consistent temperature distribution both along the shelves and across the refrigerator, and a more incremental cooling profile. In addition, the heat pipe shelves produced lower temperature gradients between the various product locations inside the cabinet compared to traditional shelves. The average daily electrical power usage of the cabinet fitted with heat pipe shelves was about 7.6 kWh, whereas the sufficient electrical power consumption for the traditional configuration was higher by 1 kWh per day (8.6 kWh). Finally, the cabinet fitted with the heat pipe shelves showed an overall energy saving of 12%, which translates to an economic benefit of almost £44 per cabinet per year.

Table 4.2 Maximum and minimum product temperatures and positions achieved on the heat pipe and conventional shelves.

		Traditional shelves temperature		Heat pipe shelves Temperature	
	Product	**Lowest**	**Highest**	**Lowest**	**Highest**
Top shelf	Lasagne	3.7 °C Position B	4.7 °C Position C	2.7 °C Position B	4.0 °C Position C
	Soft cheese	3.7 °C Position B	5.3 °C Position C	2.9 °C Position B	5.0 °C Position C
	Coleslaw	3.6 °C Position B	4.7 °C Position C	3.6 °C Position B	5.0 °C Position C
Middle shelf	Taramasalata	2.2 °C Position B	4.7 °C Position C	3.0 °C Position B	4.2 °C Position C
	Lasagne	2.8 °C Position B	4.7 °C Position C	3.0 °C Position B	3.8 °C Position D
	Butter	2.6 °C Position B	3.7 °C Position C	3.0 °C Position A	4.0 °C Position D
Base shelf	Soft cheese	2.8 °C Position B	4.1 °C Position C	2.4 °C Position B	4.1 °C Position D
	Taramasalata	2.6 °C Position B	3.8 °C Position C	3.6 °C Position A	3.8 °C Position D
	Coleslaw	3.5 °C Position B	4.7 °C Position C	4.5 °C Position B	6.0 °C Position D

4.2 Ventilation

Heat recovery ventilators (HRVs) are air-to-air heat exchangers that perform effective waste heat recovery in domestic, commercial, and industrial applications. In this way, pre-conditioning of the incoming air supply to a building delivers a heating benefit by using the otherwise wasted temperature gradient between air supply and exhaust.

It is widely recognised that HRVs are therefore effective in saving energy whilst preserving the required rates of ventilation, and the sort of HRVs that reuse expelled indoor heat have been very successfully used in buildings and for which, a number of studies have been undertaken to investigate the effect of heat recovery systems on energy building efficiency.

Studies have demonstrated that the use of HRVs can save energy for heating purposes. However, additional cooling energy is required to deal with high outdoor temperatures in summer. Beneficially though, other experiments have shown that HRVs capable of transferring both latent and sensible heat have effectively decreased both heating and cooling energy together.

Energy invested in maintaining heating, ventilation, and air-conditioning (HVAC) systems amounts to roughly 60% of global overall building energy consumption, and this segment will only increase with population growth and prosperity, hence the development and deployment of cost-effective innovations that increase the performance of HVAC, including specific waste heat recovery methods, continues to be pursued.

By way of example, the US Department of Energy published a review stating that the energy usage of buildings could be decreased by up to 50% by the use of currently available energy recovery/management technology, but it also further suggested that advanced waste heat recovery/ventilation technology would substantially reduce the energy required to condition ventilation air by utilising the frequently ignored temperature potentials between cold/hot air streams in HVAC duct systems. This temperature resource could, for example, push heat transfer by air-to-air heat exchangers, thereby facilitating pre-cooling and/or pre-heating of air to minimise the power requirement for air-conditioning (AC) and heating units.

4.2.1 Applications

Researchers have recently made significant and important advances in waste heat recovery for ventilation technologies by investigating the properties, architecture, construction, and thermal efficiency of heat pipes. Conventional heat pipe (CHP) technologies have been developed for use of low-temperature sources (15–55 °C), and Noie-Baghban and Majideian [1] investigated the recovery of waste heat using CHPs for surgery rooms in hospitals. The study showed that the efficacy of CHP was 16%, but it relied on the diameter and the fin difference and so the heat quality is limited since the CHP has been configured for low-temperature operating conditions.

Another investigation [10] implemented CHPs for heat recovery between two sources of fresh air and the return air in an AC facility where the incoming fresh air was to be cooled. To confirm the heat transfer and the temperature shift of the fresh air, the ratios of the mass flow rate between the return and fresh air were examined. During the experiments, the temperature of the fresh air inlet was monitored between 32 and 40 °C, while the temperature of the inlet return air was maintained at approximately 26 °C.

Martínez et al. [2] designed a mixed-energy recovery system consisting of two CHPs and indirect evaporative regenerators for AC and using experimental modelling approaches, the energy characterisation of the mixed-energy recovery mechanism was carried out. The key finding was that part of the energy from the return airflow could be extracted by adding a mixed-energy recovery system to AC installations consisting of two CHPs and indirect evaporative systems, thereby increasing energy efficiency, and reducing the environmental effects.

Lukitobudi et al. [3] developed, fabricated, and fully tested a two-phase closed thermosyphon (TPCT) heat exchanger for medium-temperature heat recovery in bakeries for TPCT applications. The TPCT had efficiency of 65%, although the authors commented that the TPCT could be harmed by the overload pressure during the processes. By recovering the heat from the exhaust gas of the engine, Yang et al. [9] analysed the potential use of TPCT for passengers on a large bus. The study determined that TPCTs can be used efficiently as a heat recovery system, and the findings of the tests agree well with the numerical results.

The efficacy of the TPCT heat exchangers for naturally ventilated buildings was investigated [4, 5], and in this investigation, a two-zone chamber with a horizontal partition examined the efficiency of three types of TPCT heat recovery units. These heat recovery units were just three of a set of seven externally finned heat pipes, but included firstly, a TPCT heat exchanger, a second with cylindrical spine fins, and a third consisting of two rows of spaced TPCTs. For the pressure loss characteristics of the systems, a CFD simulation was used. According to experimental findings, the performance of the TPCT heat recovery units was greatly affected by the air velocity, but potentially the heat recovery was 16% to 17% more effective for the same speed, using two TPCT banks with simple fins instead of using one bank.

The predictive pressure loss coefficient for a two-row, parallel configuration of six pipes was 3.3 compared to 4.2 for the staggered pipes and 3.7 for seven smaller parallel TPCTs, all based on the CFD modelling findings at a velocity of 1 m/s. The study proposed that any planned mean air velocity should be less than 1 m/s in naturally ventilated low-rise buildings without the impact of wind. This style of heat exchanger can be an innovative substitute for traditional reheat coils, leading to energy savings and improving the cooling capacity of cooling coils with limited or no external energy being necessary.

For the applications of an oscillating heat pipe (OHP) as a heat exchanger, Rittidech et al. [5] used the air preheater of the closed-end oscillating heat pipe (CEOHP) for a drying process. The system used, attached the condenser section to the fresh-air section and the evaporator section was in contact with the gas burner heat source. The dryer bath chosen configuration being ideal for the CEOHP air preheater. The heat transfer rate and efficiency were calculated and compared to the expected values. The study showed that energy can be saved by using a CEOHP heat exchanger for pre-heating air as part of a drying system.

4.3 Solar Energy

In spite of extensive research conducted in improving the functionality of photovoltaic (PV) solar systems, there is still much resistance when it comes to the application of this technology due to the limitations offered by the solar conversion efficiency. In this regard, many studies have been performed which encouraged the development of different technologies to improve the utilisation efficiency of solar energy. Particularly, the effect of temperature, and the generation of waste heat and overheating are noted as major problems associated with significant efficiency reduction and the poor usability of photovoltaic modules. To put this into context, the efficiency of photovoltaic cells can drop by nearly 0.5% as the surface temperature rises by every 1 °C above the normal operating condition, and this value can affect the overall productivity of a system by a large margin, especially since the efficiency of PV modules typically range from 10% to 20% [6].

This has therefore resulted in many studies which investigated different technologies and systems to absorb and recover the waste heat from solar modules. Employing a waste heat recovery system can bring several advantages to the system in terms of output efficiency improvement and generation of thermal energy, and the waste heat recovery technologies used for solar application are, in the main, based on passive or active techniques. Active techniques usually require energy from an external source to function, whereas passive technologies do not need an input from other components and operate as a stand-alone system.

For example, Kabeel et al. [7] investigated the viability of cooling photovoltaics panels in the hot Egyptian climate and discovered that by using a water-cooling technique, nearly 10 Wh/day of waste heat could be absorbed. In this same study, other methods for cooling the panels such as forced air were also examined. However, the technique that included water as cooling medium was shown to deliver the greatest amount of energy gain. Niztic et al. [8] used water as the cooling fluid and sprayed this medium both over and under the surface of the panel to absorb generated waste heat. The experiment, which was tested in a Mediterranean climate, showed that the electrical power output can increase by more than 16% from a reduction in the overall panel temperature of nearly 30 °C.

Alternatively, and as shown by Golzari et al. [11], the thermal efficiency of a PV/T system can also be increased by an electro-hydrodynamic single pass air-cooled technique, and the results noted that the thermal efficiency of the panel was increased by nearly 29% when the generated waste heat was absorbed and dissipated through secondary flows and vortices of the developed winds. Pang et al. [12] examined the performance of a PV/T system and discovered that by increasing the water mass flow rate through the panels, the absorption of the waste from the system can be increased. The experiment was conducted for water at different mass flow rates and demonstrated that the electrical and thermal efficiency of the system were in the range of 11% and 57%, respectively.

A multi-concept cooling technique was proposed by Idoko et al. [13], where different water, air, and conductive cooling methods were combined with each other to passively absorb the waste heat from photovoltaic panels. The experiment involved mounting the panels in a configuration at a height to facilitate natural air cooling, whilst, at the same time, allowing water to flow through the surface. The panels were also attached to aluminium fins which acted as a heat sink and cool water was supplied in the form of blocks of ice. The study, based on the combination of the mentioned methods, showed nearly 20 W increase in power output from their system, which corresponds to almost 5% in output efficiency improvement. The use of fins was also examined by Elbreki et al. [14], who tried to dissipate the waste heat from the panels using lapping fins and planer reflectors. The results demonstrated, that by using a specific fin configuration, slight electrical efficiency improvements of almost 2% can be achieved and that the technique, although passive, can still help in reducing the overall temperature of the module. Figure 4.6 shows the lapping fins and planar reflector arrangement developed by Elbreki et al. [14].

Having considered the mentioned passive technologies, it should be noted that other techniques, such as the use of phase change or phase change materials (PCMs), are also viable solutions to absorb and dissipate the waste heat from the panels to improve the system output efficiencies. For instance, Hasan et al. [15] studied the effect of PCMs on reducing a photovoltaics module temperature and used two different variations of PCMs – namely a salt hydrate and a eutectic mixture. The results from the experiments indicated that both materials were successful in reducing the temperature of the photovoltaics modules. However, the salt hydrate was shown to provide better performance and achieved nearly 4% increase in power output and almost 3% higher temperature drop compared to the second PCM – the eutectic mixture of fatty and capric palmitic acid. The investigation nevertheless concluded that the PCMs used have better functionality in locations where the solar radiation absorption is more stable, and the temperature are higher.

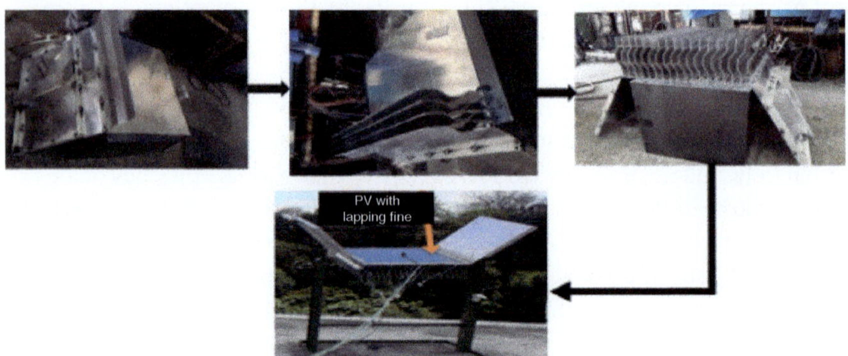

Figure 4.6 Lapping fins with planar reflector. Source: Elberki et al. [14].

A similar study was performed by Hachem et al. [16] where white petroleum jelly, pure PCM, and a mixture in the form of white petroleum jelly combined with graphite and copper was used to provide thermal management for a photovoltaics panel. The experimental results showed that the temperature of the cells can be reduced using all of the mentioned materials, however, the PCM mixture provided the highest amount of cooling and increased the efficiency of the panel by almost 6%. Salem et al. [17] performed an experiment using a PCM mixture comprising a number of nanoparticles and discovered that by passing water through straight aluminium channels placed beneath a photovoltaic panel can achieve a substantial amount of waste heat recovery from the cells. The study concluded that increasing the number of nanoparticles in the PCM mix could subsequently reduce the cells temperature even further, so making further improvements to the electrical, thermal, overall, and exergy efficiencies. Figure 4.7 shows the arrangement of channels proposed for cooling of the panel by Salem et al. [17].

Moreover, Abdelrazik et al. [18] performed an experimental investigation on a combined PV/PCM module and discovered that depositing nanoparticles in the PCM at different loadings can profoundly enhance the waste heat recovery absorption by the material. In this study, graphene nanoplatelets (GNP) were mixed in with paraffin wax (PW) which was used for heat absorption from a photovoltaic panel. The experiment was conducted in a hot location and the results indicated that significant electrical output efficiency can be obtained from the cells, even in the warmer summertime. Karthikeyan et al. [19], nevertheless, managed to dissipate the generated waste heat from a photovoltaic panel by using a radiation-based composite PCM. In their study, a thermal heat transfer network was utilised which predominantly eliminated the PCM reconduction mode and resulted in an improvement in the thermal energy absorption by the material. This then resulted in overall efficiency improvements from the panel even at higher cell temperatures.

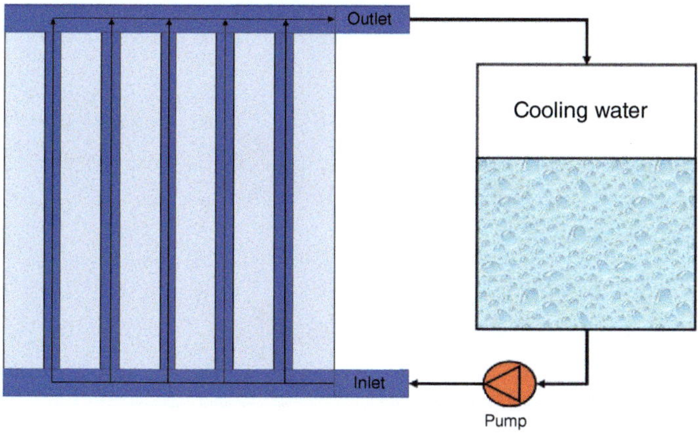

Figure 4.7 Schematic representation of cooling channel configurations. Source: Salem et al. [17].

Notwithstanding the benefits of using PCMs, it should be noted that the use of water- or air-based heat exchangers are, as yet, the most predominant of all the methods of waste heat recovery from photovoltaic panels. For instance, Elminshawi et al. [20] forced the ambient air through a buried heat exchanger that was connected to several channels at the bottom surface of a photovoltaic panel and by doing so, managed to dissipate the waste heat from the cells. At the time of the tests, the ambient air was at three temperatures of about 35, 40, and 45 °C and pre-cooled air at various flow rates were supplied to the back of the panel to perform waste heat dissipation from the cells. The technique allowed the panel to be kept below the reference cell temperature which, in return, improved the average daily efficiency by a minimum of 18%. It was also suggested, that by recovering this waste heat from the panel, the thermal efficiencies of the whole system could also be increased by almost 40%. Figure 4.8 shows the configuration of the buried heat exchanger proposed by Elminshawi et al. [20].

Baloch et al. [21], on the other hand, used a similar approach and employed a converging channel heat exchanger to absorb and dissipate the waste heat from the bottom surface of a photovoltaic panel and demonstrated the dependence on the particular shape of the converging channels, so by optimising the same, the photovoltaic cells can be effectively cooled by almost 10 °C, which can result in significant power output and conversion efficiency improvements.

Soliman and Hassan [22] investigated the potentiality of using microchannels and heat spreaders to exchange the waste heat generated from photovoltaic cells to a flow of cooling water and discovered, through this technique, the net power efficiency from the panel can be increased by nearly 4%. In the study, several experiments were conducted using different configurations of microchannel and heat spreader systems with coolant water examined at different concentration ratios and Reynolds numbers. Hossain et al. [23] used a similar idea and managed to recover and utilise the generated waste heat by introducing a novel parallel serpentine pipe

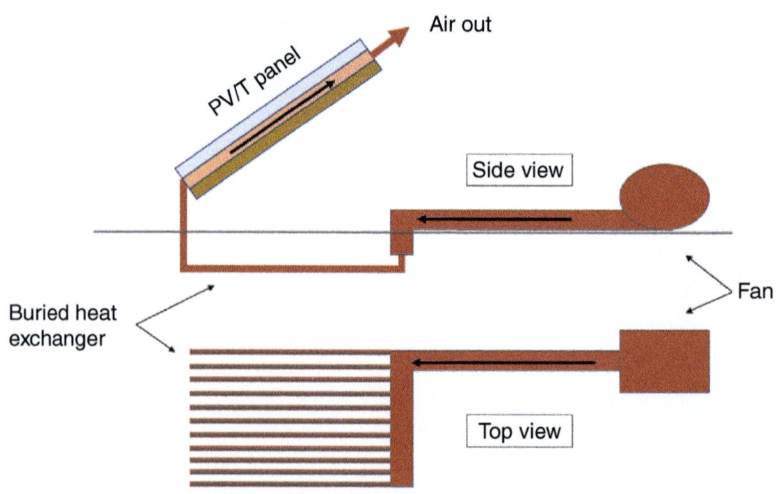

Figure 4.8 Buried heat exchanger PV cooling system. Source: Elminshawy et al. [20].

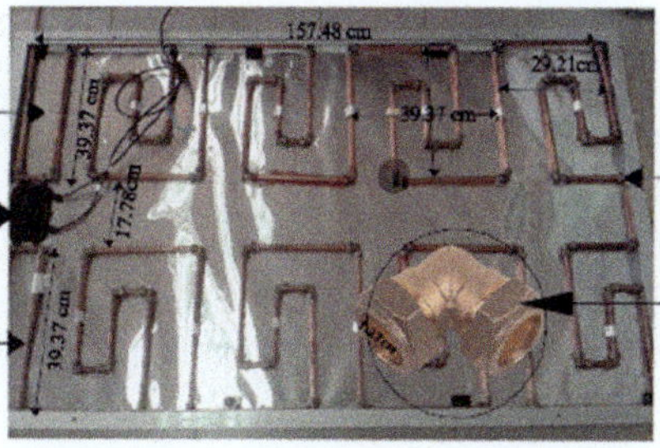

Figure 4.9 Parallel serpentine configuration. Source: Hossain et al. [23].

flow configuration to the bottom surface of a photovoltaic panel. In this study, water, as the coolant, was passed at different volume flow rates through the pipe system and it was discovered that the technique, although simple, could improve the energy conversion efficiency of the panel by nearly 0.23% per 1 °C. The results indicated that the electrical output efficiency of the photovoltaic panel can therefore reach a maximum of almost 9%. Figure 4.9 shows the parallel serpentine-type flow configuration.

Having recognised the importance of the above state-of-the-art thermal management technologies, it is important to recall that the use of heat pipes as a means of waste heat recovery from photovoltaic panels has also been shown to be just as an effective solution by several studies. For instance, Jouhara et al. [24] conceived the innovative idea and design of a unique *flat* heat pipe heat exchanger – namely a heat mat – to efficiently and effectively remove and transfer the waste heat from different types of photovoltaic panels. Their study showed that by attaching the heat mat to the bottom surface of photovoltaic modules, the waste heat generated by panel can be absorbed and transferred to a cooling manifold, where a coolant such as water is used to manage the temperature of the heat pipe. Thanks to the specific properties of the heat pipe, the study demonstrated that the photovoltaic panels could be cooled and kept at a relatively constant temperature and the experimental outcomes showed that this resulted in an improvement of electrical energy output efficiency by almost 15%. In addition, Jouhara et al. [25], in another study, furthermore demonstrated that the technology can be also used as a substitute for building envelope material, and the performance of the heat mat was examined in terms of its contribution to a district heating application. In this, an experiment was performed to investigate the functionality of the developed heat pipe-based photovoltaic-thermal panel under different solar irradiation conditions. The results from the study showed that the system is capable of delivering the thermal requirements of a single-family house on days with high solar radiation, and up to nearly 60% of the dwelling's hot water needs on days with low radiation [24]. Figure 4.10 shows the heat mat PV/T setup and Figure 4.11 presents the experimental roof installation by Jouhara et al. [25].

Other technologies such as pulsating heat pipes (PHPs) have also been utilised as effective waste heat recovery solutions that can be used to thermally manage photovoltaic panels. Thanks to the two-phase heat transfer properties, and as explained by Alizade et al. [26], PHPs are rather compact heat transfer devices that offer high heat transfer thermal performance. Based on this and in a study, it was

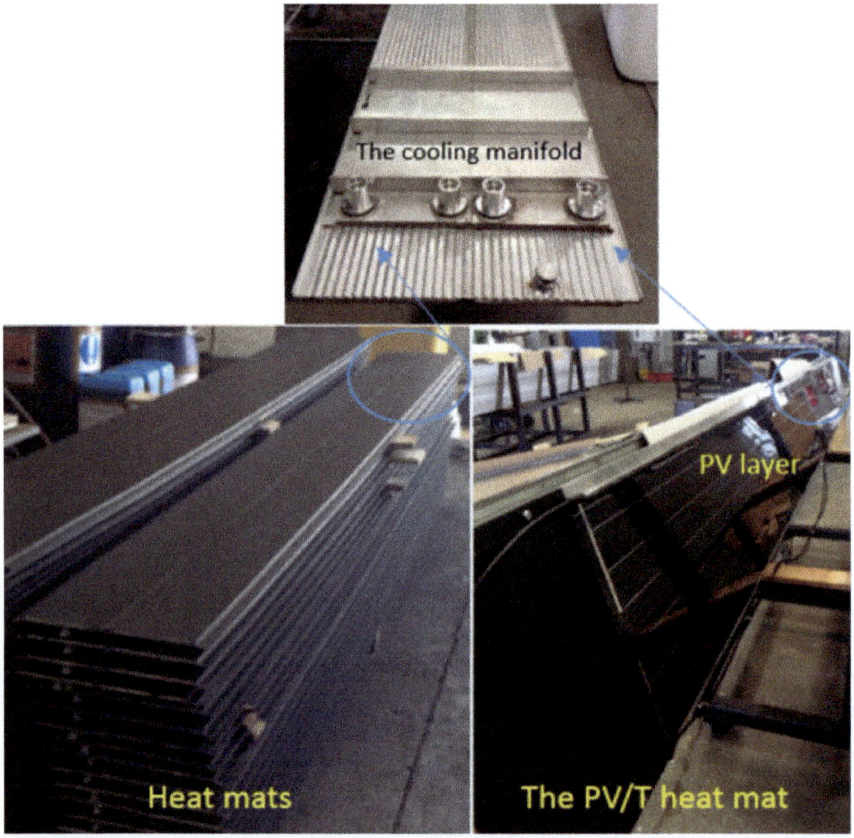

Figure 4.10 Heat mat PV/T setup.

Figure 4.11 Heat mat solar roof installation.

demonstrated that by attaching a PHP to the underneath of a photovoltaic panel, the panel can be effectively cooled and nearly an 18% increase in electrical output efficiency can be obtained.

Also of interest here, Yu et al. [27] investigated the functionality of a novel solar microchannel loop-heat-pipe photovoltaic/thermal system (MC-LHP-PV/T), where, in this study, a coaxial tubular heat exchanger was used as a condenser for a loop heat pipe (LHP) which was attached to the underneath of a photovoltaic panel. This particular configuration allowed the photovoltaic panel to act as the evaporator of the heat pipe, so when viable waste heat is available, the working fluid absorbs and transfers that to the condenser section where it is delivered to a stream of coolant. The study furthermore discovered that the output solar efficiencies largely depend on several factors, such as the inlet temperature and flow rate of the coolant, the ambient temperature, and the actual distance between the evaporator and the condenser section. To explain this further – the higher the mass flow rate and the lower the inlet temperature of the coolant; the higher the ambient temperatures and the larger the distance between the condenser and the evaporator of the system, the higher the solar thermal output efficiencies.

Based on this, the study demonstrated that a peak solar thermal efficiency of nearly 70% can be achieved at an optimal system operating condition when the water coolant inlet temperature is at about 18 °C and the flow rate is at 0.17 m^3/h. The height distance between the condenser and the evaporator was indicated to be at 1.3 m and the ambient temperature was recorded as approximately 30 °C. Relating this to tests under actual solar radiation, the system managed to obtain a maximum peak solar thermal efficiency of around 62% whilst the electrical efficiency delivered was shown to be at a minimum of about 16%. Figure 4.12 shows the design of the MC-LHP-PV/T system by Yu et al. [27].

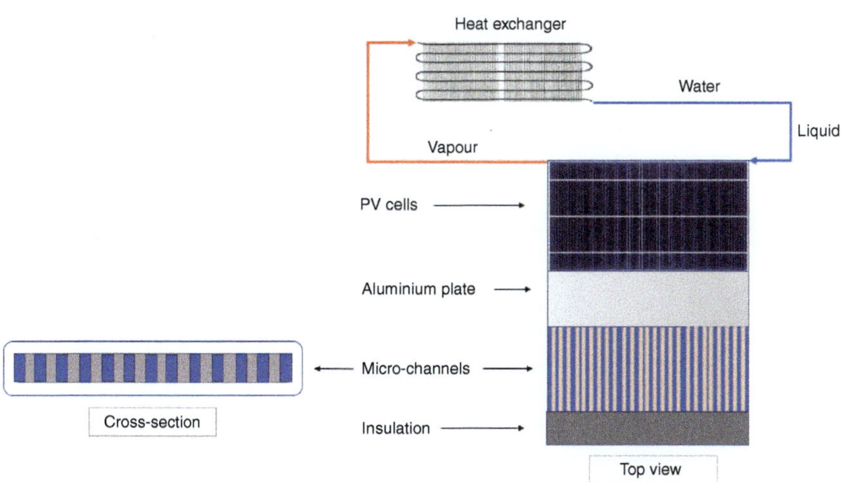

Figure 4.12 Design configuration of the MC-LHP-PV/T system. Source: Yu et al. [27].

4.4 Geothermal Energy

Geothermal energy is the heat contained beneath the Earth's crust within the sub-surface of the Earth that stimulates numerous volcanic processes across the globe. The overall heat content of the planet is estimated to be 12.6 × 1024 MJ. Part of this energy is available at high concentrations near the surface in a variety of geologically active areas [28]. This energy supply is sustainable, green, and can be extracted for a number of uses, including the generation of electricity. The schematic of a geothermal power plant is shown in Figure 4.13.

Typically, the geothermal source is accessible at a depth of 2–3 mi (3–5 km) below the surface of the Earth, where rocks can reach temperatures of over 150 °C. Holes are then drilled to allow the water to be pumped into these porous rocks to extract the energy. As the water interacts with the porous bed, it is converted to steam which is then used to power a turbine on the surface to generate electricity. The path through which the water is injected into the rocks is called the well of injection, and the path through which the steam rises is called the well of production.

Since geothermal resources are environmentally sustainable, significant advances have been made in both the utilisation technologies and the geothermal resource drilling technologies, and the use of geothermal energy has been rapidly growing worldwide, especially in the field of geothermal district space heating for buildings.

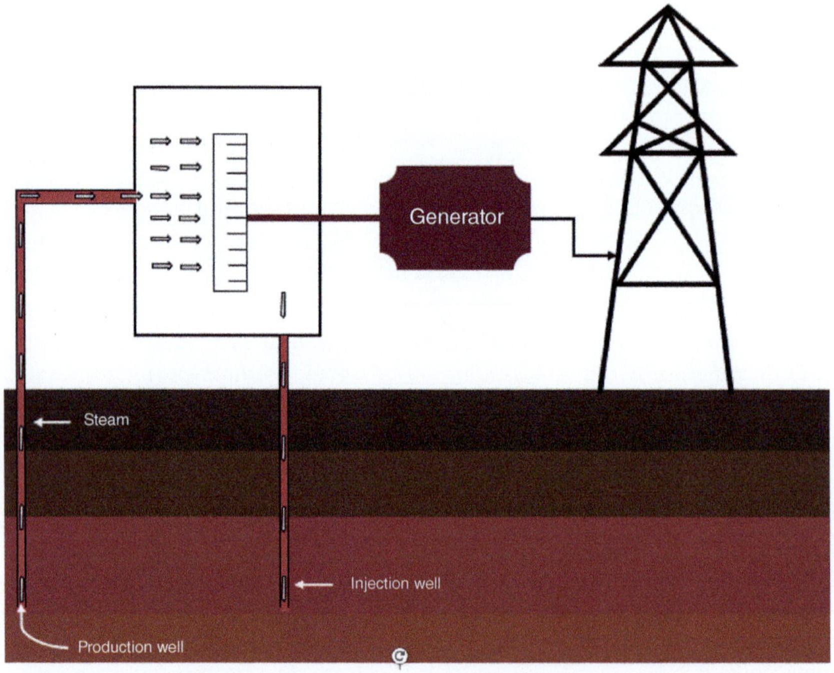

Figure 4.13 Schematic of a geothermal power plant.

There are three basic types of geothermal power plants [29]:

1. *Dry steam power plant*: In this type of power plant, the steam produced when the groundwater reaches the hot porous rocks is directly used to operate a turbine to produce energy. This power stations demand fluids at a temperature of at least 150 °C [30].
2. *Flash steam power plant*: This power plants are typically installed in geothermal reservoirs with temperatures above 182 °C. Within the separator, the hot water that emerges from the output wells flashes into steam, where the pressure reduces. This vapour is isolated from the liquid water and is then used to power the turbine. This process is the most common form of power plant used for the production of electricity.
3. *Binary-cycle power plants*: In a binary-cycle power plant, the heat from the water that comes out of the production well is exchanged with another fluid that is then used to run the turbine. Using this method (low-grade heat recovery), energy from geothermal sources can be recovered at a lower temperature, typically between 107 and 182 °C. Currently, low-grade heat recovery (based on binary cycle) options are gaining prominence.

A binary-cycle power plant is primarily used for heat recovery from low-temperature geothermal sources [31]. It allows the use of geothermal energy that cannot otherwise be used to produce electricity. Typically, this plant works on the ORC or Kalina cycle, and the working fluid is a low boiling point organic fluid [32]. New Zealand's Ngatamariki geothermal power plant uses a binary cycle to collect energy from a high-enthalpy geothermal source and produces 100 000 kW of electric power [33]. The production wells have a high temperature of about 290 °C, and no water is used to draw energy from the source.

The performance of the geothermal power plant can be enhanced by recycling and using waste heat from the plant in combination with thermodynamic cycles [34]. Wang et al. [35], evaluated the prospective geothermal infrastructure in China's Tianjin City and concluded that introducing further phases to the power generation system could result in higher net power output and enhanced geothermal resource utilisation. Growing the number of phases, however, also increases expenditure and maintenance costs. In most cases, given the expense and economic advantages, it is advised that the number of steps be reduced to two, and that potential schemes for integrating these cycles be pursued to produce or recover more energy [36].

The fluids leaving the geothermal power plants are usually over 60 °C and, in most situations, this available heat is not retrieved. The heat from the fluid can be extracted for a variety of uses and can be used for either additional power generation or direct heating applications. The hot water remaining after generation in a geothermal power plant can be used for numerous heating purposes, which are typically low-temperature applications [37], but if the hot water flowing from the power plants is immediately fed into appropriate applications, or the heat exchanger collects heat from the hot water, it can then be used in more complex applications.

Some of the forms in which the heat recovered from the power plants can be used are described as follows:

- *Space heating*: Waste heat can be fed into a district heating system and used for space heating purposes. Usually, it is suggested that geothermal resources as low as 20 to 30 °C can be used for space heating, especially in northern regions [37]. For instance, in Tianjin, China, about 620×10^6 m^2 of space heating was generated by geothermal wells in the region [36].
- *Aquaculture applications*: To control the water temperature for sustaining water-based flora and fauna [38].
- *Horticulture applications*: In 2008, Dutch developers used geothermal waste heat at 60 °C to provide 60–90% of the energy requirements for growing tomatoes [38].
- Heating the outdoor and indoor pools, which is now a tourist destination in Iceland and Croatia.

The heat from the geothermal resource can be used 'in series' for more than a few purposes. Dickson and Fanelli [28] explored the scenario of cascading applications in a location where the atmosphere is cold and snowy and a geothermal source at a temperature of 150 °C is available. A binary power plant operating between 80 and 150 °C may use the brine from a geothermal source to generate electricity. The fluid that leaves the turbine would then be at a temperature of 80 °C.

This fluid would then enter the district heating system at 80 °C and exit at 40 °C to supply heat to residential buildings. The fluid leaving the district heating system could then be used to heat swimming pools and public baths, which, it must be said, normally demand the fluid to be at a temperature of 50 °C. This fluid, from which the energy is recovered, then exits at 25 °C.

On snowy days, the de-icing of roads and parking lots allows the fluid to be at 30 °C and the fluid that leaves the balneological unit can fulfil this demand. The final outflow fluid is then estimated to be at a temperature of 10 °C. So, by such methods of cascading energy use, geothermal resources can be well utilised, and their applications maximised.

The Nesjavellir high-enthalpy geothermal system, sited in the southwest of Iceland, is a co-generation power plant that generates power and operates as a district heating system [39]. The surplus heat from the power plant is used to heat water up to 75 °C and supports about 180 000 residents in the district of Reykjavik. This district heating system is an example of a cascading system since the water, after space heating, is between 25 and 40 °C and is then used for de-icing applications. The overall length of the pipeline in the project is 2157 km, and the success of this district heating project has contributed to the discovery of other geothermal sources for district heating systems in Iceland.

The Altheim Geothermal Project in Austria derives heat from a low enthalpy geothermal source, where the water from the output well is heated to a temperature of up to 104 °C [40]. This water is subdivided into two sources, one part of which goes into the ORC to generate electrical electricity, and the other part into the district heating system to service 1500 households consuming approximately 10 000 Kilowatt-thermal (kWth) of electricity. From the ORC, the water (at about 70 °C) is cascaded to a regional school and swimming pool to provide approximately 1000 kWth of heating.

Summaries of numerous waste heat applications of geothermal systems are presented in Table 4.3.

4.5 Automotive Industry

Amongst the big problems that face humankind in the coming years are the problems of reducing greenhouse gas emissions and minimising our global footprint, but, the same time, our energy needs are still rising every day. Hence the recovery of wasted thermal energy for conversion to electricity, although part of the solution, is still seen as a major challenge for researchers and industry. Thermoelectricity, in the form of thermoelectric generators, has a strong potential for waste heat recovery, and has been researched and demonstrated in a variety of experimental and theoretical works; considering how to use thermoelectric generators to recover part of the energy that is usually lost during the production operation and then convert this into electricity.

In this respect, the transport sector might be the most desirable sector in which to use thermoelectric generators to recover lost heat, because until now, there have been few solutions to recycle the excess heat from engine exhaust gas. Unquestionably, one of the most successful fields for energy recovery is, and will be, the automobile market, where competition for ever-cleaner automobiles is quite dynamic and government-sponsored. In addition, the aeronautics field is also looking at the use of hot gases from aircraft or helicopter engines, and, of course, maritime transport provides unique possibilities due to the availability of a free cold sink (fresh water or seawater).

Automotive industrialists are projecting a great deal of interest in thermoelectric generators in order to turn the heat wasted within the exhaust gas produced by an internal combustion (IC) engine into electrical energy [43]. In a motor car, only about 25% of the fuel-fired capacity is used for vehicle mobility and operating accessories, whilst 40% is wasted within the exhaust gas [44]. On the other hand, the amount of heat generated from the exhaust system is very high and could range from 100 to 800 °C with a thermal output of up to 10 kW depending on the vehicle speed and the fuel category, making this amount of heat a valuable source of sustainable and adequate energy.

The transfer of emitted heat would also be the key solution for enhancing engine efficiency and supplying additional electronic equipment such as guidance systems, electronic brakes, additional powertrain/body controllers, stability sensors, telematics, and accident prevention systems for conventional and hybrid vehicles, and would also help reduce atmospheric emissions and energy cost.

Given the wide range of IC engines, it is essential to develop more than one form of thermoelements, all of which would contribute to improving the efficiency of conversion. As such, segmented thermoelectric materials have been used for this form of application due to their broad range of the temperature extraction [45].

Table 4.3 Waste heat applications of geothermal systems.

Location	Application	Description	Features	Drawbacks
Hengill, Iceland [39]	Geothermal, space heating, and water heating	Rankine cycle where waste heat from a cycle and geothermal brine used to produce hot water for district heating	Power generation: 120*103 kWe Heat production: 290*103 kWth	District heating requires a 27 km transmission pipe
Altheim, Austria [40]	Geothermal, space heating, and water heating	Production wells supply heat for district heating, and organic Rankine cycle and the waste heat from the heat cycle supplied to a district heating	Power generation: 2.21*106 kWh/yr Heat Production: 30.653*106 kWh/yr	Capital expenditure: 17 million EURO (US$ 19.55 million) Operating expense: 350 million EURO (US$ 402.48 million)
Egat, Thailand [41]	Geothermal and water heating	Organic Rankine binary power plant and geothermal brine using directly in industrial hot water applications	Power generation: 1.2*106 kWh/yr Heat Production: 920*103 kWth/yr	One of the production wells in an artesian well and requires chemical-scale removal
Bad Blumau, Austria [41]	Geothermal, water heating, and space heating	Organic Rankine cycle with degassing of geothermal fluid, then the fluid is fed into district heating after the power cycle	Net power production: 180 kW Thermal capacity: 5.1*103 kWth	Total investment: 75 million EURO (US$ 86.25 million)
Washoe County, Nevada [42]	Geothermal and water heating	Organic Rankine cycle with geothermal fluid for pre-heating and vapourising, and then waste fluid used as hot water for industrial process	Net power output: 3.8*103 kWe Water production: 1000 gpm (141 °C)	Total cost: US$ 2 585 000 (estimate)

In addition to the segmented thermoelectric generator (TEGs), cascaded TEGs have also been investigated for vehicle applications [46], because the independent mechanical structure of this method enables the incompatibility of the segmented arrangement to be avoided. Wilbrecht and Beitelschmidt [47] introduced two phases of cascaded thermo electric (TE) materials for railway vehicles. The cascaded TEG was made of Bi_2Te_3 (220 °C) and $Mg_2SiSn/MnSi$ (410 °C) and provided 2.5 kW of electrical power.

As discussed within the literature, two TEG positions have been used within an IC engine system – within the cooling mechanism (radiator) and within the exhaust heat exchanger. The first method was reported by Crane et al. [48], in which they presented the modelling of the integrated TEG radiator and demonstrated its potential to supplement the power output from the cooling system to the alternator.

The generators were mounted between the radiator surface and the fins. The estimated temperature difference between the radiator and the fins system was approximately 80 °C, but in addition, the engine load also influences the heat transfer to the radiator and the output power of the TEG. Hence a unit employing a Bi_2Te_3-based TEG even at low engine loads (25%) was able to produce more than 1 kW of electrical power, this could supplement the power output from the alternator and result in a major increase in fuel efficiency.

The second described method of positioning a TEG within the exhaust heat system extracts energy from the exhaust gases as they flow from the catalytic converter to the muffler/silencer. Heat exchangers or this type of application are available in two forms: flat and cylindrical, and the placement of flat TEGs on the surface of the exhaust heat exchanger has been analysed [49] for this high temperature source. For the study, the TE modules were attached to the exhaust device surface in a matrix shape – heat flowing through the exhaust inlet to the exhaust outlet giving thermal energy to the hot sides of the TEG. The study addresses the placement of 32 TEG units, each with 24 TCs, on to a flat exhaust gas. For cold TE units, coolant heat exchangers – often liquid heat exchangers – are the most commonly used to sustain cold temperatures.

Orr et al. [43] presented the use of gas heat exchangers with liquid heat pipes connected to both the cold and hot sides of the TEG, the hot waste gas stream entering the exhaust conduit, where eight water pipes are integrated into the channel fins and connected to the hot side of the eight TEG units. A fan was mounted in the inlet of the cool air channel, and the cold channel also included eight water heat pipes, which were built into the fins and connected to the cold side of the TEGs. Hussain et al. [50] used water-cooled heat sinks to have a minimum temperature at cold junctions for six modules of 96 Bi_2Te_3-based TCs each and to produce 14 W at $\Delta T = 200\,°C$ [51]. Hussain et al. also proposed a direct loop to the coolant fluid radiator to retain cold junctions at lower temperatures.

4.5.1 Industrial Processes

Industry is an area where heat is always a by-product of most operations, and in certain cases, some of this heat is reused to supply heating networks or converted into electrical energy by steam turbines – Rankine or Stirling engines – but much of the time this heat is simply wasted to the atmosphere. Accordingly, various waste heat recycling projects using TEGs have been studied.

A field test of a thermoelectric generation system was began [52] at the carburising furnace of a factory. Residual carburising gas containing CO, H_2, and N_2 is burnt at the site, and this resulted in a burning capacity of 20 kW, which continuously heated the hot side of the TEG. The TEG consisted of 16 Bi_2Te_3 modules and a heat exchanger that received about 20% of the heat (4 kW). The highest electrical output was about 214 W, which reflects an efficiency of 5%, however the electricity used to cool the cold side of the modules was not included with these results.

Aranguren et al. [53] performed laboratory analyses and mathematical tests for a TEG configured to recover waste heat from a combustion chamber. In their investigation, they included the energy used by the TEG cooling system and addressed various heat exchangers. They concluded that their prototype had a theoretical output of 100 W/m^2, and applying this finding to a large industrial chimney on a ceramic tile furnace with a flue gas mass flow of 18 400 Nm^3/h and a temperature of 187 °C, they forecasted an expected average electrical output of 136 MWh annually [54].

Likewise, the steel industry also produces a lot of waste heat, particularly radiant heat from steel products and so TEGs are possible and reliable candidates for the recovery of radiant heat from molten metal, and one company has introduced a 10 kW TEG method (about 4 m × 2 m) using radiant heat from continuous casting slabs [55]. The TEG is attached to the grid and consists of 896 TE modules (56 TEG units of 16 Bi_2Te_3 TE) and produces approximately 9 kW when the slab temperature is about 915 °C.

Cement manufacturing is also an energy-consuming industrial process, and Luo et al. [56] have researched the feasibility of applying TEGs to Portland cement production. Approximately 10–15% of the energy is dissolved directly into the atmosphere through the outer surface of the rotary turbine, but due to the permanent movement of the furnace, this kind of waste heat is difficult to recover. For a furnace with a 4.80 m outer diameter and 72 m in length, the gross heat loss without the TEG recovery system is approximately 10 000 kW. A coaxial shell was considered on the inside of the rotary kiln filled with Bi_2Te_3–PbTe hybrid TE modules and a theoretical model developed which estimated that for 20 units with 3480 TE modules (30 × 30 mm^2) each, the total TE system will generate 210 kW and save 3280 kW because of the insulation shell. The impact of thermoelectric generation being roughly 2%.

These numerous examples and industrial investigations have shown that TEGs have two key opportunities for industrial manufacturing: either in situations where it is challenging to recycle excess heat from traditional systems (radiated heat energy) or in circumstances where modern TEG materials are able to provide maintenance-free and low-cost electrical power, even considering the low performance.

4.6 Aviation

Waste heat recovery in the aviation and aerospace industry is a topic which has not been extensively researched, primarily because of certain design parameters such as weight limitations associated with aircraft. Yet, ever-escalating fuel prices and the harmful products emitted by aircrafts has resulted in many studies to investigate different theoretical and experimental methods to recover heat, mainly from the aircraft engines, and utilise that recovered heat in other segments of an aircraft system for overall efficiency improvements.

For example, Saadon [57] proposed a one-dimensional analysis method and investigated the functionality of using an ORC for waste heat recovery utilisation in turbofan engines. The paper included a numerical model and predicted the performance of a turbofan engine integrated to an ORC. The result of the study showed that the thrust-specific fuel consumption (TSFC) using the waste heat recovery configuration reached a value of almost 0.9 lbm/lbf h when the thrust force was at about 7000 lbf. Having taken into consideration weight factor of the system, the results translated into a 2% reduction of fuel consumption, indicating that the implementation of the ORC waste heat recovery system to an aircraft engine could provide substantial economic and environmental benefits to the aerospace and aviation industry. Figure 4.14 shows an ORC system integrated into an aircraft engine.

That said, the use of other state-of-the-art technologies such as TEGs are still recognised as the most promising area of investigation for waste heat recovery in aerospace and aviation industries and for which, for example, Ziolkowski et al. [59]

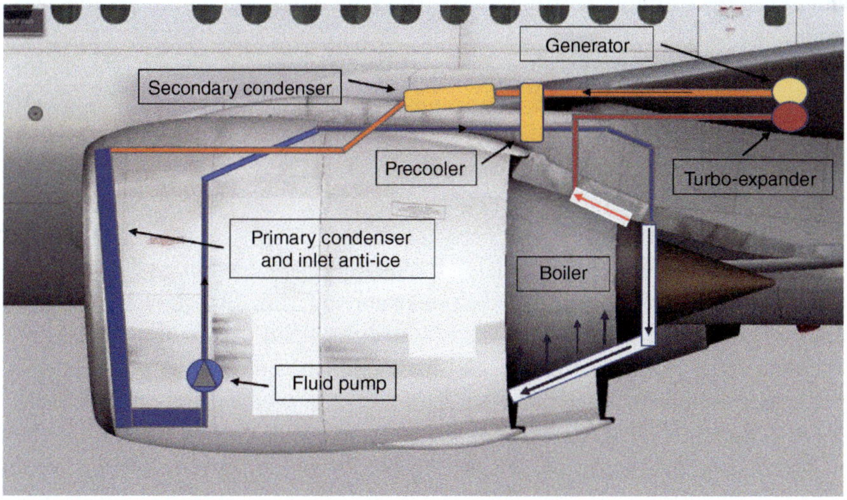

Figure 4.14 Aircraft engine integrated with an ORC system. Source: Adapted from Saadon and Saiful Islam [58].

used finite element modelling simulations to analyse the performance of a TEGs in recovering waste heat to generate power. The modelled device was placed between the cold bypass flow and the hot core stream of a turbofan engine nozzle, and the result from the study showed that, in order to obtain a positive effect on the specific fuel consumption, a direct balance between the TEG design specification (relating to the element length and filling factor) needs to be implemented. The work indicated that achievable and viable power output can be obtained by optimising the thermoelectric material and varying the conductive heat transfer coefficient between the flow stream and the hot nozzle surfaces. Through this method, the study also showed that by extending the TEG cover around the whole surface of the nozzle, the power output from the system can reach nearly 1.65 kW for each engine of the aircraft. The study therefore demonstrated the positive effect on fuel consumption reduction that can be achieved by varying the design of the device, indicating its potential for a cost-efficient and high-performance waste heat recovery technology.

As a further example, the effect of integrating a TEG device inside an airliner engine nozzle was demonstrated by Büchter et al. [60]. In the study, a rudimentary Breguet range-equation model was developed and fuel savings potentials were investigated by taking into consideration the engine specification. The model considered a simple off-design mission model and compared this against a more complex mission-based model in respect of the aircraft size and mission specification, and the study concluded that the output delivered by the technology for the mentioned models are rather comparable. Nevertheless, in order to reach a break-even fuel efficiency, a specific power density of about 180 W/kg of weight is required to be delivered by the device.

Having noted that the study further investigated the relationship between the heat flux limitations and fuel consumption reduction and explored this with a specific TEG device that presents 270 W/kg of power density, thereby showing that fuel savings of nearly 0.05% can be achieved; although, as might be expected, this result categorically depends on the size and type of engine being used. For instance, and based on further investigations, it has been demonstrated that by using a lighter TEG device that delivers a specific power density of around 100 W/kg, it is possible to achieve higher fuel saving potentials of about 0.1%.

As has already been inferred, enhancing the heat transfer coefficient is an important factor in increasing the electrical output efficiency from TEG devices, because the heat flux applied through the device is largely affected by the boundary layer around the device. Hence, placing the TEG device into the hotter areas of the engine nozzle – where higher temperature gradients and waste heat are available – can potentially increase the heat flux through the device and hence improve the specific power density and TEG efficiency.

Going forward, Bode et al. [61] studied the potential of future thermoelectric energy recovery for the aviation industry and showed that for a conceptual aircraft engine starting commence service by the year 2030, considerable energy efficiency improvements can be achieved through the use of TEG technology. In this regard, an advanced engine design called GasTurb was selected as a reference engine and an evolutionary TEG device was used to perform analytical performance simulations. The results showed through a design that includes the use of heat pipes, significant efficiencies of up to 7% can be achieved. Nevertheless, it must be remembered that this study is entirely futuristic, and any such figures delivered will require advancements in both the aircraft engine, heat exchanger, and thermoelectric generator technologies.

A preliminary analysis undertaken by Boeing Science & Technology also found a decrease in fuel consumption of 0.5% (or more) can be obtained with TEGs. The study [62] on turbine nozzles with Bi_2Te_3 modules shows that the electrical power generated under real operating conditions is substantial, but is currently inadequate if the weight of the cold exchanger is included.

Heat can be stored on the turboshaft on the compressor section, combustion chamber, and exhaust pipe. TEGs could therefore be attached to these various positions, but they must not have any effect on that turboshaft working stage which therefore prevents the use of heat exchangers. In addition, heat travelling through the TEG must be emitted elsewhere. Of course, fuel, oil, air, or compressed air are all available for this purpose, but if these must be added to the TEG, it will raise the total weight of the system. Hence and currently, the incorporation of TEGs (0.05 kW/kg-module as state-of-the-art) would not fulfil the criteria for aircraft power density (0.5 kW/kg), and these findings were subsequently confirmed in the Thetagen [63] European review. The road map for the future is therefore likely to come up against the limitations of either adding TEGs to the configuration of the aircraft engine or to add them to an environment where there is already a hot heat flow and a cold heat flow. Of course, new, lightweight TE materials will also lead, but it should be noted that the weight of the equipment must also be include the overall weight of the heat exchanger.

References

1 Noie-Baghban, S.H. and Majideian, G.R. (2000). Waste heat recovery using heat pipe heat exchanger (HPHE) for surgery rooms in hospitals. *Appl. Therm. Eng.* 20: 1271–1282. https://doi.org/10.1016/S1359-4311(99)00092-7.

2 Martínez, F.J.R., Álvarez-Guerra, P.M., Velasco Gómez, E. et al. (2003). Design and experimental study of a mixed energy recovery system, heat pipes and indirect evaporative equipment for air conditioning. *Energy Build.* 35: 1021–1030. https://doi.org/10.1016/S0378-7788(03)00056-2.

3 Lukitobudi, A.R., Akbarzadeh, A., Johnson, P.W., and Hendy, P. (1995). Design, construction and testing of a thermosyphon heat exchanger for medium temperature heat recovery in bakeries. *Heat Recovery Syst. CHP* 15: 481–491. https://doi.org/10.1016/0890-4332(95)90057-8.

4 Riffat, S.B. and Gan, G. (1998). Determination of effectiveness of heat-pipe heat recovery for naturally-ventilated buildings. *Appl. Therm. Eng.* 18: 121–130. https://doi.org/10.1016/s1359-4311(97)00033-1.

5 Rittidech, S., Dangeton, W., and Soponronnarit, S. (2005). Closed-ended oscillating heat-pipe (CEOHP) air-preheater for energy thrift in a dryer. *Appl. Energy* 81: 198–208. https://doi.org/10.1016/j.apenergy.2004.06.003.

6 Ceylan, I., Erkaymaz, O., Gedik, E., and Gurel, A.E. (2014). The prediction of photovoltaic module temperature with artificial neural networks. *Case Stud. Therm. Eng.* 3: 11–20. https://doi.org/10.1016/j.csite.2014.02.001.

7 Kabeel, A.E., Abdelgaied, M., and Sathyamurthy, R. (2019). A comprehensive investigation of the optimization cooling technique for improving the performance of PV module with reflectors under Egyptian conditions. *Sol. Energy* 186: 257–263. https://doi.org/10.1016/j.solener.2019.05.019.

8 Nižetić, S., Čoko, D., Yadav, A., and Grubišić-Čabo, F. (2016). Water spray cooling technique applied on a photovoltaic panel: the performance response. *Energy Convers. Manage.* 108: 287–296. https://doi.org/10.1016/j.enconman.2015.10.079.

9 Jouhara, H., Nannou, T., Ghazal, H. et al. (2017). Temperature and energy performance of open refrigerated display cabinets using heat pipe shelves. *Energy Procedia* 123: 273–280.

10 Abd El-Baky, M.A. and Mohamed, M.M. (2007). Heat pipe heat exchanger for heat recovery in air conditioning. *Appl. Therm. Eng.* 27: 795–801. https://doi.org/10.1016/j.applthermaleng.2006.10.020.

11 Golzari, S., Kasaeian, A., Amidpour, M. et al. (2018). Experimental investigation of the effects of corona wind on the performance of an air-cooled PV/T. *Renewable Energy* 127: 284–297. https://doi.org/10.1016/j.renene.2018.04.029.

12 Pang, W., Cui, Y., Zhang, Q. et al. (2019). Experimental effect of high mass flow rate and volume cooling on performance of a water-type PV/T collector. *Sol. Energy* 188: 1360–1368. https://doi.org/10.1016/j.solener.2019.07.024.

13 Idoko, L., Anaya-Lara, O., and McDonald, A. (2018). Enhancing PV modules efficiency and power output using multi-concept cooling technique. *Energy Rep.* 4: 357–369. https://doi.org/10.1016/j.egyr.2018.05.004.

14 Elbreki, A.M., Sopian, K., Fazlizan, A., and Ibrahim, A. (2020). An innovative technique of passive cooling PV module using lapping fins and planner reflector. *Case Stud. Therm. Eng.* 19: 100607. https://doi.org/10.1016/j.csite.2020.100607.

15 Hasan, A., McCormack, S.J., Huang, M.J. et al. (2015). Increased photovoltaic performance through temperature regulation by phase change materials: materials comparison in different climates. *Sol. Energy* 115: 264–276. https://doi.org/10.1016/j.solener.2015.02.003.

16 Hachem, F., Abdulhay, B., Ramadan, M. et al. (2017). Improving the performance of photovoltaic cells using pure and combined phase change materials – experiments and transient energy balance. *Renewable Energy* 107: 567–575. https://doi.org/10.1016/j.renene.2017.02.032.

17 Salem, M.R., Elsayed, M.M., Abd-Elaziz, A.A., and Elshazly, K.M. (2019). Performance enhancement of the photovoltaic cells using Al_2O_3/PCM mixture and/or water cooling-techniques. *Renewable Energy* 138: 876–890. https://doi.org/10.1016/j.renene.2019.02.032.

18 Abdelrazik, A.S., Al-Sulaiman, F.A., and Saidur, R. (2020). Numerical investigation of the effects of the nano-enhanced phase change materials on the thermal

and electrical performance of hybrid PV/thermal systems. *Energy Convers. Manage.* 205: 112449. https://doi.org/10.1016/j.enconman.2019.112449.

19 Karthikeyan, V., Sirisamphanwong, C., Sukchai, S. et al. (2020). Reducing PV module temperature with radiation based PV module incorporating composite phase change material. *J. Energy Storage* 29: 101346. https://doi.org/10.1016/j.est.2020.101346.

20 Elminshawy, N.A.S., El Ghandour, M., Gad, H.M. et al. (2019). The performance of a buried heat exchanger system for PV panel cooling under elevated air temperatures. *Geothermics* 82: 7–15. https://doi.org/10.1016/j.geothermics.2019.05.012.

21 Baloch, A.A.B., Bahaidarah, H.M.S., Gandhidasan, P., and Al-Sulaiman, F.A. (2015). Experimental and numerical performance analysis of a converging channel heat exchanger for PV cooling. *Energy Convers. Manage.* 103: 14–27. https://doi.org/10.1016/j.enconman.2015.06.018.

22 Soliman, A.M.A. and Hassan, H. (2018). 3D study on the performance of cooling technique composed of heat spreader and microchannels for cooling the solar cells. *Energy Convers. Manage.* 170: 1–18. https://doi.org/10.1016/j.enconman.2018.05.075.

23 Hossain, M.S., Pandey, A.K., Selvaraj, J. et al. (2019). Thermal performance analysis of parallel serpentine flow based photovoltaic/thermal (PV/T)system under composite climate of Malaysia. *Appl. Therm. Eng.* 153: 861–871. https://doi.org/10.1016/j.applthermaleng.2019.01.007.

24 Jouhara, H., Szulgowska-Zgrzywa, M., Sayegh, M.A. et al. (2017). The performance of a heat pipe based solar PV/T roof collector and its potential contribution in district heating applications. *Energy* 136: 117–125. https://doi.org/10.1016/j.energy.2016.04.070.

25 Jouhara, H., Milko, J., Danielewicz, J. et al. (2016). The performance of a novel flat heat pipe based thermal and PV/T (photovoltaic and thermal systems) solar collector that can be used as an energy-active building envelope material. *Energy* 108: 148–154. https://doi.org/10.1016/j.energy.2015.07.063.

26 Alizadeh, H., Ghasempour, R., Shafii, M.B. et al. (2018). Numerical simulation of PV cooling by using single turn pulsating heat pipe. *Int. J. Heat Mass Transfer* 127: 203–208. https://doi.org/10.1016/j.ijheatmasstransfer.2018.06.108.

27 Yu, M., Chen, F., Zheng, S. et al. (2019). Experimental investigation of a novel solar micro-channel loop-heat-pipe photovoltaic/thermal (MC-LHP-PV/T) system for heat and power generation. *Appl. Energy* 256: 113929. https://doi.org/10.1016/j.apenergy.2019.113929.

28 Mary, H. and Dickson, M.F. (2005). *Geothermal Energy: Utilization and Technology* (eds. M.H. Dickson and M. Fanelli) Google Books. Paris, France: UNESCO.

29 US Energy and Information Administration. (2017). Geothermal Power Plants.

30 European Geothermal Energy Council. (2009). Geothermal Electricity and Combined Heat & Power The European Geothermal Energy Council.

31 Setel, A., Gordan, M., Antal, C., and Bococi, D. (2015). Use of geothermal energy to produce electricity at average temperatures. *2015 13th International Conference* *on Engineering of Modern Electric Systems (EMES)*, : 1–4. https://doi.org/10.1109/EMES.2015.7158398.

32 Obi, J.B. (2015). State of art on ORC applications for waste heat recovery and micro-cogeneration for installations up to 100kWe. *Energy Procedia* 82: 994–1001. https://doi.org/10.1016/j.egypro.2015.11.857.

33 Legmann, H. (2015). *The 100-MW Ngatamariki Geothermal Power Station a Purpose-Built Plant for High Temperature, High Enthalpy Resource*. World Geothermal Congress: NV, USA.

34 Leff, H.S. (2012). Thermodynamics of combined-cycle electric power plants. *Am. J. Phys* 80 https://doi.org/10.1119/1.3694034.

35 Wang, Q., Liu, W., Yuan, X. et al. (2018). Environmental impact analysis and process optimization of batteries based on life cycle assessment. *J. Cleaner Prod.* 174: 1262–1273. https://doi.org/10.1016/j.jclepro.2017.11.059.

36 An, Q., Wang, Y., Zhao, J. et al. (2016). Direct utilization status and power generation potential of low-medium temperature hydrothermal geothermal resources in Tianjin, China: a review. *Geothermics* 64: 426–438. https://doi.org/10.1016/j.geothermics.2016.06.005.

37 International Energy Agency. (2011). Technology Roadmap Geothermal Heat and Power.

38 DeLovato, N., Sundarnath, K., Cvijovic, L. et al. (2019). A review of heat recovery applications for solar and geothermal power plants. *Renewable Sustainable Energy Rev.* 114: 109329. https://doi.org/10.1016/j.rser.2019.109329.

39 Lund, J.W. (ed.) (1996). *Hitaveita Reykjavikur and the Nesjavellir Geothermal Co-generation Power Plant*. US Department of Energy, Office of Scientific and Technical Information.

40 Geotherm Workshop Veli Lošinj. MND. (2015). Altheim in Upper Austria – an example of cascaded geothermal energy use. International School of Geothermal Development.

41 Lund, J.W. and Chiasson, A. (2007). Examples of Combined Heat and Power Plants Using Geothermal Energy. Proceedings European Geothermal Congress 2007. Unterhaching, Germany.

42 R. Gordon Bloomquist. (2004). Empire Energy, LLC – A Case StudyGHC Bulletin, 25: 11–13.

43 Orr, B., Akbarzadeh, A., and Lappas, P. (2017). An exhaust heat recovery system utilising thermoelectric generators and heat pipes. *Appl. Therm. Eng.* 126: 1185–1190. https://doi.org/10.1016/j.applthermaleng.2016.11.019.

44 Tang, Z.B., Deng, Y.D., Su, C.Q. et al. (2015). A research on thermoelectric generator's electrical performance under temperature mismatch conditions for automotive waste heat recovery system. *Case Stud. Therm. Eng.* 5: 143–150. https://doi.org/10.1016/j.csite.2015.03.006.

45 LaGrandeur, J., Crane, D., Hung, S., et al. (2006). Automotive waste heat conversion to electric power using Skutterudite, TAGS, PbTe and BiTe. *2006 25th International Conference on Thermoelectrics*: 343–348. https://doi.org/10.1109/ICT.2006.331220.

46 Krzywaniak, A., Proficz, J. and Czarnul, P. (2018). Analyzing energy/performance trade-offs with power capping for parallel applications on modern multi and many core processors. *2018 Federated Conference on Computer Science and Information Systems (FedCSIS)*: 339–346.

47 Wilbrecht, S. and Beitelschmidt, M. (2018). The potential of a cascaded TEG system for waste heat usage in railway vehicles. *J. Electron. Mater.* 47: 3358–3369. https://doi.org/10.1007/s11664-018-6094-z.

48 Crane, D., Jackson, G., and Holloway, D. (2001). Towards optimization of automotive waste heat recovery using thermoelectrics. https://doi.org/10.4271/2001-01-1021.

49 Liu, C., Pan, X., Zheng, X. et al. (2016). An experimental study of a novel prototype for two-stage thermoelectric generator from vehicle exhaust. *J. Energy Inst.* 89: 271–281. https://doi.org/10.1016/j.joei.2015.01.019.

50 Hussain, Q.E., Brigham, D.R., and Maranville, C.W. (2009). Thermoelectric exhaust heat recovery for hybrid vehicles. *SAE Tech. Pap.*: 1132–1142. https://doi.org/10.4271/2009-01-1327.

51 Haidar, J.G. and Ghojel, J.I. (2001). Waste heat recovery from the exhaust of low-power diesel engine using thermoelectric generators. *Proc. ICT2001. 20 Int. Conf. Thermoelectr.* (Cat. No.01TH8589), p. 413–8. doi:10.1109/ICT.2001.979919.

52 Kaibe, H., Makino, K., Kajihara, T. et al. (2012). Thermoelectric generating system attached to a carburizing furnace at Komatsu ltd., Awazu plant. *AIP Conf. Proc.* 1449: 524–527. https://doi.org/10.1063/1.4731609.

53 Aranguren, P., Astrain, D., and Pérez, M.G. (2014). Computational and experimental study of a complete heat dissipation system using water as heat carrier placed on a thermoelectric generator. *Energy* 74: 346–358. https://doi.org/10.1016/j.energy.2014.06.094.

54 Champier, D. (2017). Thermoelectric generators: a review of applications. *Energy Convers. Manage.* 140: 167–181. https://doi.org/10.1016/j.enconman.2017.02.070.

55 Kuroki, T., Murai, R., Makino, K. et al. (2015). Research and Development for thermoelectric generation technology using waste heat from steelmaking process. *J. Electron. Mater.* 44: 2151–2156. https://doi.org/10.1007/s11664-015-3722-8.

56 Luo, Q., Li, P., Cai, L. et al. (2015). A thermoelectric waste-heat-recovery system for Portland cement rotary kilns. *J. Electron. Mater.* 44: 1750–1762. https://doi.org/10.1007/s11664-014-3543-1.

57 Saadon S. *Computational Modelling of an Organic Rankine Cycle (ORC) Waste Heat Recovery System for an Aircraft Engine* n.d. doi:https://doi.org/10.1051/matecconf/201815102001.

58 Saadon, S. and Md Saiful Islam, S. (2020). A recent review in performance of organic Rankine cycle (ORC). Org. Rank. Cycles waste heat recover. *Anal. Appl.*, IntechOpen https://doi.org/10.5772/intechopen.89763.

59 Ziolkowski, P., Zabrocki, K., and Müller, E. (2018). TEG Design for Waste Heat Recovery at an aviation jet engine nozzle. *Appl. Sci.* 8: 2637. https://doi.org/10.3390/app8122637.

60 Büchter, K.D., Kling, U., Bode, C., and Friedrichs, J. (2018). Estimation of mission fuel savings potential using thermoelectric recuperation in aero-engines.

Transp. Res. Procedia 29, Elsevier B.V.: 23–33. https://doi.org/10.1016/j.trpro.2018.02.003.

61 Bode, C., Friedrichs, J., Somdalen, R. et al. (2017). Potential of future thermoelectric energy recuperation for aviation. *J. Eng. Gas Turbines Power* 139: 101201–101202. https://doi.org/10.1115/1.4036527.

62 Kousksou, T., Bédécarrats, J.-P., Champier, D. et al. (2011). Numerical study of thermoelectric power generation for an helicopter conical nozzle. *J. Power Sources* 196: 4026–4032. https://doi.org/10.1016/j.jpowsour.2010.12.015.

63 Commission E. *Thermoelectric Generator for Engine Control System*. 2015.

5

High-Temperature Applications

Previous chapters, in particular Chapter 2, have described, in detail, the various waste heat recovery (WHR) technologies available. This section examines case studies of WHR applied to high-temperature applications – categorised as energy recovery at temperatures over 400 °C. It must be stressed that a higher temperature is not necessarily synonymous with increased energy recovery. Temperature is only one parameter. The energy content is also determined, for example, by the mass flow rate of the heat source and whether the entire waste heat content can be recovered. Many high-temperature heat streams contain corrosive gases that would condense if these gases were to drop below their dew point and so the temperature is usually kept above this value, thereby limiting the potential energy recovery. Hence an ideal scenario would be a clean, high temperature, high flow rate heat source that can be reduced to ambient temperature in order to maximise energy recovery.

To a large extent, applications of WHR technology on an industrial scale are in the private sector and as such a large number of case studies are not available for publication. In addition, many companies are not willing to disclose process information so as to retain their commercial advantage. Whilst understandable, this does create a barrier to disseminating information on WHR technologies and cloaks information about its application to real industrial settings rather than sourced from research bodies or equipment suppliers. For these reasons, simulation studies have also been included in addition to case studies of real-world examples.

High-temperature applications are usually associated with the recovery of energy from combustion processes, but other applications also include nuclear, electrical resistance, or electrically induced waste heat sources. With this in mind, as broad a range of examples of possible applications have been compiled so as to provide a more extensive representation of WHR opportunities. Table 5.1 highlights the technologies discussed within this chapter.

5.1 Steel Industry

The steel industry processes a large amount of material to produce usable products for other sectors such as automotive, informatics, construction, and the like. And

Waste Heat Recovery in Process Industries, First Edition. Hussam Jouhara.

Table 5.1 High-temperature recovery technologies.

High temperature range (>200 °C)
• Recuperators • Heat pipe heat exchangers • Material preheating • Decorating • Regenerative burners • Recuperative burners • Rotary regenerators • Direct electrical conversion devices (still in the research stage as an emerging technology)

because of the significant volume processed, the amount of energy consumed is also amongst the largest of the secondary sector industries. The potential therefore for heat recovery is also very large and several studies have been conducted to reduce the amount of energy wastage throughout the various processes – the main recovery methods being focused on the optimisation of burners and heat exchangers.

5.1.1 TEG Modules

A thermoelectric generator (TEG) can be used to generate electrical power from a heat source and generate hot water. Ghosh et al. [1] numerically investigated the use of a TEG module to recover heat and generate electrical power during the cooling process of hot steel slabs. The TEG module is placed above the hot steel slab with the heat collection plate of the TEG operating by natural convection and heat transfer via radiation. Thermoelectric legs will then generate electrical power whilst the heat sink will be cooled by water. An illustration can be seen in Figure 5.1.

In the study, the thermoelectric module was exposed to the slab at a distance of 2 m with a thermoelectric leg thickness of 3 mm and the results show that the TEG module was able to recover up to 4.6% for a heat source at 1.5 kW/m^2. The generated electrical output is estimated to be at approximately 0.2 US$/W. The main issue highlighted by this study is the temperature of the slab being quite high for this type of material. This technology is still in development, and the cost of a full-scale installation would be high. Currently, research into TEG modules is quite extensive with multiple studies being conducted to improve the cost, efficiency, and the size of TEGs.

5.1.2 Heat Exchangers

5.1.2.1 Application 1: Slag Particles Blast Furnace Retrofit

A potential for process improvement was identified by Liu et al. [2] utilising the slag particles produced during a smelting process. Slag is a by-product of metal production because when processing an ore, any impurities are separated from the molten metal and these impurities are referred to as slag. The hot slag is usually discarded to disposal facilities, however the energy contained within this slag offers a huge

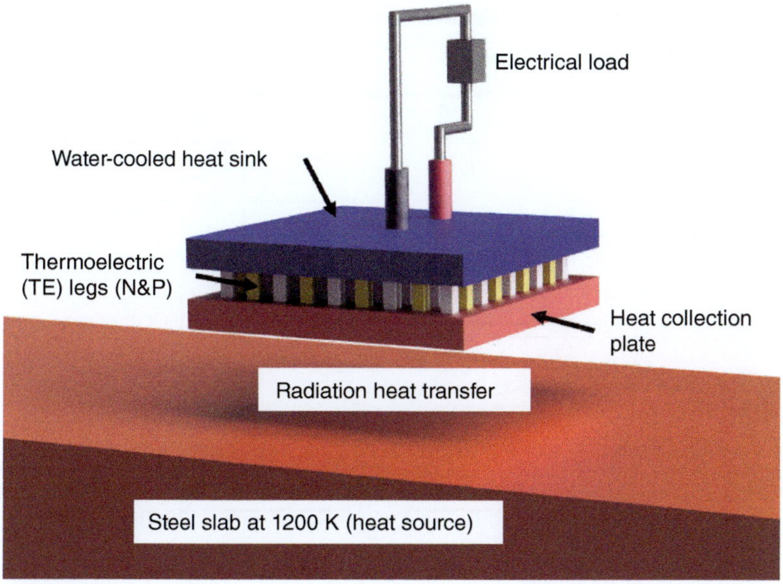

Figure 5.1 Schematic of the TEG placed above the steel slab. Source: Ghosh et al. [1].

potential for WHR. In the investigation, the authors presented a combination of two WHR technologies: a waste heat boiler and a heat exchanger as shown in Figure 5.2. The system consisted of a bed and a serpentine heat exchanger connected to a steam pipe network and a steam turbine/condenser system. The waste slag enters from the top of the bed and when passing through the coil its heat is released to a working fluid, which vapourises to generate steam.

The experimental investigation presented by the authors considered the addition of slag particles ranging between 300 and 700 °C and the outcome of the study highlighted varying efficiencies ranging between 60% and 95%, depending on the Reynolds number, tube arrangement, and temperature of the slag particles. The application of this technology is promising because it offers both high efficiencies and high heat transfer rates, but it is currently limited to laboratory-scale applications. As with many emerging technologies, scalability can be a limiting factor, causing detrimental effects to the final efficiencies. However, it is important to note that one of the key products from this technology is the production of steam. The steam and high-temperature water can be repurposed for other technologies within a processing plant, to preheat rollers and the like.

5.1.2.2 Application 2: Flat Heat Pipe Heat Exchanger

One large contributor to the inefficiency of a steel processing plant is the cooling of their different manufactured products, and this presents multiple opportunities and a vast untapped potential for heat recovery from these cooling processes to thereby increase the efficiency of metal production. For example, Jouhara et al. [3] developed and installed a radiative heat pipe to recover heat from the cooling conveyor in a metal wire production line. The heat pipe heat exchanger (HPHE) consists of a row

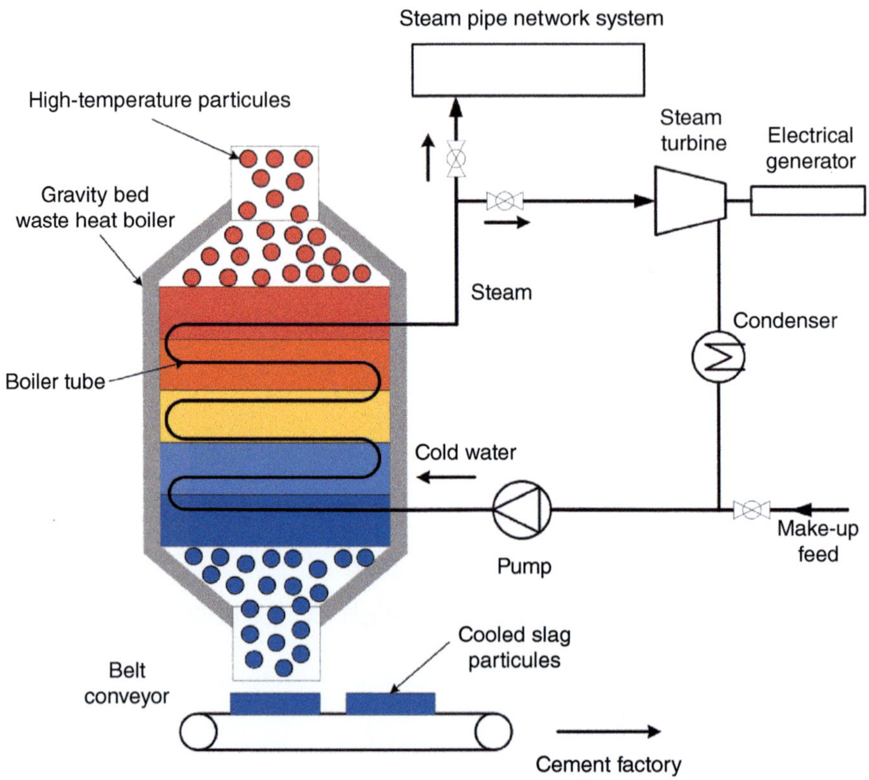

Figure 5.2 Illustration of the gravity-assisted waste heat boiler–heat exchanger system. Source: Liu et al. [2].

of tubes connected by a bottom collector and a shell and tube condenser section with an additional feature of a back panel to increase the WHR. Water passes through the cooling pipe to the condenser to extract the heat from the heat pipe system and which can then be used to preheat other systems within the processing plant. An illustration of the system can be seen in Figure 5.3.

The heat pipe is placed above the conveyor as shown in Figure 5.4, and heat is transferred from the wire to the evaporator section of the heat pipe via radiation and forced convection.

Even though the experimental setup covered only a small surface area of the conveyor, the heat pipe still managed to recover up to 15.6 kW at a water flow rate of 0.38 kg/s when the hot steel wire was at 450 °C. So, by applying this system to the entirety of the conveyor (approximately 100 m long with a heat pipe only 1 m in width) the recovered waste heat could significantly increase the process efficiency. In addition, it is also important to note that the presented flat heat pipe (FHP) does not impact the temperature of the wire, however, the initial cost of the installation could be problematic. Advantageously, heat pipe technology has a relatively low maintenance cost, and providing high efficiencies, the FHP is still a good candidate for emerging waste heat technologies within the steel manufacturing process industry.

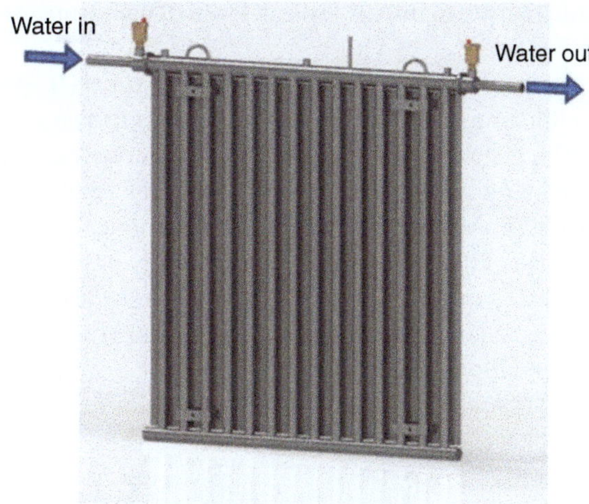

Figure 5.3 Flat heat pipe heat exchanger.

Figure 5.4 Flat heat pipe installed on the conveyor.

5.1.3 Recuperators

5.1.3.1 Application 1: Heat Recuperator for Steel Slag

As noted earlier, WHR from the slag produced by smelting is becoming one of the main areas of research and innovation for WHR within the steel industry, and in relation to which a new type of heat recuperator was developed by Trashorras et al. [4]. The heat recuperator is located after the blast furnace, and the slag produced by the furnace is transferred by conveyor to a slag tank which then feeds a heat recuperator conveyor. The heat recuperator conveyor is fully enclosed in an alumina brick lining, inserted into the internal wall of which are the heat exchanger tubes as shown in Figure 5.5.

Molten salt flows through the tube channels to extract the heat from the slags and an illustration of the process can be seen in Figure 5.6.

The system was studied through a computer simulation, the results of which indicated a recovery of 23.2 MW of heat. This result points the way to the development of a highly efficient system for successful heat recovery from metal slags, but due to the size of the recuperator, the scalability of the technology could be challenging. It is also important to note that the conditions used for any simulation may not always reflect real life because a significant number of assumptions need to be made for any computational study.

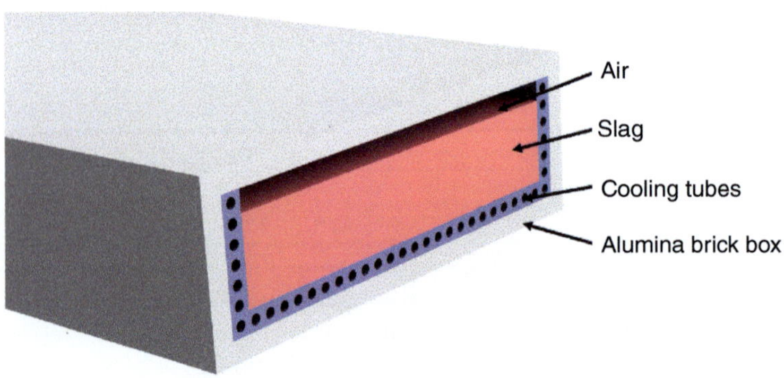

Figure 5.5 Cross section of recuperator unit. Source: Gutiérrez Trashorras et al. [4].

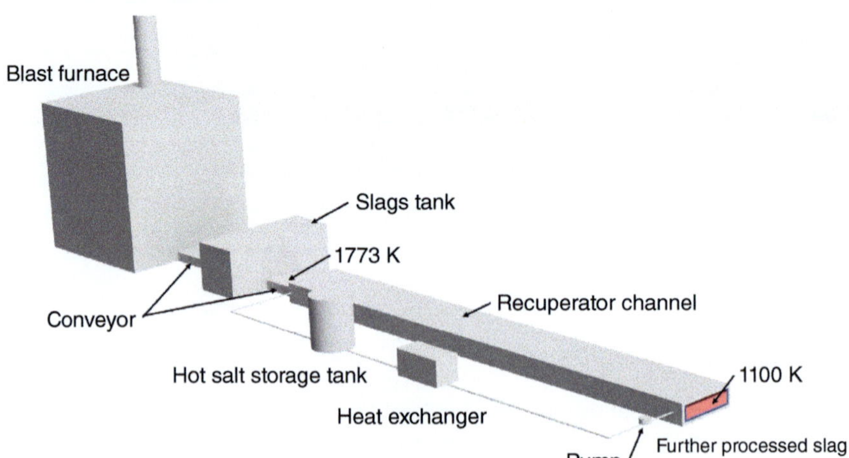

Figure 5.6 Overall heat recovery process. Source: Gutiérrez Trashorras et al. [4].

5.2 Ceramic Industry

5.2.1 Introduction

The ceramic industry is one of the key energy consumers within the secondary processing industries, but due to the nature of the process not many WHR technologies can address the corrosive exhaust stream. Unlike the metallurgy-based industries introduced in this chapter, the ceramic industry presents its own unique set of problems, revolving mainly around the final products produced. The brittleness of ceramics is the main downfall because any changes in production temperature can structurally compromise the integrity of the final product. Nonetheless, the ceramic industry is still a key candidate for WHR, with the main focus being on the development of heat exchangers, and over the past few years a small number of investigations have been conducted utilising HPHEs to improve the overall efficiency of ceramic kilns.

5.2.2 Heat Exchangers

5.2.2.1 Application 1: Radiative Heat Pipe

A significant amount of energy in a ceramic tile production line is dissipated in the cooling section of the kiln. The wall of the kiln will radiate heat to its surroundings and in the process will waste energy to the environment and cause an unnecessary generation of corresponding CO_2 emissions – all in order to regulate the cooling profile. So to help address this issue, a preliminary investigation was conducted by Delpech et al. [5]. A heat pipe system was designed to recover the heat from the tile whilst it is within the slow cooling section of the kiln as shown in Figure 5.7. The radiative heat pipe acts as an active cooling device for the tile in order to achieve a uniform temperature over the tile during its cooling period.

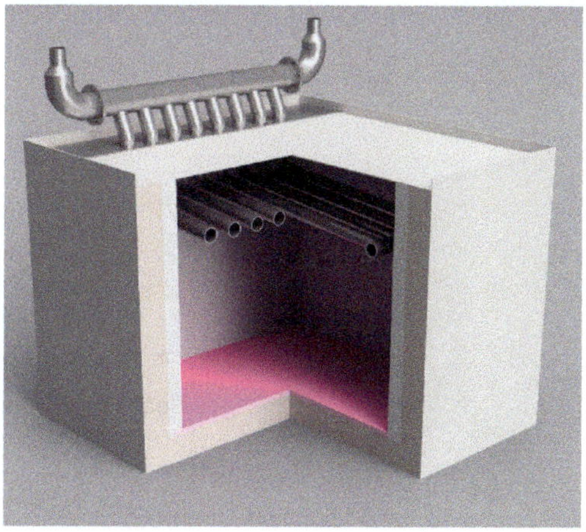

Figure 5.7 Radiative heat pipe in a laboratory ceramics kiln.

The heat pipe system is separated into two sections: an evaporator section – located within the kiln – which extracts heat from the tile, and then a condenser section – arranged in a shell and tube configuration – extracts heat from the heat pipe. Finally, the recovered heat is then transported to dryers or used to preheat the incoming air for the gas burners. The developed technology allows for full integration into existing ceramic cooling sections, but it does mean that the technology is intrusive and a section of the kiln needs to be removed for installation.

5.2.2.2 Application 2: Multi-Pass Heat Pipe

A laboratory-scale HPHE was developed by Brough et al. [6] to investigate the potential for the use of a multi-pass HPHE in the ceramics industry as shown in Figure 5.8. The HPHE consists of a tube bundle with baffles in the evaporator section to maximise the available heat transfer area. Similarly, the condenser section was also arranged with baffles.

The HPHE was connected to the exhaust stack of a ceramic kiln and Figure 5.9 shows the configuration of the HPHE and its control system. To control the outlet temperature and prevent overheating, the evaporator section is connected to a bypass system, water acting as a heat sink for the condenser section of the system.

The system was tested at a range of temperatures from 135 to 270 °C and the flow rate of the exhaust stream was also modified to allow for a full characterisation of the HPHE. The system was able to recover up to 63 kW for an exhaust temperature of 270 °C at a flow rate of 1298 kg/h and a water flow rate of 1320 kg/h. This configuration of the HPHE is vertical and baffled, which both increases the compactness of the system and reduces its footprint, but the main limitations of HPHEs are their initial cost. Initial costs for traditional shell and tube heat exchangers are far lower and so more attractive; however, the benefits of a low maintenance HPHE often outweigh the higher maintenance, more traditional heat exchangers, as discussed in Chapter 2.

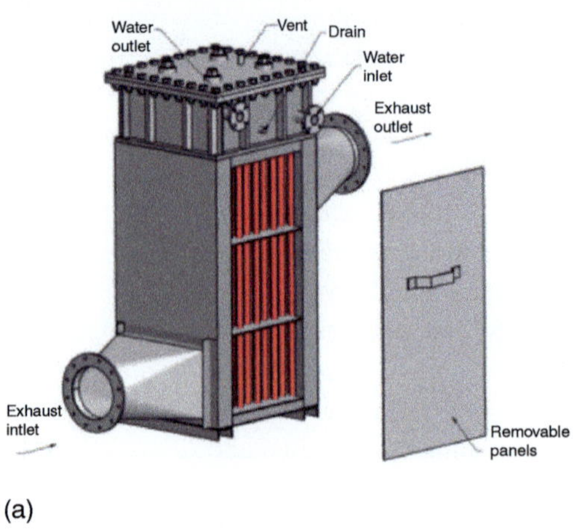

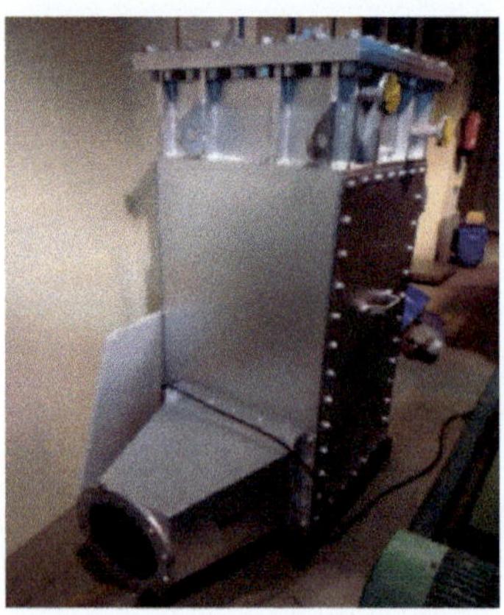

Figure 5.8 Heat pipe heat exchanger installed on the laboratory-scale kiln.

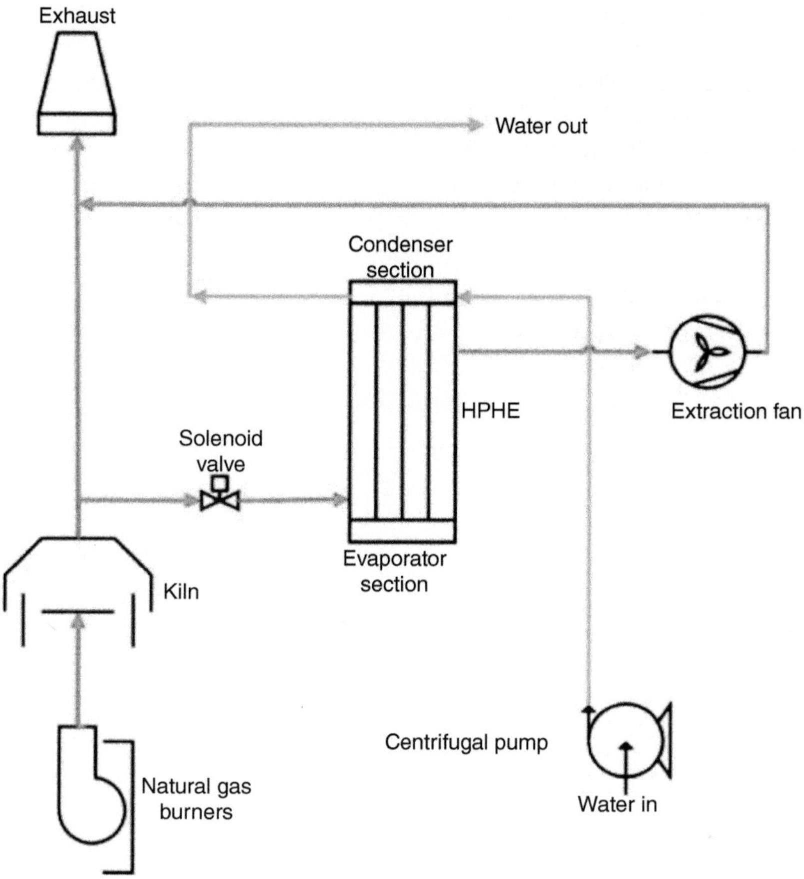

Figure 5.9 P&ID control system.

5.2.2.3 Application 3: Forced Convection Heat Pipe

As a third example of heat recovery applied to the ceramics industry, Delpech et al. [7] designed and installed a HPHE on a ceramics roller earth kiln. This was designed to cope with the high-temperature and heavy-particle-laden flow of the exhaust stream from a ceramic processing plant as shown in Figure 5.10. The HPHE designed consists of an evaporator section – for the hot exhaust gas, and a condenser section which draws in cool ambient air. The heated air leaving the HPHE unit then be used to supply clean hot air to the kiln burners and dryers.

The HPHE was tested at different working temperatures and conditions to determine the overall performance for the system. Table 5.2 shows the maximum operating conditions for the unit.

The total heat recovered by the installation and the total fuel and CO_2 saving per year are presented in Table 5.3.

The benefit of this installation are the low maintenance costs for the unit and the smaller footprint in relation to the energy recovered.

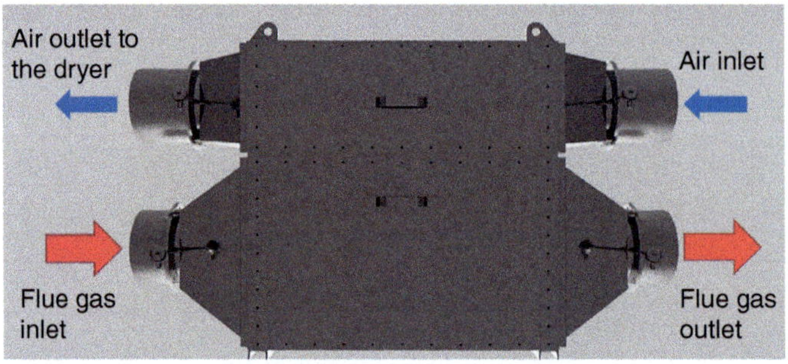

Figure 5.10 Heat pipe illustration.

Table 5.2 Heat pipe parameters.

Exhaust air mass flow rate	6000 kg/h
Fresh air mass flow rate	2640 kg/h
Exhaust air inlet temperature	204 °C
Exhaust air outlet temperature	145 °C
Fresh air inlet temperature	20 °C
Fresh air outlet temperatures	164 °C

Table 5.3 Total energy recovered and energy saving.

Parameter	Value	Units
Recovered energy from HPs	863.9	kWh
Additional fan power requirement	5.41	kW
Fuel savings	110 599	Sm^3/y
Avoided fuel cost	22 120	Euro/y
Avoided CO_2	164	tonnes/y

5.3 Cement Industry

The cement industry is widely used across the construction industry and plays an important role in the development of infrastructure. In 2019, approximately 4086 MT of cement was produced globally and, quickly increasing, this rapid expansion accounts for 0.5% increase in carbon dioxide emissions annually, which by 2030 will increase to 0.8%. Studies conducted by the IEA suggest a significant amount of this is caused by the high thermal energy intensity combined with the amount of the clinker ingredient used. The decomposition of limescale needed to form clinker is a significant contributor for CO_2 production. It has been noted that an increased efficiency in clinker production and reducing the thermal intensity of the process will actively reduce the amount of CO_2 produced. An increased efficiency has allowed the development of alternative materials to form a clinker substitute and the use of biofuel-powered processes. These changes sound promising but the structural integrity of the final product needs to be considered. An alternative method to improve the efficiency of the cement industry is to adopt WHR methods to reduce the amount of fuel consumed.

5.3.1 Gas Suspension Preheaters

Cement clinker is an intermediary solid material produced in the manufacture of Portland cement and, although rotary kilns make up approximately 20% of clinker production in the United States, worldwide it is now looked upon more favourably to use a dry process kiln that uses heat from exhaust gas to dry the powder. Clinker production makes up approximately 90% of delivered energy consumption and 74% of total energy consumption. The main fuel used is coal, but natural gas and oil-fired kilns are also available. Exhaust gases from power generation clinker cooling or stack exhaust pass through the dry meal in a series of cyclones with 4–6 cyclones being typical, although using fewer is possible depending on the scale of the plant [8]. The temperature of the waste heat is modified to suit the design of the plant. Currently a range between 204 and 1000 °C has been noted in literature as a waste heat temperature, although the influence of flow rate has not been discussed [9].

5.3.1.1 Application 1

In the United Kingdom, there are two plants with gas suspension preheaters. By installing gas suspension preheaters, the energy consumption was significantly reduced from 5.65 to 3.5 GJ/t by replacing the existing kilns with preheater kilns. A saving of 36% was reported [10].

Potentially these technologies can save up to 3 GJ/t by a reduction in operating fuel. However, to counter this, there is an increased electricity consumption of 9 kWh/t per clinker. Table 5.4 provides the energy content per tonne of clinker

Table 5.4 Process type with corresponding energy consumption per tonne of clinker.

Technology	Fuel consumption (GJ/t)
Wet process	5.9–6.7
Long dry process	4.6
1st stage preheater	4.2
2nd stage preheater	3.8
3rd stage preheater	3.3
4th stage preheater	3.1
5th stage preheater	3.0–3.1
6th stage preheater	2.9

Source: Adapted from Ref. [10].

Table 5.5 Emission reductions by country.

Country	Emission reduction (CO_2/ t clinker)
China	284
India	297
US	233

Source: Adapted from Ref. [10].

corresponding to the size of the preheater and, in comparison, if no preheating is undertaken. Although the technology increases electricity consumption, the direct emission reduction is also significant, as shown in Table 5.5.

5.3.1.2 Application 2

Caputo et al. [11] modelled a radiant heat recovery exchanger applied to a rotary kiln, the authors discussing how 8–15% of heat input is lost through the exterior surface of the kiln to ambient. To offset some of these losses, innovations in insulation were installed but these were not able to provide drastic savings, leading, more recently, to the use of WHR being considered. For this application, the chief approach being studied is to pass water through tubes within a secondary shell and then use the waste heat to preheat this water from the hottest sections of the kiln which, in practise, are the sections closest to the infeed where the majority of combustion takes place. The authors showed that 2.5 MW could be recovered (1.75 MW actual) from a tube bundle 20 m long and with a kiln wall temperature of 400 °C. However, the authors also noted that there would be diminishing returns in the subsequent cooler sections of the kiln because, for these stages, the water would already be at an elevated temperature and this would reduce the heat transfer coefficient. However, because the cost per unit length of the heat exchanger remains the same, a cost analysis was also conducted (as shown in Figure 5.11), the revenue from energy recovery per section being plotted against cost. This shows that reasonable returns are only made in the earlier sections of the kiln

5.3.2 Heat Pipe Thermoelectric Generator

One emerging technology of WHR is the use of TEGs (as discussed in Chapter 2). This technology has been coupled with heat pipe technology and Wang et al. [12] proposed the heat pipe thermoelectric generator (HPTEG) (shown in Figure 5.12), which uses potassium as a working fluid to recover waste heat from high-temperature industrial processes, and recognised cement production as an example.

The following results were observed:

- At 630 °C, the HPTEG achieved a thermal conductivity of roughly 35.8 kW/m K.
- When the TEG part of the HPTEG reaches 200 °C, the thermal conductivity is approximately 3 W/m K.

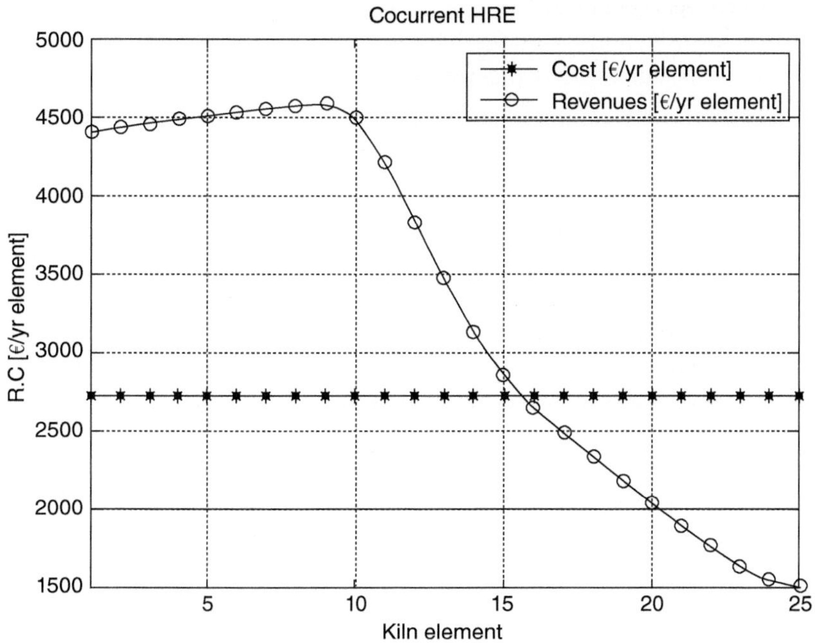

Figure 5.11 Cost and revenue of a radiative heat recovery element applied to a rotary kiln. Source: Caputo et al. [11].

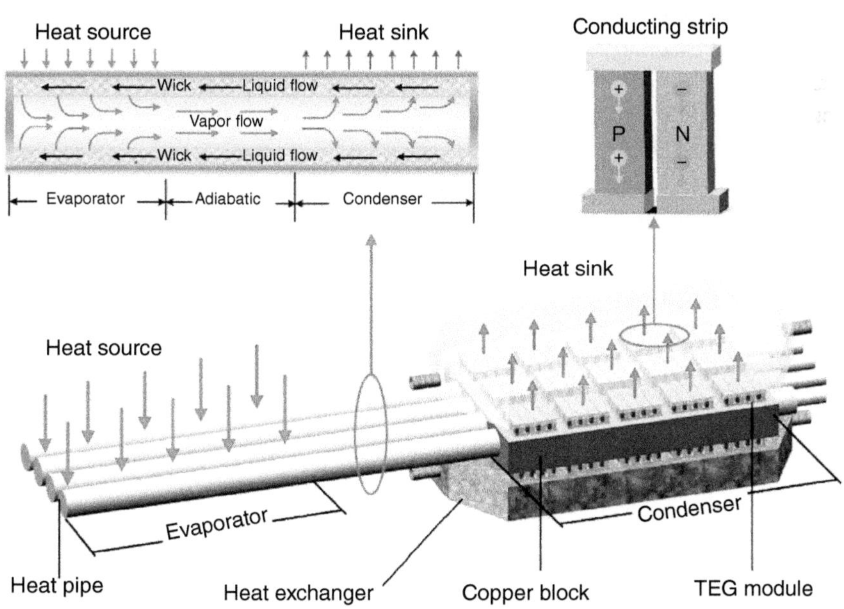

Figure 5.12 A heat pipe thermoelectric generator. Source: Wang et al. [12].

- A temperature difference of 600 °C induced a voltage over 40 V.
- A current of up to 20 A could be produced leading to an experimental performance of up around 130 W, but less than theoretically predicted value of up to 250 W.
- The efficiency peaked at 7.5%, low, but acceptable for thermoelectric conversion.

5.4 Aluminium Industry

The aluminium industry is one of the largest metallurgy sectors, second only to the steel industry, and both industries face similar issues from an environmental perspective. As aluminium is used for a wide range of applications and alloys, the use of metallurgical processes is increasing by 3.6% annually, which is also contributing to a 1% annual increase in anthropogenic greenhouse gases and 2.5% increase of carbon dioxide (CO_2) being emitted into the atmosphere. A reduction in fuel and electricity consumption is extremely desirable in energy-intensive industries, and a significant amount of this can be achieved by WHR methods [13]. Table 5.6 shows a list of extensive technologies applicable to the aluminium industry [14].

5.4.1 Rotary Regenerator

Rotary regenerators developed by Jasper GmbH aim to reclaim heat from furnace exhaust gases. Crucially, high temperatures up to 1600 °C are targeted by this technology to reduce the exhaust gas temperature to between 140 and 310 °C. The technology highlights a 62% fuel reduction, and the recovered heat was used to reheat combustion air for burners up to 1400 °C [15, 16]. (As discussed in Section 2.5.)

5.4.2 Heat Exchangers

Figure 5.13 shows a 528 kW exhaust-to-gas HPHE unit which was installed for the aluminium industry in the United States. The evaporator inlet temperature (from the exhaust stack) was approximately 400 °C and reduced to 266 °C; meanwhile within the condenser section, cold ambient air at 30 °C, acting as a heat sink, was drawn into the system. Lastly, the outlet temperature of the condenser section was 293 °C, and this was used for air preheating. As a result of the installation, approximately £155 k of equivalent energy was recovered per annum [17].

5.4.3 Heat Pumps

Nowicki and Gosselin [18] studied the application of heat pumps to Alcoa's Deschambault-Grondines primary aluminium smelter plant in Quebec province, Canada, to recover waste heat for space heating purposes. The studies of the cast house showed that approximately 4.9 MW of heat was available from the cast house cooling water, with a further 2.5 MW being identified in the compressor room, all using a water/glycol mixture. The study highlighted the potential if

Table 5.6 Sources of waste heat and applications of WHR technology [14].

Sector	Equipment	Heat loss source	Temperature range	Existing applied WHR technology	Potential applicable WHR technology
Primary	Reduction cell pots	Exhaust gases	Medium	Not available in literature	Thermoelectric devices, HPHEs, Rankine cycles
		Cell walls	High	Not available in literature	Heat pumps, thermoelectric devices, Rankine cycles
	Anode baking furnaces	Exhaust gases	Medium–High	Not available in literature	Regenerators, economisers, HPHEs, recuperators, thermoelectric devices, waste heat boilers, burner technology
		Furnace walls	Medium–High	Not available in literature	Thermoelectric devices
	Compressor rooms	Cooling medium	Low	Not available in literature	Recuperator, heat pumps
Refining	Flash tanks	Steam	Medium	Thermo-compressors, shell and tube exchangers	Other heat exchangers
	Rotary kiln calciners –	Exhaust gases, walls	Medium–High	Review use in favour of CFB technology	
Primary and secondary cast houses	Melting and holding furnaces	Exhaust gases	High	Regenerators, economisers, HPHEs, recuperators, waste heat boilers, burner technology, Material preheating, decoating	Thermoelectric devices
		Furnace walls	High	Not available in literature	Thermoelectric devices
		Slag or dross	High	Not available in literature	Radiative heat pipes
	Vertical direct chill casting or conveyors	Aluminium product to atmosphere	High	Not available in literature	Radiative heat pipes
		Cooling water	Low	Not available in literature	Recuperators, heat pumps

Source: Brough and Jouhara [14].

Figure 5.13 Heat pipe heat exchanger used to preheat combustion air. Source: Taken from Echnotherm Limited [17].

a heat pump with a coefficient of performance of 4 were used. In addition, the study also highlighted that the corresponding reduction in emissions would save approximately 6 kt of $CO_{2eq.}$ if this waste heat was used for space heating purposes, equating to a saving of around $ 300 000 per annum

5.4.4 Recuperators

5.4.4.1 Radiative Recuperator

Traditionally in the primary sectors operating at over 850 °C, radiative recuperators were used to recover heat from exhaust gases with minimal pressure drop [19]. There is some economic justification for their use with 40–60% efficiencies being seen, but they require high maintenance and frequent cleaning due to fouling by condensable vapours and particulates [20]. In addition, they also require dilution to cool the exhaust gas if metallic radiation recuperators are used; plus, other technologies now available which are more effective at transferring heat.

Beltrán [21] studied this smaller radiative recuperator (applied to a lead glass [crystal] furnace) to preheat the combustion air prior to the burner. The recuperator was 1393 mm long with an internal diameter of 520 mm. The temperature of the exhaust gas inlet averaged 880 °C but peaked at 1150 °C with the mass flow rate of 0.0869 kg/s. The combustion air mass flow rate was similar at 0.084 kg/s and averaged 37 °C. The exhaust gas was lowered to 515 °C and the combustion air temperature was raised to 197 °C. However, the effectiveness was calculated to be 0.195 – which is extremely low – and this was attributed to the low heat transfer coefficient of the radiation mechanism and to similar specific heat capacities and mass flow rates for both streams.

To their advantage, radiative recuperators have a very simple construction, consisting of only two concentric tubes, and as such they tend to require lower capital expenditure and are easy to install. They can be fitted into existing ducts (so do not require the rerouting of ductwork) which reduces the system footprint. Also, due to their simplicity of design and almost non-existent flow impedance, these exchangers have one of the very lowest pressure drops (if not the lowest) compared to other technologies and so they tend not to require additional power input from fans. Furthermore, for these reasons, a high mass flow rate is achievable and the simple design and low-pressure drop also help reduce the risk of fouling. Some designs can include finning within the recuperator to increase the heat transfer coefficient and heat transfer surface area, but then these can only be used for clean streams where deposits will not build up on the fins

By way of disadvantages in comparison to other technologies, radiative recuperators have a very low efficiency. Plus, during operation they suffer thermal stresses from heating and cooling leading to fractures. Now this can be overcome by using specialist materials, such as specialist alloyed steels or ceramics, but this will naturally increase their capital cost.

5.4.4.2 Convective Recuperator

Case study examples of installations using convective recuperators are not readily available. However, significant simulation work has been published regarding their installation and material improvement [22]. One example of which, was the modelled design for a convective recuperator applied to an aluminium furnace for preheating combustion air from exhaust gases. The real-world example was a 5 × 20 row of tubes. Case studies for six applicable steel construction recuperator designs were simulated with varying air and exhaust velocities. The temperature of the air was assumed to be ambient (25 °C) and the exhaust was modelled at 1050 °C. The outlet results of the simulations for both the air and exhaust gases are shown in Table 5.7.

The exhaust flue gas temperatures were reduced by just under 300 °C and the air was preheated by between 135 and 290 °C, depending on the air and exhaust inlet velocity, hence the mass flow rate.

Table 5.7 Six design cases for the 20 × 5 tube convective recuperator.

	Air inlet velocity (m/s)	Flue gas inlet velocity (m/s)	Maximum air outlet temperature (°C)	Minimum flue gas outlet temperature (°C)
Case 1 (steel)	48	1.2	160.7	776
Case 2 (steel)	12	1.2	284.5	786
Case 3 (steel)	6	0.6	317	773.8
Case 4 (HE steel)	6	0.6	316	773.9
Case 5 (HE steel + air)	14.4	1.2	171	753
Case 6 (rev.) (HE steel + ir)	10.5	1.2	184.5	760

Source: Gorog et al. [22].

In operation, because of the high temperatures and material choices, tube lifespans can be low – they suffer from bending, warping, and corrosion causing thinning of the walls. Any point of failure within the system leads to contamination and because the operation is not isothermal, cold spots can be present, which can give rise to corrosion or thermal stresses leading to cracks which can then cause failure. In addition, since it is not possible to view the installation during operation, regular checks must be made to ensure structural integrity, and monitoring systems must be in place to detect failure. The use of tube bundles causes a restriction in the flow of the fluid streams resulting in a high-pressure drop with additional fan power being required to overcome this – all at an increased cost. Plus, the high-pressure drop, the change in stream direction, and the challenging stream composition means that this technology is susceptible to fouling, thus reducing the effectiveness over time. This leads to high maintenance costs as the units have to be treated as consumables because cleaning the pipes is difficult which, even without replacement, would still require their disconnection from the system.

5.4.4.3 Hybrid Recuperator

Hybrid recuperators are used when either a radiative or a convective recuperator alone do not give the required performance. Typically, the exhaust stream is first passed through the radiative recuperator in order to reduce its temperature prior to the convective recuperator stage. As you would imagine, the advantages and disadvantages of both individual radiative and convective recuperators apply to a hybrid unit, but a hybrid recuperator has a higher effectiveness than either recuperator technology used in isolation because both radiation and convection thermal transfer mechanisms apply. Hence, more energy can be recovered and there is greater control over the heat recovery process because the system can be designed to suit the needs of the facility. However, the design and installation are more complicated and, as such, result in higher maintenance costs. It has also been noted that pressure drops are higher compared to isolated systems, which can lead to fouling and a requirement for additional fan power.

5.4.5 Thermoelectric Device

Yazawa and Shakouri [23] modelled water-cooled low profile thermoelectric devices in a 54 cm thick refractory lining for a melting furnace. The furnace modelled was utilised to melt glass pellets and furnace gases at 1500 °C used to maintain the molten glass at 1000 °C. 10 kW/m^2 heat flux thermoelectric devices were chosen and power generation of up to 1.72 kW/m^2 was estimated for the 500 t per day facility.

5.4.6 Regenerative Burner

Hassan and Al Kindi [24] conducted a feasibility study using regenerative burners on aluminium holding furnaces by developing a thermodynamic model and validating their results at a working foundry. Interestingly, the authors found that regenerative burner furnaces were not profitable in saving energy and their use could shorten the lifespan of the furnace. The main reason for this was that an increase in combustion air temperature resulted in a higher flame temperature, which then led to thermal shocks within the refractory lining, increased dross generation, and metal alloy degradation due to superheating of the surface of the molten metal bath. In fact, the study found that the most significant factors for increasing the efficiency of operation were (i) reducing the amount of time the furnace doors are open and (ii) reducing the overall holding time.

Two further examples are:

(1) Bloom Engineering Co., Inc. installed a regenerative burner system within an aluminium remelting furnace, and this involved six burners with two regenerators, additional valving, and a new exhaust fan [25]. The unit preheated the air temperature to approximately 525 °C from a furnace temperature of 600 °C. This gave an additional 12% fuel saving and the overall furnace efficiency was 80.1%.

(2) Manatura and Tangtrakul [26] combined both a regenerative burner system with a recuperator applied to a steel reheating furnace to preheat combustion air. The furnace had a capacity of 30 T/h and preheated the steel to between 1100 and 1250 °C. The furnace was split into three sections: preheating, heating, and soaking zones. The regenerative burners were rated at 2326 kW and applied to the heat zone.

    Originally, the recuperator could preheat the combustion air only to 300 °C due to the material limits of the recuperator, but adding the regenerators increased the combustion air temperature to 1000 °C. The energy balance analysis then showed an additional 43% saving in energy consumption in comparison to the conventional recuperative system (Table 5.8).

Table 5.8 Pros and cons of available regenerative systems.

	PROS	CONS
Rotary	Continuous/no cycling hardware needed	Leakage at moving bed seals
	Fewer burners than coupled	Loss of POC gases according to wheel design/number of segments
		Bed rotation mechanism maintenance
		Temperature limited by hot-side valving
Coupled	No hot-side valves Temp limit at least 1500 °C	Double number of burners required Cost and space required increase Special design/housing options can help to offset this
	No hot-side ducting	Cycling hardware required
LTR™	Fewer burners than coupled Retrofit may be easier	Cycling hardware required
	Minimized amount of cycling equipment	Temperature limited by hot-side valving

5.4.7 Preheating Scrap

Johansen and Strømhaug [27] measured energy input to a typical furnace cycle where charged metal was preheated and then molten metal was added prior to casting. They calculated the relative energy consumption used in the process and found that around 75% of the energy was used to preheat the cold metal and only 18% during casting to prevent freezing and maintain metal temperature.

Albert [28] conducted a study on heat recovery systems including preheating systems. The study showed that 1.65 GWh per annum less energy would be required if preheated aluminium charges were raised to 300 °C from a 20 °C ambient temperature. The tonnage considered was 55 T per charge for 365 charges. It was concluded that the initial investment and additional handling did not justify this method of WHR over others.

By preheating scrap, the melting time is decreased which decreases the cycle times and subsequently increases productivity. It was found that preheating aluminium to only 100 °C prior to casting reduced cycle times by eight minutes [29]. A further study has shown a 25.5% increase in productivity by tonnage produced if exhaust gases are used to preheat the charge [30]. This equates to a 21.1% energy saving for the process. An invaluable health and safety benefit is achieved because the removal of water content from scrap prevents superheated steam explosions.

However, even though the concept is relatively simple, the capital cost can be high and the installation can involve major rerouting of ductwork if the preheating area is located at a distance from the heat source. Furthermore, not only must the design be sturdy to hold the weight, but it also has to be resistant to mobile plant impacts. Additionally, extensive controls are required to prevent contact between operators and the hot exhaust when loading and unloading the chamber. Additional care also needs to be taken to maintain the integrity of the seals of the doors to the chamber to avoid leakage of hot exhausts or of particulate laden gases.

Other problems include the following:

(1) The possibility of cooling the exhaust below their corrosive gas dew point temperatures which should be avoided to prevent condensing these gases. Furthermore, chimney design requires that there is enough velocity and buoyancy to ensure the cleaned gases are punched into the atmosphere and do not prematurely ground.
(2) If the exhaust temperature is too hot, the edges of the material can heat more rapidly than conduction can pass energy through to the centre of the mass being heated, and this can lead to the development of hot spots on the metal rather than uniform melting. This is of particular concern in the aluminium industry where aluminium and aluminium alloys have a low melting temperature. The material being heated usually has a large thermal mass, so it can take a long period of time to increase the temperature of the material. And Migchielsen [31] reported that preheating a charge takes approximately the same amount of time as it would do if melted in a melting furnace. So, this should also be taken into account during the design process of the installation.

Table 5.9 De-coating savings balance.

Metal loss saving	US$ 7 840 000
Salt usage saving	US$ 1 630 000
Salt disposal cost saving	US$ 400 000
Total savings	**US$ 9 870 000**
Electricity cost	US$ 320 616
Labour cost	US$ 87 600
Capital equipment cost	US$ 350 000
Land/building cost	US$ 12 800
Total costs	**US$ 771 016**
Overall saving	**US$ 9 098 984**

Source: Evans and Guest [32].

5.4.8 De-coating

Case studies on de-coating systems are limited and so only predicted savings from manufacturers are provided in this section. Reference [32] provides a breakdown of the savings for a 40 kT aluminium recycling facility, summarised in Table 5.9. Material, in this case aluminium, is preheated to 400–500 °C, below the melting temperature and enough to volatilise the organic content.

The organic material coating the scrap has calorific value and this can also be used as a fuel to supply energy for de-coating the material or it can be returned to the furnace [32, 33]. The advantage of de-coating can be seen in that the presence of contamination from coatings can increase the oxidisation that occurs with the metal during melting and, in fact, there is a direct correlation between percentage coating present and amount of melt loss. For example, in the case of used beverage cans, the melt loss is between 15% and 30% if melted with a coating compared to less than 6% if the material is first pre-treated to remove the coating. Hence when the coating is removed, although there is an increase in gross production time, to offset this there is an increase in metal recovery. Also, the material melts over a shorter period of time so there is less exposure for oxidation to occur and overall, this also results in an increase in the recovery yield [34].

References

1 Ghosh, S., Margatan, K., Ahn, C.H., and Bahk, J.H. (2018). Radiant heat recovery by thermoelectric generators: a theoretical case-study on hot steel casting. *Energy Convers. Manag.* 175: 327–336. https://doi.org/10.1016/j.enconman.2018.08.106.

2 Liu, J., Yu, Q., Peng, J. et al. (2015). Thermal energy recovery from high-temperature blast furnace slag particles. *Int. Commun. Heat Mass Transf.* 69: 23–28. https://doi.org/10.1016/j.icheatmasstransfer.2015.10.013.

3 Jouhara, H., Almahmoud, S., Chauhan, A. et al. (2017). Experimental and theoretical investigation of a flat heat pipe heat exchanger for waste heat recovery in the steel industry. *Energy* 141: 1928–1939. https://doi.org/10.1016/j.energy.2017.10.142.

4 Gutiérrez Trashorras, A.J., Álvarez, E.Á., Río González, J.L. et al. (2013). Design and evaluation of a heat recuperator for steel slags. *Appl. Therm. Eng.* 56: 11–17. https://doi.org/10.1016/j.applthermaleng.2013.03.019.

5 Delpech, B., Axcell, B., and Jouhara, H. (2019). Experimental investigation of a radiative heat pipe for waste heat recovery in a ceramics kiln. *Energy* 170: 636–651. https://doi.org/10.1016/J.ENERGY.2018.12.133.

6 Brough, D., Mezquita, A., Ferrer, S. et al. (2020). An experimental study and computational validation of waste heat recovery from a lab scale ceramic kiln using a vertical multi-pass heat pipe heat exchanger. *Energy* 208: 118325. https://doi.org/10.1016/j.energy.2020.118325.

7 Delpech, B., Milani, M., Montorsi, L. et al. (2018). Energy efficiency enhancement and waste heat recovery in industrial processes by means of the heat pipe technology: case of the ceramic industry. *Energy* 158: 656–665. https://doi.org/10.1016/j.energy.2018.06.041.

8 Agico Group. (2018). Cyclone Preheater.

9 Johnson, I., Choate, W.T., and Davidson A. (2008). Waste Heat Recovery: Technology Opportunities in the US Industry – prepared by Bcs, Incorporated. US Department of Energy – Industrial Technology Program, 1–112.

10 Industrial Efficiency Technology Database, Institute for Industrial Productivity. (n.d.)

11 Caputo, A.C., Pelagagge, P.M., and Salini, P. (2011). Performance modeling of radiant heat recovery exchangers for rotary kilns. *Appl. Therm. Eng.* 31: 2578–2589. https://doi.org/10.1016/J.APPLTHERMALENG.2011.04.024.

12 Wang, C., Tang, S., Liu, X. et al. (2020). Experimental study on heat pipe thermoelectric generator for industrial high temperature waste heat recovery. *Appl. Therm. Eng.* 175: 115299. https://doi.org/10.1016/j.applthermaleng.2020.115299.

13 Migchielsen, J.M. and De Groot, J.D. (2006). Design consideration for charge preheating ovens. TMS (The Minerals, Metals and Minerals Society), Annual Meeting & Exhibition, 737–740.

14 Brough, D. and Jouhara, H. (2020). The aluminium industry: a review on state-of-the-art technologies, environmental impacts and possibilities for waste heat recovery. *Int. J. Thermofluids* 100007 https://doi.org/10.1016/j.ijft.2019.100007.

15 Jasper GmbH. (n.d.). EcoReg, Rotary bed regenerator.

16 Henning, B. and Jasper, R. (2004). Multimelter – The new generation of aluminium melting furnaces. Heat Process.

17 Econotherm Limited (n.d.). Heat Pipe Heat exchanger. https://www.econotherm.co.uk/project-7.

18 Nowicki, C. and Gosselin, L. (2012). An overview of opportunities for waste heat recovery and thermal integration in the primary aluminum industry. *JOM* 64: 990–996. https://doi.org/10.1007/s11837-012-0367-4.

19 Incropera, F.P., Prescott, P.J., and Voelkel, D.D. (1985). Hybrid systems for furnace waste heat recovery: I. use of a radiation recuperator with a Rankine cycle. *J. Heat Recover. Syst.* 5: 321–330. https://doi.org/10.1016/0198-7593(85)90006-2.

20 Thekdi, A. and Nimbalkar, S.U. (2014). Industrial Waste Heat Recovery – Potential Applications, Available Technologies and Crosscutting R&D Opportunities. https://doi.org/10.2172/1185778.

21 Beltrán, R.A.P. (2017). Study of Radiative Recuperators in High Temperature Furnaces, 112.

22 Keiser, J.R., Sarma, G.B., Thekdi, A. et al. (2007). *Final Report, Materials for Industrial Heat Recovery Systems, Task 1 Improved Materials and Operation of Recuperators for Aluminum Melting Furnaces.* U.S. Department of Energy, Office of Scientific and Technical Information.

23 Yazawa, K. and Shakouri, A. (2016). Thermal optimization of embedded thermoelectric generators in refractory furnaces. 15th IEEE ITHERM Conf., 1489–1493.

24 Hassan, M.I. and Al, K.R. (2014). Feasibility study of regenerative burners in aluminum holding furnaces. *JOM* 66: 1603–1611. https://doi.org/10.1007/s11837-014-1110-0.

25 Schalles, D. (2012). Bloom Engineering Co. Regenerative heat recovery /burner system as a replacement for conventional recuperators. *American Flame Research Fall Symposium*, 1–8. University of Utah, Salt Lake City, Utah, USA.

26 Manatura, K. and Tangtrakul, M. (2010). Energy conservation in a reheating furnace using regenerative burners combined with recuperator. *Silpakorn Univ. Sci. Technol. J.* 4: 7–13. https://doi.org/10.14456/sustj.2010.6.

27 Johansen, I. and Strømhaug, S.I. (2013, 2016). Energy control in primary aluminium casthouse furnaces. In: *Light Metals* (ed. B.A. Sadler), 935–939. Cham: Springer International Publishing https://doi.org/10.1007/978-3-319-65136-1_159.

28 Albert D. (2012). Design of heat recovery system in an aluminium cast house. NTU – Trondheim (Norwegian University of Science and Technology).

29 Arink, T. and Hassan, M.I. (2017). Metal scrap preheating using flue gas waste heat. *Energy Procedia* 105: 4788–4795. https://doi.org/10.1016/J.EGYPRO.2017.03.945.

30 U.S. Department of Energy (2007). Energy tips – process heating. *Combust. Technol. Man.*

31 Migchielsen, J., Gräb, H., and Schmidt T. (2010). Retrofitting aluminium melting furnaces. TMS (The Minerals, Metals & Minerals Society), Annual Meeting & Exhibition, pp. 663–668, San Francisco.

32 Evans, R. and Guest, G. (2002). *The Aluminium Decoating Handbook.* Wolverhampton, UK: Stein Atkinson Stordy Ltd.

33 Capuzzi, S. and Timelli, G. (2018). Preparation and melting of scrap in aluminum recycling: a review. *Metals (Basel)* 8: 24. https://doi.org/10.3390/met8040249.

34 Schlesinger, M.E. (2013). *Aluminium Recycling.* New Edition. Taylor & Francis Inc.

Index

a

actual Rankine cycle 17–19
adiabatic efficiency 18, 19, 24, 58
adiabatic index 56
aerospace industry 237
air preheaters 67–79, 160, 166
air source heat pump system 103
aluminium industry
  convective recuperator 261–262
  de-coating 265
  heat exchangers 258
  heat pumps 258
  hybrid recuperator 262
  preheating scrap 264
  radiative recuperator 260–261
  regenerative burner 262–263
  rotary regenerators 258
  thermoelectric device 262
ammonia 191
  based refrigeration cycle 192
  lithium nitrate 194
  low capacity heat pump system 196
  sodium thiocyanate 194
  water 194
ammonia–water solution 48
aquaculture applications 232
assisted circulation HRSGs 99, 169
automotive industry 233–236

b

back-work ratio (BWR) 57
binary-cycle power plant 231
binary heat pump fluids 194
biomass energy generation 45
Biot number 155, 156
blackbody model 164
bottom-up technique 132
Brayton cycle
  conservation of mass 54
  pressure ratio 56
  regenerative 57–59
  schematic of 54
  temperature entropy graph 54
  thermal efficiency 54
burner technology 71, 76–78, 191

c

carbon dioxide-based heat pump system 192
Carnot cycle 10–12
Carnot power cycle 10
cement industry
  gas suspension preheaters 255–256
  heat pipe thermoelectric generator 256–258
ceramic industry 251–253
chlorofluorocarbons (CFCs) 44
Clapeyron diagram 5
classical Rankine cycle 16
closed-end oscillating heat pipe (CEOHP) 222
closed FWH cycle 22
coefficient of performance (COP) 102, 172, 194
cogeneration 26–29
CO_2 heat pump system 193

cold heat exchanger 162
cold storage method 203
combined cycle gas turbine (CCGT) power plant 26
combined gas and steam turbine system (COGAS) 88
combined heat and power (CHP) 26
combustion air 69–75, 154
compact heat exchanger 85, 86
compressor analysis 58
condensers 146
convective recuperator 261–262
conventional heat pipe (CHP) 221
coolant fluid 37, 191, 235
copper thermosyphon 198, 199
counter-flow heat exchanger 145
cross-flow heat exchanger 145
cryogenics 198–204
cryogenic thermosyphon 198

d

data acquisition (DAQ) system 217
de-coating 265
D-frame evaporator 96, 170
diesel cycle 11, 54
dimensionless coefficient 18
direct contact heat exchangers 145
direct electrical conversion devices 107, 173, 176
distillation and condensation subsystem (DCSS) 48
double-pipe heat exchanger 141
double-pressure system 95
drain cascaded backward technology 24
drain pump forward technology 24
dry steam power plant 231
dry working fluid type 42

e

economiser 96
effectiveness-number of transfer units (ε-NTU) method 140, 151, 152
enthalpy 7, 14, 18, 19, 29, 35, 57
eutectic PCMs 128
evaporator 95
exfoliated graphite 134
exhaust air system 106
extended Surface heat exchangers 144
external forced convection 138

f

fatty acids 127
feed water heater (FWH) 22
feed water preheating effect 88–89
Fermi level 113–114
first-generation Kalina cycles 51
fixed bed regenerators 75
flash steam power plant 231
flat heat pipe 247–248
floating production storage and offloading facility (FPSO) 89
flow boiling 157
food industry, energy use 213
forced convection 138
forced convection heat pipe 253
fossil fuel power plants 31–32
fouling factor 142
Fourier's law 137
freezing 191
fuel jet 71

g

gas suspension preheaters 255–256
GasTurb 238
gas turbine cycle 53
geothermal energy 46, 230, 233
global warming potential (GWP) 44, 48, 192
graphite nano-platelets (GNPs) 134
ground source heat pump system 103–105

h

heat capacity ratio 18, 56
heat exchangers 79, 191
  aluminium industry 258
  ceramic industry 251–253
  cold 162
  compact 85–86

double-pipe 141
flat heat pipe 247–248
horizontal tube 98
hot 162
HPHE 83–85
PHE 82–83
shell and tube 79–82
spiral plate 83
steel industry 246–248
heating, ventilation, and air-conditioning (HVAC) 204–209
heat pipe-based heat exchangers 147
heat pipe heat exchanger (HPHE) 83–85, 204
heat pipe thermoelectric generator (HPTEG) 256–258
heat pumps 170, 173, 191, 206
air source 103
aluminium industry 258
cryogenic liquification unit 202
cycles 1
exhaust air system 106
fundamental principles of 100–102
ground source 103–105
hybrid 106
reverse Rankine cycle 102
solar-assisted 106
variation 102
water loop 105–106
water source 105
heat recovery steam generators (HRSG) 94–100, 166, 170
heat recovery vapour generator (HRVG) 48
heat recovery ventilators (HRVs) 221
heat transfer 79
local 140–147
mechanics 216
Hirn cycle 16
horizontal tube heat exchangers 98
horticulture applications 232
hot heat exchanger 162
hybrid heat pump 106
hybrid recuperator 67, 69, 262

i

ideal Rankine thermodynamic cycle 13
I-frame evaporators 98
indirect contact heat exchangers 144
inorganic PCMs 128
intercooling 60
irreversibility 18
isentropic compression 12
isentropic expansion 13
isentropic-type working fluid 41
isobaric heat addition 13
isobaric heat rejection 13

j

Joule cycle 53
Joule heating 111–112

k

Kalina-based plant (KCS-34) 53
Kalina cycles
advantages 49–50
applications 50–53
case studies 53
drawbacks 50
fundamentals 46

l

laminar flow 138
latent heat 139
latent heat storage 120–123
local heat transfer 140–147
log mean temperature difference (LMTD) method 140, 147, 151
loop heat pipe (LHP) 199
low-grade waste heat recovery 192
low-temperature applications
cryogenics 198–204
HVAC 204–209
refrigeration 191–198

m

microchannel loop-heat-pipe photovoltaic/thermal system (MC-LHP-PV/T) 229

multi-pass heat exchanger 145
multi-pass heat pipe 252

n
nanoparticle diffusion 132
nano-PCM 132
nano-sizing 114–115
natural circulation HRSGs 98
Newton's law of cooling 137, 140, 150
nitrogen-charged heat pipe 199
non-azeotropic mixture 47
non-paraffin PCMs 127
NO_x formation 71, 77–78
nuclear power plants 32
Nusselt number 154

o
O-frame evaporator 98, 170
ohmic heating 112
open FWH cycle 22
optimising thermodynamic cycles 89–90
organic material coating 265
organic materials 125–127
organic Rankine cycle (ORC) 215, 237
  basic configuration 35–36
  mass conservation 34
  net mechanical work 35
  organic working fluids 40–42
  with preheating 36–38
  recuperative 38–39
  recuperative, with preheating 39–40
  renewable energy 45–46
  typical parts 34
  waste heat recovery 45–46
  waste heat recovery system 44
  working fluid selection 42–45
organic working fluids 40–42
oscillating Heat Pipe (OHP) 201, 222
Otto cycle 11
ozone depletion potential (ODP) 44

p
parallel flow heat exchanger 145
Peltier effect 109–110
Peltier module 197
phase change materials (PCMs) 123, 126
  encapsulation 129, 130
  eutectic 128
  heat exchanger systems 208
  inorganic 128
  macroencapsulation 132
  materials 226
  microencapsulation 129–132
  nanomaterials 132–135
  non-paraffin 127
  shape stabilisation 135
photovoltaics/PCM (PV/PCM) module 225
piezoelectric power generation 174
pinch point 100, 168
plate heat exchanger (PHE) 82–83, 143
plugging 74
pool boiling 157
Prandtl number 154, 158
preheating scrap 264
pressure ratio 56
pressure–temperature diagram 5
P-T diagram 3, 7, 8
pulsating heat pipes (PHPs) 200, 201, 227
P-V diagram 3–6, 31

r
R22 196
R123 44
R600 44
R134a 196
R404a 196
radiative heat pipe 251, 252
radiative recuperator 68, 260–261
Rankine cycle
  application of 1
  Carnot cycle 10–12
  case studies 33–34
  fossil fuel power plants 31–32
  ideal and actual 12–29
  nuclear power plants 32
  overall efficiency of a power plant 32–33

with reheat 21–22
reverse 30–31
supercritical Rankine cycle (SRC) 29–30
thermodynamic diagrams 2–10
Rankine–Hirn cycle 16
R407c 196
recuperative burners 68, 70, 73, 152, 154, 156, 157
classification 71–72
fuel savings 72–74
recuperative organic Rankine cycle (ORC) 38–39
recuperators 57, 67, 249–250
refrigeration 191–198
cycles 1
regenerative Brayton cycle 57–59
regenerative burners 75, 76, 152, 154, 156
advantages/disadvantages 76
aluminium industry 262–263
regenerative heat exchangers 144
regenerative Rankine cycles 22
regenerator efficiency 57
reheating 60
renewable energy 45–46
re-radiation 70
reverse absorption cycle 46
reverse Rankine cycle 30–31, 102
reversibility 18
reversible water-to-water heat pump 206
Reynolds number 138, 141, 154
rotary air preheaters 74
rotary regenerators 74, 75, 258
run-around coil (RAC) 78, 79
R1234ZE (E) 193

s

Seebeck effect 109
selective catalytic reactors (SCRs) 91
self-recuperative burners 70
sensible heat storage 117–120
SG-2a cycle 51
shape stabilisation 135
shape-stabilised phase change material (SSPCM) 135
single-pass heat exchanger 145
single-phase heat exchanger 145
slag particles 246, 247
solar-assisted heat pump 106
solar energy 45, 223, 229
solar irradiation 227
soloendash gel method 132
sorption chiller 28
space heating 232
specific enthalpy 7
specific entropy 7
spiral plate heat exchanger 83
spray drying method 131
static regenerators 75
steam drums 96
steel industry
heat exchangers 246–248
recuperators 249, 250
TEG modules 246
steel slag, heat recuperator 249–250
Stefan–Boltzmann constant 166
Stefan–Boltzmann law 139
sub-zero cooling 191
sub-zero refrigerators 191
supercritical fluid 4
supercritical Rankine cycle (SRC) 15, 29–30
superheater component 96
supplementary firing 100

t

temperature–volume diagram 5
thermal resistance 142
thermal storage 116, 176, 178
latent heat storage 120–123
organic materials 125–127
phase change materials 123–126
sensible heat storage 117–120
thermochemical storage 123–124
thermionic generators 175
thermochemical storage 123–124
thermodynamic diagrams, Rankine cycle 2

thermoelectric generators (TEGs) 107, 197, 235, 236
  domestic refrigerator 198
  efficiency of 115
  Fermi level 113–114
  Joule heating 111–112
  merit 112–113
  modules 246
  nano-sizing 114–115
  Peltier effect 109–110
  Seebeck effect 109
  Thomson effect 110–111
  working principle 108
thermoelements 233
thermophotovoltaic (TPV) generators 176
thermosiphons 216
Thomson effect 110–111
top-down technique 132
transitional flow 138
trichlorofluoromethane 44
trigeneration 28
TRNSYS 206
tubular heat exchanger 143, 229
turbine analysis 58–59
T-V diagram 3
two-phase closed thermosyphon (TPCT) 222
two-phase heat exchangers 145

V
vapourisers 146
ventilators, heat recovery 223

W
waste heat boilers (WHB) 86, 157, 160
  design 87–88
  failure modes 92–93
  feed water preheating effect 88–89
  optimising thermodynamic cycles 89–90
  water spray systems 91–92
waste heat recovery (WHR) 45–46, 86
waste heat recovery boilers (WHRB) 89
waste heat technologies 197, 213, 248
water loop heat pump system 105–106
water source heat pump system 105
water spray systems 91–92
wet steam 15, 20
wet-type working fluid 41
wraparound loop heat pipe (WLHP) 205